D0557762

STREAMS

Their Ecology and Life

*In **Streams**, Cushing and Allan present a comprehensive blueprint for preserving and restoring rivers and streams. Using clear language and specific examples, the authors show the interrelationships among waterways with the flora and fauna that have evolved over time. In doing so, they demonstrate that the maintenance of this complex web of life is vital to the health of our planet. **Streams** represents a remarkable literary achievement. To present such complex and comprehensive information in a form the average person can grasp is truly an accomplishment. Kudos to Cushing and Allan!*

Bob Madgic, Ph.D., and author of *A Guide to California Freshwater Fishes*

***Streams** is an exceptionally welcome addition to the public literature that explains science to the general reader. It will provide much pleasure as well as knowledge to an increasing number of persons who are coming to appreciate both the joy and the importance of the nation's river resources. The color photographs, depicting principles and stream organisms, are superb. Authors Cushing and Allan are among the tops in the river science business.*

Thomas F. Waters, professor emeritus of fisheries and stream ecology, University of Minnesota, and author of *Wildstream: A Natural History of the Free Flowing River*

STREAMS

Their Ecology and Life

COLBERT E. CUSHING
Department of
Fisheries and Wildlife Biology
Colorado State University

J. DAVID ALLAN
School of Natural
Resources & Environment
University of Michigan

ACADEMIC PRESS
A Harcourt Science and Technology Company

San Diego San Francisco New York Boston London Sydney Tokyo

Cover photo credit: Used with permission from C.E. Cushing.
Front cover, Selway River, Idaho. Back cover, stream on
Mt. Rainier, Maryland.

This book is printed on acid-free paper. ♾

Academic Press
A Harcourt Science and Technology Company
525 B Street, Suite 1900, San Diego, California 92101-4495, USA
http://www.academicpress.com

Academic Press
Harcourt Place, 32 Jamestown Road, London NW1 7BY, UK
http://www.academicpress.com

Library of Congress Catalog Card Number: 2001086448

International Standard Book Number: 0-12-050340-9

PRINTED IN HONG KONG
01 02 03 04 05 06 TPC 9 8 7 6 5 4 3 2 1

Contents

Preface

A number of personal experiences prompted us to write this book. Foremost was our awareness of the increased concern for rivers and streams shown by conservation organizations, adopt-a-stream programs, and individual citizens who are taking an active interest in the well-being of the flowing waters of our country. From talking to lay groups, teaching summer field courses, and providing scientific advice to conservation organizations, we became aware of a desire for a book on stream ecology written for the nonprofessional, but seriously interested, individual. This is not a book filled with scientific jargon, statistics, and other information appreciated only by the professionals in the field. Instead, we have attempted to provide a straightforward explanation of how river ecosystems function as well as to describe the rich diversity of organisms that make up the biological communities of rivers and streams.

Both authors have spent their entire careers conducting research in stream ecology. C.E.C is now retired, with an affiliate faculty appointment at Colorado State University; J.D.A is a professor at the University of Michigan. Scientific understanding of the ecology of flowing waters has made great strides over the past 30-plus years, and it has been exciting to be part of this. Equally important, for us as with countless others, rivers are wonderful places for recreation and quiet contemplation. Each of us has favorite rivers to fish, to canoe, and to visit. And it is likely that any of our readers, with similar interests and of an age to look back over 20, 30, or more years, can tell of places that once were special, and now may be less so. For us, again as with many others, it is clear that it is not enough to study rivers and to love them. We must also work to protect them.

Flowing water seems to hold a special fascination for everyone. Of course there are practical reasons that so many towns and cities are built along rivers—transport, drinking water, waste removal, and water power. But rivers also provide recreation and aesthetic appreciation to great numbers of people, and these benefits are likely to become even more valued in the future because they are in limited supply. As the demand for access to attractive, healthy rivers increases, we appear to be entering a positive era in which restoration and recovery are increasingly called for and, in a growing number of cases, accomplished. We think that our chances of success can only improve if more citizens are aware of what constitutes a healthy, functioning river ecosystem.

Our intent, then, is to provide an introduction to stream ecosystems and a field guide to their biota, written for the nonspecialist. In Part I (Chapters 1–5), which describes the river ecosystem, we begin with an examination of the physical nature of rivers—the dynamics of flow, the intricacies of habitat, and the physical and chemical factors that determine the suitability of the environment for different life forms. We then turn to the ecology, considering first the sources of energy for the stream's biota and second the feeding roles and food webs. These elements are linked together in a fascinating and powerful model of how stream ecosystems function and how they change along their length, known as the River Continuum Concept.

In Part II (Chapters 6–11), we undertake to describe streams and rivers of different geographical regions of the United States. We visit a number of different ecosystems, ranging from trout streams to great rivers of the West to often-neglected Midwestern rivers. We try to acquaint readers with the variety of flowing waters and their idiosyncrasies in terms of ecosystem structure and function, using the River Continuum Concept as a framework for comparisons.

In Part III (Chapters 12–21), we describe the major groups of organisms that can be found in rivers and streams. Some of these groups—fishes and other vertebrates—are included in field guides, but the plants and invertebrates, which constitute most of the food web, are not. We include a number of photographs and drawings, but we have not attempted to construct a pictorial guide. Rather, our emphasis is on describing the ecological roles that these organisms serve, and introducing the reader to what we think are some of the fascinating aspects of their biology.

Chapter 22, the final chapter, addresses the "state of our rivers" in the United States. Although there is a need for improved regional and national assessment, scientific evidence clearly shows that much of the riverine biota is imperiled, few unaltered river segments remain, and there is an urgent need to better manage and protect rivers. On the positive side, however, there are signs that the momentum is beginning to shift in the right direction, and rivers have great powers of recuperation.

For those who would like to delve more deeply into the subject matter of each chapter, we have appended a list of recommended readings to most chapters. For those who might like to enrich their next visit to a favorite stream with a closer look at its inhabitants, we include a list of suppliers of equipment for sampling and monitoring flowing waters. We welcome communication with our readers and will do our best to respond to any queries. We hope you find something of interest herein.

Colbert E. Cushing
Estes Park, Colorado

J. David Allan
Ann Arbor, Michigan

Acknowledgments

We express our appreciation to the following people and organizations for providing photographs, information, or publications without which it would have been difficult to complete this book: Eric Bergerson, Bob Bodie, Ric Hauer, Phil Holbert, Dave Penrose, Tom Waters, Ed Van Put, and Trout Unlimited (Chapter 6); Margaret Franklin and Mark Nelson (Chapter 7); Nick Aumen, Ken Cummins, Judy Meyer, Alan Steinman, and Lou Toth (Chapter 8); Cliff Dahm, Stuart Fisher, Bill Minckley, and Jeff Whitney (Chapter 10); Alan Covich and Braley Houslet (Chapter 11); Art Benke and the North American Benthological Society (Chapter 14); Kevin Cummings and the Illinois Natural History Survey (Chapter 15); Mark Bain (Chapter 18); Jim Detterline, Stan Gregory, and Chuck Hawkins (Chapter 19); Bill Baker, Jean-Luc Cartron, Jeff Kelly, Ron Ryder, and Elizabeth Swenson (Chapter 20); Bill Baker, Martin Margulies, and Ron Ryder (Chapter 21); Paul Johnston, Larry Master, Chad Smith, and The Nature Conservancy (Chapter 22). We thank Alison Schroeer for providing the drawings that open each chapter.

The act of writing this book reminds us that many conversations, experiences, and friendships have led to the formation of our views about rivers. To all of those colleagues, fishing partners, conservationists, and others who have shared their knowledge and ideas over many years, our sincere thanks. Last, and most important, we express our deep appreciation for the support of our spouses and families.

And, of course, a special thanks to Chuck Crumly and Donna Benton James, our editor and assistant editor at Academic Press, who cajoled and helped us through the entire project.

PART I

The Ecology of Rivers and Streams

Part I contains five chapters that describe how rivers and streams function as complete, or *holistic*, ecosystems, including aspects of their channel morphology and the structure and function of the plant and animal communities living in and near them. There is much more to streams than water, fish, and insects—how many might view them—and it is our hope that Part I will help you understand and appreciate stream ecosystems as the dynamic entities that they are.

Chapter 1 explores the physical setting of stream channels and explains how the geological and climatological characteristics of a particular region interact to produce the channels characteristically found in that region, and how these affect the flow and transport of material within the channels. Chapter 2 describes the abiotic variables that influence the organisms found in the streams—factors such as water temperature, sunlight, chemical constituents, and substrate. Chapters 1 and 2 together provide the physical setting of rivers and streams, the template for their biotic communities. Before introducing the biological components, however, we need to address one other aspect—the energy sources that provide the "fuel" for these ecosystems. Chapter 3 does this, introducing both those energy sources produced within the stream itself and those produced outside of the stream in the terrestrial environment. The latter enter stream food webs as external energy subsidies. Chapter 4 introduces the wide variety of organisms that are found living in rivers and streams in terms of their functional roles as consumers and how they interact to form aquatic food webs. Finally, Chapter 5 brings all of the information from the first four chapters together and describes the structure and function of river and stream ecosystems in the context of one of the most dynamic models available for the examination of rivers and streams—the River Continuum Concept.

Truly, you will come to appreciate that streams are more than what you see from shore to shore, but are indeed integral parts of larger ecosystems and a reflection of the valleys in which they occur.

CHAPTER 1

Rivers as Dynamic Physical Entities

When we first look at a river, we usually don't see and may not even know about the complex ecosystem it supports. Rivers fascinate us as physical entities and by their variety. Who is not entranced by the complex swirling of currents that wend through protruding stones of a cobble-bottom stream or awed by the power of a mighty river as it runs swiftly downslope. Running water is exactly that—coming from somewhere, going somewhere, and interacting with its valley and the landscape along the way. As we shall see, rivers are an integral part of the water cycle, which is the balanced exchange of water among various compartments of the hydrosphere. Rivers are constantly changing, in volume, in channel shape, and over multiple time scales from momentary to millennial. Rivers transport minerals from land to sea, eroding landscapes and helping to shape continents. They have many secrets to share with us, and we continue to search these out.

Geomorphologists study rivers and their channels to try to determine the physical principles that govern behavior and form, and to see how rivers adjust to external changes, be they fluctuations in climate, geological change in landform, or man-made obstacles. Hydrologists study precipitation, groundwater supply, surface flows, and the many additional pathways by which water moves about the planet. And ecologists seek insight into the physical processes that influence the organisms and biological communities that dwell in running waters. Rivers are fluvial systems as well as ecosystems. In this chapter we will ignore life, but find plenty of dynamism. Readers wishing for an expanded treatment of these topics will enjoy *A View of the River*, by Luna Leopold.

THE HYDROLOGIC CYCLE

Water, brought to the surface of the earth by volcanic action, is essential to life as well as to many physical processes. It cycles endlessly, transferred by familiar processes such as precipitation, evaporation, infiltration, and runoff. The simplest model is one of boxes and arrows (often called *fluxes*) (Fig. 1.1). We can shrink our scale to a river valley (Fig. 1.2), or any other areal unit that interests us. Hydrologists use a mass-balance approach, similar to what many of us remember (or prefer to forget) about balancing both sides of a chemistry equation. Inputs to a watershed or a stream section must equal outputs, plus or minus any change in storage. If we measured all of the fluxes, we should find them to be in balance, and so this approach allows us to cross-check the accuracy of our measurements. If one flux is especially difficult to measure, it can be estimated by subtraction, provided the other terms are reliably estimated.

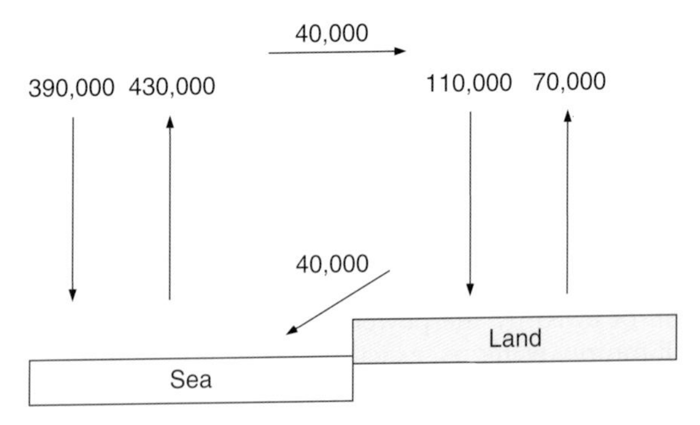

FIGURE 1.1 A simplified depiction of the global hydrological cycle. Flows, in cubic kilometers per year, are approximate. Downward arrows signify precipitation; upward arrows signify evapotranspiration. This latter term is the sum of two very distinct processes: evaporation, the physical loss of water to the atmosphere due to solar heat; and transpiration, the loss of water obtained primarily by root uptake, from plant leaves during gas exchange required for photosynthesis. The horizontal arrow represents the transfer of atmospheric moisture from sea to land and the arrow below it represents runoff from land to sea. (From Postel *et al.*, 1996.)

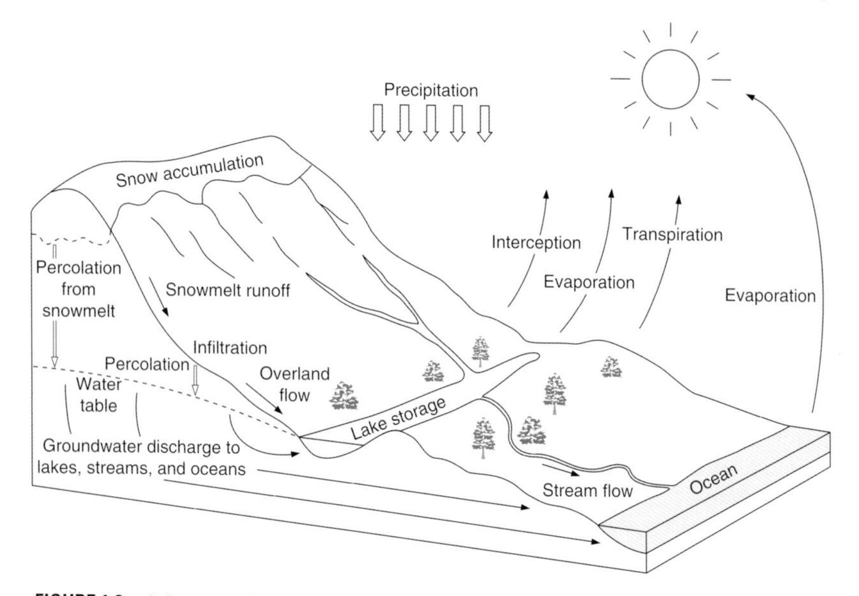

FIGURE 1.2 Schematic drawing of the water cycle for a river valley. Water moving downhill will move by overland flow when precipitation exceeds the infiltration capacity of the soil. This is rare in moist, vegetated regions, and more common in aridlands. Water that infiltrates the soil may move as shallow subsurface flow or penetrate deeper, recharging the groundwater, depending upon soil types and degree of soil saturation. Saturation of the soil can force subsurface water to rise to the surface, where, together with direct precipitation, it forms saturation overland flow. River discharge rises most rapidly when runoff follows surface and shallow subsurface pathways. (From Allan, 1995, with permission.)

Roughly speaking, about 75 cm of precipitation fall annually over the United States, and the majority, 53 cm, returns to the atmosphere via evaporation and plant transpiration. The remainder, slightly less than one-third of total precipitation, maintains river flow and contributes to groundwater. Most reaches the sea as surface water. As Fig. 1.1 illustrates, land surfaces receive more precipitation than they return to the atmosphere as evaporation, and oceans give up more water than they receive back directly as precipitation. But wind moves atmospheric water from sea to land resulting in precipitation, and rivers return that deficit back to the seas in the form of runoff, so that globally the cycle is in balance (Fig. 1.1). Locally the cycle may not be in balance, but the balance calculations can tell us that.

This surface runoff, nearly one-third of the volume of precipitation, moves from hillside into channels that converge into yet larger channels. This reminds us that when we stand on a river bank we are somewhere in a larger network that almost always connects to uplands and to the sea.

Although rivers are a critical link in the hydrologic cycle, and have influenced the development of civilizations and the locations of cities, they actually represent a miniscule fraction of the world's water supply and only a small fraction of freshwater. Of the 2.8% of the earth's water supply that occurs on land, 2.2% is locked up in ice and 0.6% is in groundwater. Lakes contain less than one one-hundreth of 1% of the world's water supply and rivers ten times less.

STREAM ORDER

When you get the chance, follow a small stream uphill to the point where it begins, perhaps as a spring or as a barely visible locale of damp earth and vegetation. This location may migrate up- and downslope depending upon the amount of recent precipitation, but at some point the stream flows year-round in a typical year. This a first-order stream. Above this point, the stream is ephemeral, flowing only during periods of ample precipitation. A first-order stream is one that lacks permanently flowing upstream tributaries. Somewhere downslope, two first-order streams merge, creating a larger, second-order stream. Other first-order streams may contribute to the river's increase in size, as can groundwater entering the channel through its bed and sides, but it takes two second-order streams to create a third-order stream, and so on (Fig. 1.3). This classification scheme is a useful way of keeping track of where we are in a stream network. It also reveals some surprising regularities in stream geometry. For example, there usually are three to four times as many streams of order n–1 as of order n, each of which is roughly half as long, and drains a little more than one-fifth of the

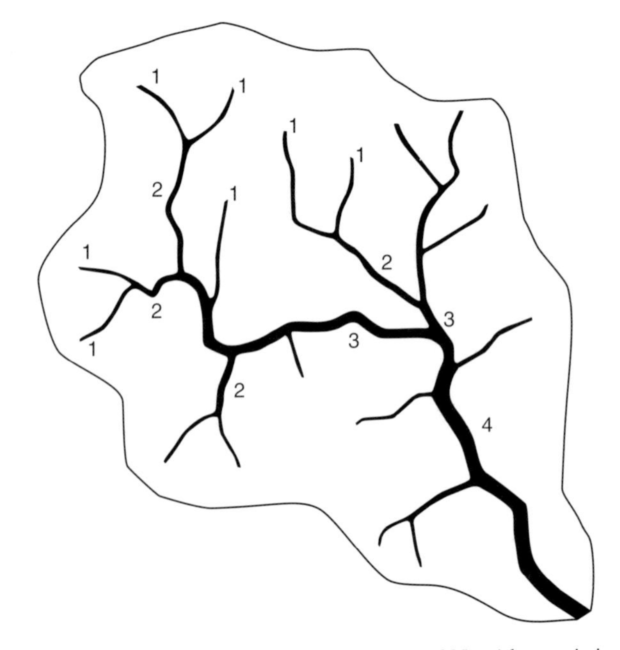

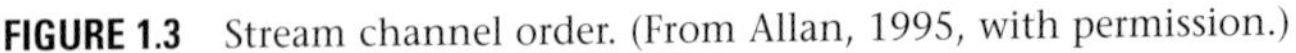
FIGURE 1.3 Stream channel order. (From Allan, 1995, with permission.)

TABLE 1.1 Number and Lengths of River Channels of Various Sizes in the United States (Excluding Tributaries of Smaller Order)[a]

Order[b]	Number	Average length (km)	Total length (km)	Mean drainage area (km²)
1	1,570,000	1.6	2,510,000	2.6
2	350,000	3.7	1,300,000	12.2
3	80,000	8.8	670,000	67
4	18,000	19	350,000	282
5	4,200	45	190,000	1,340
6	950	102	98,000	6,370
7	200	235	48,000	30,300
8	41	540	22,000	144,000
9	8	1,240	9,900	684,000
10	1	2,880	2,880	3,240,000

[a] From Leopold *et al.*, (1964).
[b] Of the approximately 5,200,000 total river kilometers, nearly 50% are first-order. The total for first-through third-order combined is just over 85%. Examples of large rivers include the Allegheny (7th-order), the Gila (8th-order), the Columbia (9th-order), and the Mississippi (10th-order).

land area. Most rivers and the great majority of river kilometers are in lower order (small- to medium-sized) streams (Table 1.1). This will influence how we think about river conservation, whereas the connectivity of small headwater stream to mighty lowland river will serve as the underpinning for later discussion of river ecosystems.

Let's imagine a selected spot on a river that we observe daily. Many changes will be apparent—over an extended period of fine, dry weather, we may see the wetted channel become narrower and its depth decrease. A period of intense precipitation can rapidly transform any river, causing it to widen and deepen, flow more swiftly, become roiled with sediments, and perhaps overflow its banks. These changes, when summarized graphically, reveal something of the dynamic character of our particular river, and may help us search for generalizations by comparing different rivers. Furthermore, the form of the channel also is not static, but adjusts dynamically to forces acting upon it and governed by variables influenced by geological and geographical location.

FLUCTUATIONS IN FLOW

Frequent measurements of the volume of water passing where we are standing on the river bank produces a graph of river discharge against time—a hydrograph (Fig. 9.1 provides two examples). Discharge usually is expressed as cubic feet per second (cfs) or cubic meters per second; conversion factors can be found at the back of the book. One can measure discharge by measuring depth and current velocity at multiple points (usually at least ten) in a transect across the stream, effectively dividing the channel cross section into a series of cells. Then one computes the discharge (width × depth × velocity) for each of the cells, and sums them to find total discharge. Of course, it gets tedious doing this every day, so hydrologists will install a gage to measure stream depth and create a graph of depth vs discharge. Thereafter, the gage provides an estimate of discharge from a measure of depth (referred to as river stage). Stream gages often are located at road crossings or other easily accessible sites and are housed in small buildings about the size of an outhouse. The gaging apparatus usually is a vertical pipe, connected by a horizontal pipe to the bottom of the stream channel. As water rises and falls in the river, a recording device measures river stage within the vertical pipe, and this is converted to discharge by an equation from the aforementioned graph. The U.S. Geological Service (USGS) maintains an extensive network of stream gages. Presently, about 2000 stream gaging stations provide daily discharge estimates, and many provide hourly data that can be rapidly and freely obtained from USGS Web servers by anyone with internet access. Annual publications record daily mean discharge and provide sums and averages for the calendar year and the water year.

To find out if your local stream has (or had–some gages are discontinued) such records, consult the U.S. Water Year Books of the USGS at your local university library, or log on to the Web at <http://waterdata.usgs.gov/nwis-w/ST> (use the two-letter state code instead of ST). The EPA's "Surf Your Watershed" site at <http://www.epa.gov/surf2/> should also get you there.

The data collected at thousands of stream gages permit a great deal of analysis of how discharge varies, on time scales of days, months, years, or decades. From a long-enough record one can estimate the magnitude of a flood that occurs on average once in 10 years, 25 years, 100 years, and so on, which is obviously worth knowing if you are building a structure near a river. One can see if a stream is relatively stable, or has a predictable period of high water, or is highly erratic. For example, the Au Sable, a premier trout stream in central Michigan, is largely groundwater-fed and highly stable throughout the year. The Colorado River has a regular peak in late spring, when snow melts in the high country. Unpredictable flash flooding characterizes many desert streams.

Human actions modify natural flow regimes in many different ways, of which dams are only the most obvious. In general, dams regulate a river's flow, making it more constant by capturing and storing flood peaks. Altered land use, such as deforestation and urbanization, usually hastens the flow of rainwater into stream channels, making them more flashy in response to storms and also more prone to drought, because that stormwater runoff might otherwise have recharged groundwater and maintained the stream during periods without rainfall. Rivers, and in fact virtually all ecosystems, are naturally variable, and that variation is critical to the physical and biological functioning of the ecosystem. In Chapter 22 we will argue that keeping the flow regime as natural as possible is a vital piece of river conservation.

Examining a hydrograph teaches us that rivers are dynamic in time. It should be no surprise that rivers are dynamic in space also. Rivers overflow their banks during large floods, generally with a frequency of once every two to three years. Often the bankfull level is easy to see, from our vantage point on the stream bank, by inspection of bank angles, and from debris pushed up on the bank. Don't be misled, however, by steep, eroded river channels such as some western arroyos. These have much higher bank edges, due to continual down-cutting. Unable to increase in width, floodwaters can swell only in depth and velocity, endangering the lives of anyone unlucky enough to be within the canyon walls during a flash flood event.

Rivers flood to bankfull on average once every two to three years. Larger floods will spill over the banks and inundate the flat area adjacent to the river channel. The inundated area is the floodplain, and its extent likely depends on flood magnitude and terrain. The extent inundated usually is greater for a 100-year flood than for a 10-year flood. A change in climate or other conditions may cause a river to degrade, in which case the old floodplain is abandoned or becomes a terrace, and a new floodplain may be formed at the river's new level.

Sometimes one can see terraces or higher landscape that is evidence of the river's floodplain when the valley was younger and before the river cut down to its present level and created its present floodplain. Between these terraces the river may have migrated laterally back and forth across the present floodplain many times, occupying different locations whenever major floods caused the river to cut a new channel. The stream acts to shape the valley as the valley helps to shape the stream, in a continuing interaction between the landscape and the river flowing through it.

THE TRANSPORT OF MATERIAL

Rocks break down by the erosive forces of nature, including transport by water, freeze and thaw, landslides, and more subtle downhill motion due to gravity. This produces materials that can be transported by water in solution, as dissolved load; as particles in suspension, or suspended load; or as particles moved by skipping and sliding along the stream bed, or bed load. In humid, warm climates, the dissolved load tends to be the greater, whereas in more arid regions the vast majority of the river's load is suspended particulate material. Worldwide, the suspended load greatly exceeds the dissolved load. Bedload usually is considerably smaller than suspended load, and is difficult to measure, requiring special traps or tubes sunk flush with the streambed.

This transport of material is of interest to fluvial scientists for many reasons. As so well put by Leopold *et al.* in their still classic 1964 book, *Fluvial Processes in Geomorphology*, "rivers are the gutters down which run the ruins of continents." Material transport is important to the shaping of landforms and to the global cycling of many elements. From our viewpoint of the river, we see that material transport strongly influences channel dynamics.

As discharge varies, rivers alternate between eroding and transporting sediments and depositing them, between scour and fill. Discharge events of progressively greater magnitude, which occur with progressively less frequency, are most effective in this erosion–deposition process and potentially may reshape the channel. However, the evidence indicates that discharges of intermediate frequency and magnitude are most important in channel formation and maintenance—big floods carry more sediment, but are infrequent and therefore do not accomplish as much work over a given time period. Figure 1.4 schematically portrays the amount of water in a channel and the frequency with which each stage occurs.

Erosion is greatest during the time when floodwaters are rising, referred to as the rising phase of a hydrograph. Once floodwaters begin to recede, material settles out of suspension due to decreasing velocity and deposition occurs. Depending on the magnitude of the flood and the composition of the sediments, a riverbed may be eroded to a surprising depth, but return to something very close to its prior state following a cycle of scour and fill. River channels usually show curvature, and it is common for erosion to occur at the outside of bends and deposition to occur on the opposite bank. Many an angler has waded out on the point bar formed by deposition on the convex bank and cast a line into the deeper water and undercut bank of the concave side of a bend in a stream. The greatest deposition occurs in the regions of the channel where flows are slowest, of course. Lateral sandbars in the Grand Canyon of the Colorado River gradually erode away under the constant flows below the Glen Canyon Dam. Occasional floods, which now must be artificially induced, scour sediments from the main channel and, as the flood recedes, these sediments are redeposited along slower sections of the river's margins as gravel and sand bars, creating critical habitat for organisms and campsites for river rafters.

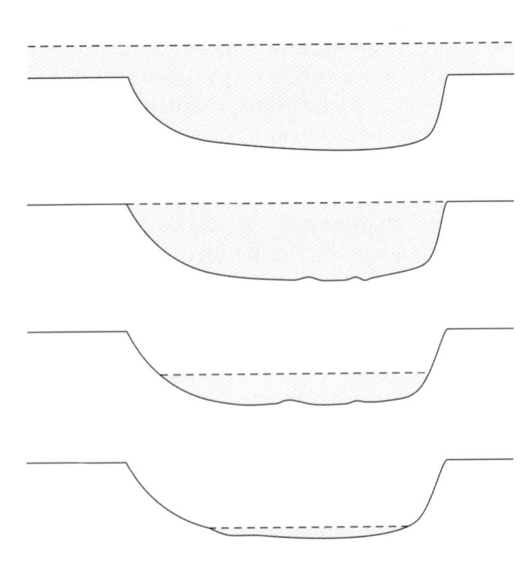

FIGURE 1.4 The amount of water in a river channel, and the frequency with which such an amount occurs. (Reprinted by permission of the publishers from A View of the River by Luna Leopold, Cambridge, MA: Harvard University Press. © 1994 by the President and Fellows of Harvard College.)

THE CHANNEL

Wading along a stream channel, we are immediately struck by the variety of conditions; habitat changes almost constantly in most small streams. In larger rivers, habitat might seem to come in larger packages, although that is probably a trick of scale, which Gulliver, wading along a sixth-order river, would not notice. Walk along an unnatural, channelized river, however, and the uniformity of habitat is both dramatic and appalling.

Variety in habitat is a consequence of the physical form of the river. Channels show curvature, vary in their cross-sectional profile, alternate between riffles, pools and other channel categories, exhibit much horizontal and vertical variation in current velocity, and contain a variety of bottom substrates, which often vary from place to place in accord with channel and velocity. It's no wonder that fisheries biologists assess habitat variability as a measure of stream health (and it's also no wonder that standardizing the protocols to quantify this variation vexes the practitioners who do so).

One of the more remarkable features of running water is its tendency to curve, tracing a sinuous path. Sinuosity can be quantified by measuring the distance between two points, following the deepest and usually swiftest channel section, or *thalweg*, which will itself shift from bank to bank as the river curves. The ratio of thalweg distance to straight-line distance is an index of sinuosity. Rivers of course will bend when they meet greater resistance, such as harder rock formations, but flowing water appears to exhibit a natural tendency to meander. Meandering simply refers to pronounced sinuosity, and it occurs at all scales, from a tiny stream of meltwater flowing over the unobstructed surface of a glacier, to the Gulf Stream current of the Atlantic Ocean. Figure 1.5 illustrates the remarkable regularity of curvature exhibited by rivers of different sizes, when drawn to the same scale.

Riffles, regions of shallow, faster water, and pools, regions of slower, deeper water, are characteristic of virtually all rivers and familiar to us all. Substrate and channel curvature play a role in the alternation of riffles and pools. In relatively straight, gravel bed rivers, a riffle–pool sequence tends to repeat at intervals of roughly 5–7 channel widths. Meandering channels show a similar regularity, although here the pools tend to be associated with the concave bank. In very steep, boulder-filled channels, step pools are separated by riffles, cascades, or small waterfalls. In streams containing substrate in a variety of sizes, finer substrate usually is found in the pools because slower currents there favor deposition of suspended materials. Coarser substrate usually characterizes riffles, where steeper gradient and faster current erode and transport finer particles, leaving behind the larger cobbles and small boulders.

Fallen trees within stream channels strongly influence channel features in regions where riparian vegetation produces large woody debris. This is especially well known in the streams of the Pacific Northwest, where towering Douglas fir, Sitka spruce, and western hemlock influence even relatively large channels. An 80-m tall tree and its root wad lying across the channel can easily shift the flow of even a good-sized river from one bank to the other. Deep pools associated with logjams provide cover for adult salmon and steelhead migrating upstream to spawn, while scour

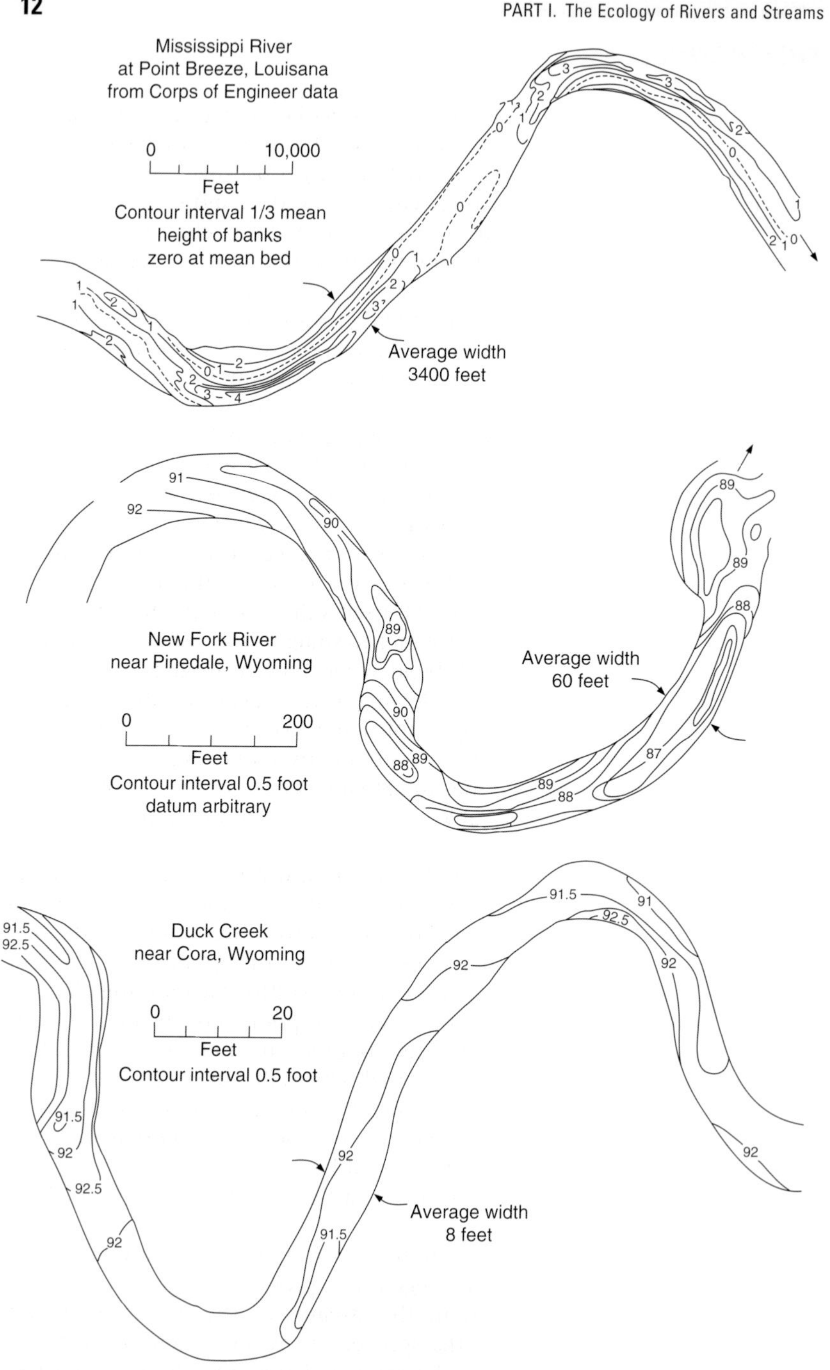

FIGURE 1.5 Plan view of the meander bends on each of three rivers that vary greatly in size. The diagrams are scaled so that the meander lengths are equal on the printed page. (Reprinted by permission of the publishers from A View of the River by Luna Leopold, Cambridge, MA: Harvard University Press. © 1994 by the President and Fellows of Harvard College.)

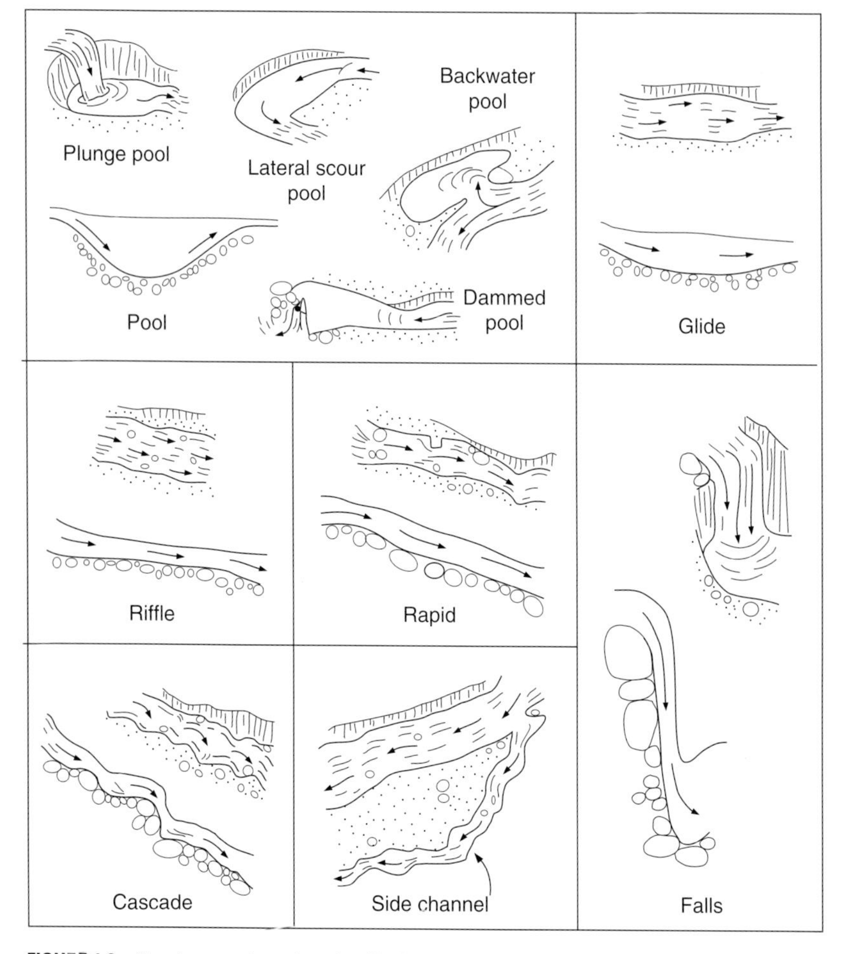

FIGURE 1.6 Fundamental pool and riffle forms used in many stream habitat survey protocols. (From Frissell *et al.*, 1986.)

pools and step pools provide valuable rearing habitat for juveniles. These same blockages cause gravel deposition and create regions of fast current, ideal for certain invertebrates, and so their contribution to habitat complexity is profound. Stream surveys often require a taxonomy of pools and riffles, such as is illustrated in Fig. 1.6.

All of this complexity in channel form results in enormous variation in current from place to place within the river channel. Canoeists and anglers know this, of course. Fly casting requires a heavy line that floats on the water surface, because one really casts the line, not the fly, and experienced fly fishers learn how to "mend" their line to compensate for the varying rates at which different sections of the line travel. Obstructions such as logs and boulders create differences in current, but even in a smooth channel, friction causes water near the sides and the bottom to travel at lower velocities. The rougher the channel bed, the greater the friction. The result is that current velocity decreases from the surface to the bottom of a stream, with the greatest decrease very near the streambed. If one wishes to measure average velocity—in order to estimate discharge, for instance—it is common to make measurements at various depths and compute the average. In shallow streams, a single measurement at 40% of depth (viewed from the bed toward the surface) usually suffices. Quantification of near-bed flows is very difficult in rough natural channels, because flows are so complex, and variation occurs on a scale of millimeters, which is smaller than the size of most current meters.

The substrate of the river channel is determined by geology, but by sorting and transport current exerts a powerful role in determining local substrate composition. Generally speaking, the larger the particle the higher the current velocity required to erode and transport that particle. And as velocity decreases, such as after the passage of a flood peak, large particles will settle out prior to smaller particles. If one examines substrate size in a gravel-bed stream, comparing riffles and pools, or across a transect from the convex to the concave side of a river bend (Fig. 1.7), the influence of current in particle sorting is apparent. It's more complicated, of course, because finer particles can be "protected" by larger particles. In streams rich in calcium carbonate from groundwater, individual particles accrete together, bound by deposits of marl, a precipitate of calcium carbonate. The finest particles, clays in particular, adhere together, and so they can resist erosion even at velocities that might move coarse sand and fine gravel, producing a slick clay substrate that can upset the unwary wader.

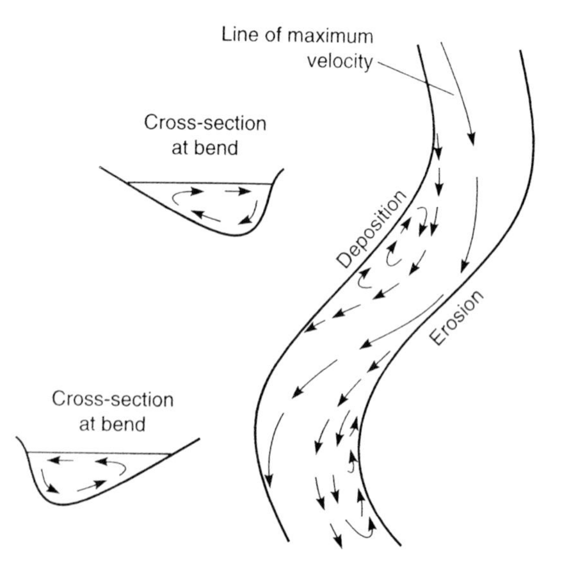

FIGURE 1.7 A meandering reach, showing the line of maximum velocity and the separation of flow that produces areas of deposition and erosion. Cross sections show the lateral movements of water at the bends. (From Allan, 1995, with permission.)

THE STREAM AND ITS VALLEY

Clearly, river channels exhibit many forms, shaped by the dynamic interaction between the landscape through which the river passes and the action of the river itself. Certain variables are "givens" for a river—its discharge is determined by climate, its substrate and sediment load by local geology, and its altitudinal extent by its highlands and sea level. But important features of channel geometry, including curvature, cross-sectional topography, and riffles and pools, are interactive and mutually adjusting, yet within limits that produce the regular patterns.

One intriguing pattern concerns the relationships among width, depth, and velocity as rivers increase in size. If we follow the main stem of a river from its origin as a first-order, headwater stream, to the point where it is sixth- or eighth-order, discharge of course increases. Because discharge is equal to width × depth × velocity, we expect some or all of these to increase as well. It turns out that velocity increases least, and width increases the most (Fig. 1.8). It may seem surprising that velocity increases at all, because slope generally decreases. One reason is that larger, deeper rivers have proportionately less of their volume in contact with bed and banks, and may have finer substrates,

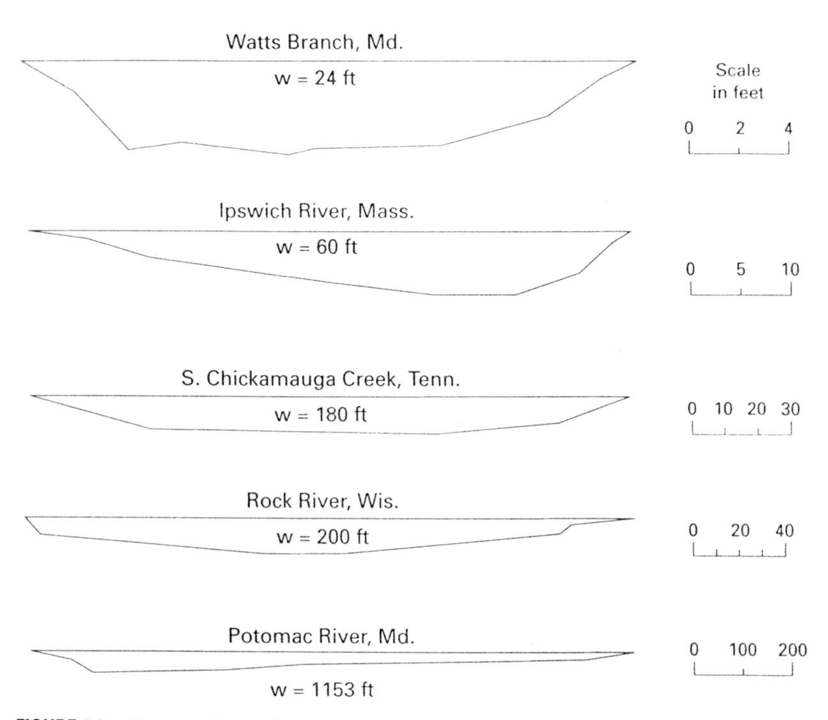

FIGURE 1.8 Cross sections of some natural rivers, scaled so that the width appears to be the same. (Reprinted by permission of the publishers from A View of the River by Luna Leopold, Cambridge, MA: Harvard University Press. © 1994 by the President and Fellows of Harvard College.)

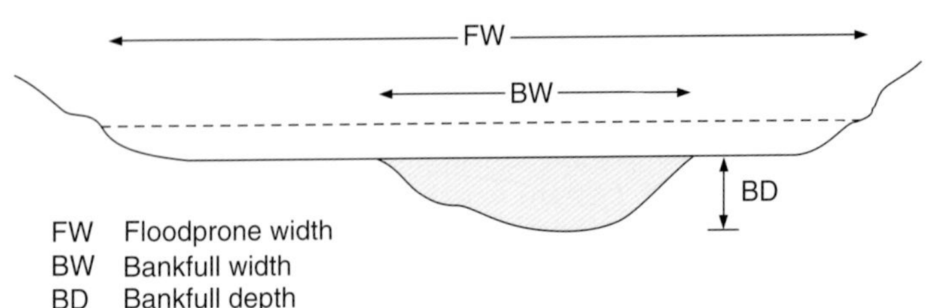

Channel type	Slope percent	Width to depth ratio, BW/BD	Entrenchment ratio, FW/BW
A	4-10	< 12	1-1.4
B	1.5-4	> 12	1.4-2
C	1.5-2	> 12	> 2.2
D	+1.5	> 50	> 2.2
E	< 2	< 12	> 2.2
F	< 1.5	> 12	1-1.4
G	2-4	< 12	1-1.4

Subtype number based on channel bed material

1	2	3	4	5	6
Bedrock	Boulder	Cobble	Gravel	Sand	Clay

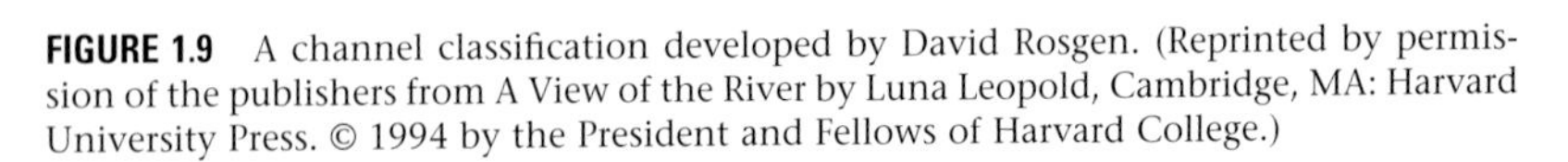

FIGURE 1.9 A channel classification developed by David Rosgen. (Reprinted by permission of the publishers from A View of the River by Luna Leopold, Cambridge, MA: Harvard University Press. © 1994 by the President and Fellows of Harvard College.)

and so there is less friction. At high flow, the Mississippi is more than 1-km wide, 20-m deep, and flowing at up to 1 m per second.

Something similar but a bit different happens when discharge increases at a location, like at your favorite riverbank. Above the floodplain, if the river channel is at all constrained by valley walls, width increases less, depth more, and velocity often most of all. Within the floodplain region, when discharge exceeds the bankfull state, width can increase greatly, more so than depth and velocity.

Channel classification represents an attempt to bring structure to the variety of channel shapes that rivers exhibit. Such a classification can be useful to scientists in characterizing rivers and to managers, if restoration approaches that prove useful for a particular river can be generalized to others of the same type. Figure 1.9 depicts a classification scheme that, while not universally accepted, is both helpful and instructive. A river type—really, a reach of some 100s of meters in length—falls into one of seven main categories, determined by a particular combination of variables: gradient or channel slope, ratio of width to depth, sinuosity, bed material, and the degree of confinement to lateral movement. This last variable is referred to as the entrenchment ratio and is defined as the ratio of floodplain width to bankfull width. The seven classes, A through G, are further subdivided according to six classes of coarseness of the bed material. This results in 42 possible channel types, but the majority fall into about 20 categories, dominated by types A to D.

Clearly, rivers are fascinating objects of study even before we consider their biota and ecology. River scientists seek to understand how channels continually adjust toward a steady state that represents a balance among counteracting forces. At each location within the channel, erosion occurs if the available stress exceeds the resisting force, and sediments are deposited if the converse occurs. In this manner, the river and its channel tends toward minimum work and equal distribution of expenditure of power. But as floods and low flows alternate, interdependent hydraulic variables mutually adjust into any of several perhaps equally probable combinations of values. Because change in forces may occur more rapidly than this adjustment, the river need not be in steady state at any instant, particularly if multiple human actions have altered erosion, regulated flows, and altered climate.

In subsequent chapters we will argue that the principles described here are of utmost importance to understanding the ecological functioning of rivers. The diversity of organisms is strongly influenced by the diversity of habitats, and this physical diversity is linked to the natural complexity and variability of flow and channel features. Many aspects of river ecosystems change predictably along a river's length, making stream order a useful reference point for position. Now to the river ecosystem, which cannot be fully appreciated separate from our just-completed consideration of the fluvial system.

Recommended Reading

Dunne, T. and Leopold, L. B. (1978). *Water in Environmental Planning*. Freeman, San Francisco.
Leopold, L. B. (1994). *A View of the River*. Harvard University Press, Cambridge.

CHAPTER 2

Abiotic Factors

Life will flourish or be limited, depending upon the physical environment. Just as rivers exhibit enormous physical variation in size, slope, discharge, and channel features, flowing waters also vary greatly in physical and chemical factors—collectively called abiotic factors. These include sunlight, necessary for photosynthesis; current, which may bring food to a trout or cause an insect to be swept from the streambed and into the trout's vision; temperature, so critical to activity levels and growth rates; substrate, which creates specific habitat; and a variety of chemical factors from nutrients to contaminants. The possible combinations are infinite, lending a uniqueness to every stream and posing a formidable challenge to ecologists in search of patterns and generalities. In this chapter we continue to develop the physical setting for life in running waters by exploring the principal abiotic factors which, together with hydrologic and geomorphological variables, create the physical template for river ecology.

A useful, albeit simplified, model for considering abiotic factors is shown in Fig. 2.1. This model envisions an optimal range for the particu-

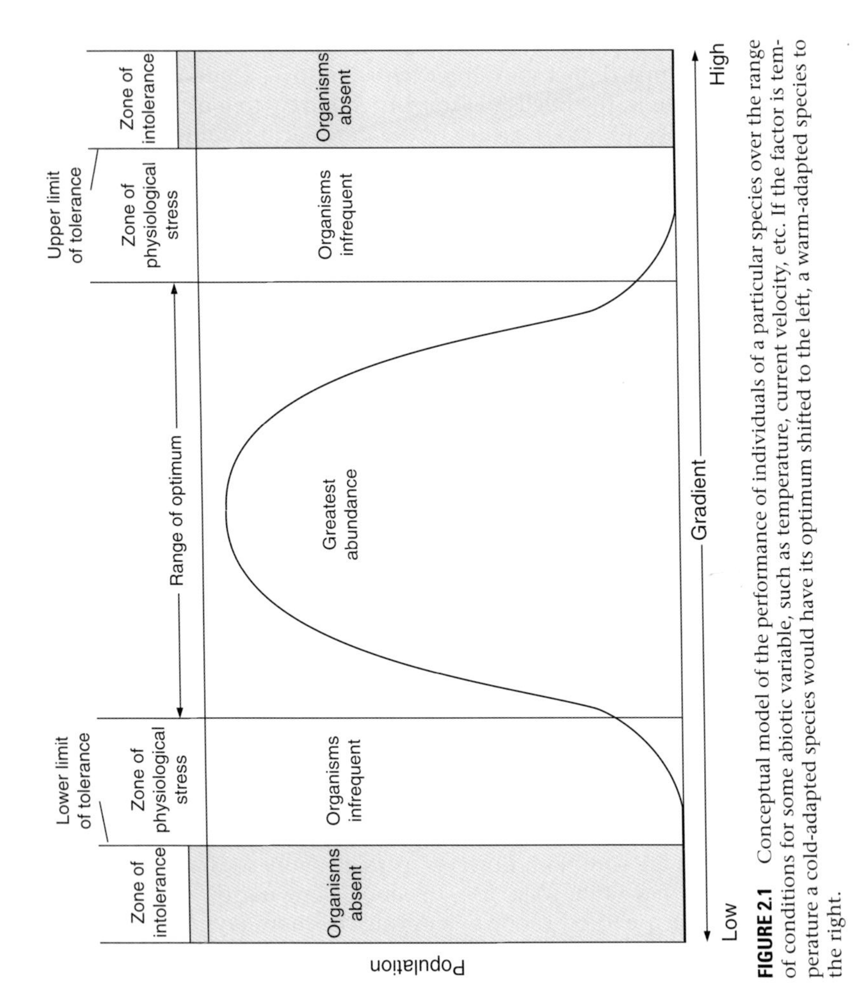

FIGURE 2.1 Conceptual model of the performance of individuals of a particular species over the range of conditions for some abiotic variable, such as temperature, current velocity, etc. If the factor is temperature a cold-adapted species would have its optimum shifted to the left, a warm-adapted species to the right.

lar variable (and it also depends on the particular species, because organisms may adapt physiologically to perform best within different ranges of a factor—as do cold- and warm-water fishes). Outside the optimal range one encounters the range of tolerance, increasing stress, and eventually, lethality. The model is too simple in many ways—the middle part of the curve can be more flat than bell-shaped, the curve may not be symmetrical, and organisms in nature must cope with many variables at once. If any one factor is extreme, we might expect to find few species able to tolerate ambient conditions; diversity and productivity will be low. One of us remembers his first dipnet encounter with the River Duddon, a glorious mountain stream in the English Lake District. Lovely to look at, it is virtually devoid of life, the consequence of too much acid from Europe's contaminated rain. In most natural ecosystems it is likely that no single factor is perennially limiting, as is the acidity of the Duddon. More frequently we see seasonal and spatial variation in the importance of various abiotic factors. Temperature may limit activity during winter, light may be abundant in early spring but limiting after the canopy leafs out, and so on. When all abiotic factors are within tolerable limits for most species, the virtual infinity of possible combination of factors may favor different combinations of species, influence productivity, and help to explain the marvelous ecological variety that makes each stream to some degree unique.

LIGHT

Of all the abiotic factors, light is one of the most straightforward to describe. It is necessary for photosynthesis. Light may also be a habitat factor. Fishes or invertebrates may avoid sunny spots within the stream, perhaps because they are more visible to predators. Sometimes mayfly nymphs congregate in sunny spots, probably to graze on abundant algae. But the influence of light on plant growth is its primary role, one well demonstrated by cave streams. In the unending darkness of a cave, life subsists on a very small amount of imported organic debris and its associated microbes; plants are absent. One finds very few species, often amphipods and springtails, at very low densities. Caves of course are an extreme, used here to demonstrate how severely life is limited by the absence of light. Low light levels are common, however, perhaps more so than most of us realize. We may not often walk, fish, or canoe along narrow stream channels beneath a dense forest canopy, or during the short days of winter, but to do so is a reminder that stream channels can be surprisingly dark.

The availability of light for photosynthesis varies seasonally, of course, and is very much affected by whatever features might prevent its reaching the stream surface and the streambed. Small streams in forested landscapes are shaded by the forest itself, or sometimes by valley walls. If the forest is deciduous, algae may grow vigorously in the spring, and less so after leaf-out, and perhaps again briefly in the autumn, after leaf fall. If the forest is dense and coniferous, such as the rain forests of the Pacific Northwest and southeastern Alaska, small streams can be very dark at all times. Some algae, such as diatoms, which are adapted to low light, nonetheless exhibit some growth under quite gloomy conditions. As we describe in more detail in Chapter 3, autumn-shed leaves and other sources of nonliving organic matter often supplant photosynthesis as the energy base for aquatic food webs in these highly shaded streams. In many small streams, light limitation is a variable factor, of greater significance in some locations or season than in others, and partially offset by the adaptation of some algae to allow for photosynthesizing in low light.

Larger streams tend to be less shaded by forest and valley walls. At a width of 10 or 20 m, a substantial amount of sunlight will reach the stream surface during at least part of the day. These circumstances can favor very high photosynthetic rates. But as the river deepens, light must penetrate deeper water to reach the streambed. Particles in the water column attenuate light; in very turbid rivers this can occur in half a meter or less. Neither algae nor rooted plants will grow well except in shallow embayments, and even plankton (suspended algae) can be seriously disadvantaged. Large rivers are fairly turbulent, and so these plankton are continuously swirled into shaded depths. From time to time they may be swept into shallow well-lit waters and briefly photosynthesize, almost likely a swimmer gasping for air, before again being swept into darkness. As rivers increase in size, plenty of light reaches their surfaces, but the river itself shades its depths, and so light reaching the streambed may become as limiting as in a densely forested headwater stream.

CURRENT

Current is the defining physical variable of rivers and streams. It seems to fit well with the model of Fig. 2.1; many organisms occupy a certain range of water velocities and appear stressed by extreme values. But current is a complex physical variable because it interacts with many other abiotic and biotic factors. Organisms must shelter from the current, develop attachment devices, or perform energetically costly work to hold their positions. The flow of water transports oxygen to organisms, which may aid their respiration. The same flow carries carbon dioxide and nutrients to plants, refreshing their requirements for photosynthesis. As Chapter 6 describes and every fly fisher knows, current conveys food to waiting organisms, such as a trout sheltered behind a rock, ready to dart out when a sufficiently enticing morsel drifts by. Water velocity also shapes the streambed habitat, eroding substrate here and depositing it there, creating riffles, glides, and pools. The speed of the current determines if fine particles will settle to the surface of the channel, causing a form of habitat degradation where silt fills the interstices and coats the surfaces, referred to as *embeddedness*. The importance of current, coupled with the complexity of its influence, has led to much study of this key variable of running waters.

It may seem simple enough to measure current, and then study its effects, but, unfortunately, the more we study water movements the more challenging the subject becomes. Water velocity increases when rain or snowmelt delivers more water to the channel, which may occur with seasonal regularity or episodic unpredictability. Current differs between the inside and the outside of bends, between riffles and pools, and with every log, rock, or other obstruction. These subjects are further discussed in Chapter 1. The benthic ecologist, one who studies the ecology of the streambed where virtually all algae and invertebrates, and many fishes, are found, is most interested in the velocity at or very near the surface of the streambed. Unfortunately, this is exceedingly difficult to measure because robust instruments capable of distinguishing current at the scale of millimeters simply do not exist. In addition, we know from using dyes and making such measurements as we can, that fine-scale current velocity is incredibly heterogeneous. Theory tells us that the layer of water molecules in contact with the surface of a stone or other fixed object does not move, whereas subsequent layers do; the friction between moving and nonmoving molecules creates turbulence and shear, which affects organisms. Measurement of flow at the millimeter scale can only be approximated in carefully controlled laboratory settings without the structural complexity that comes with stones of varying size and texture. But progress is being made, and the adaptations of the organisms tell us much about the ways in which current influences their lives.

Stream-dwelling animals exhibit a wide variety of adaptations to withstand current, and perhaps use it to advantage. Diverse attachment devices are found among invertebrates, including silk and other sticky secretions, hooks, claws, and suckers. A flattened profile also may enable organisms to avoid some shear stress, although this is uncertain. Black fly larvae (Fig. 2.2) tolerate very high velocities. Often they form dense aggregations in very

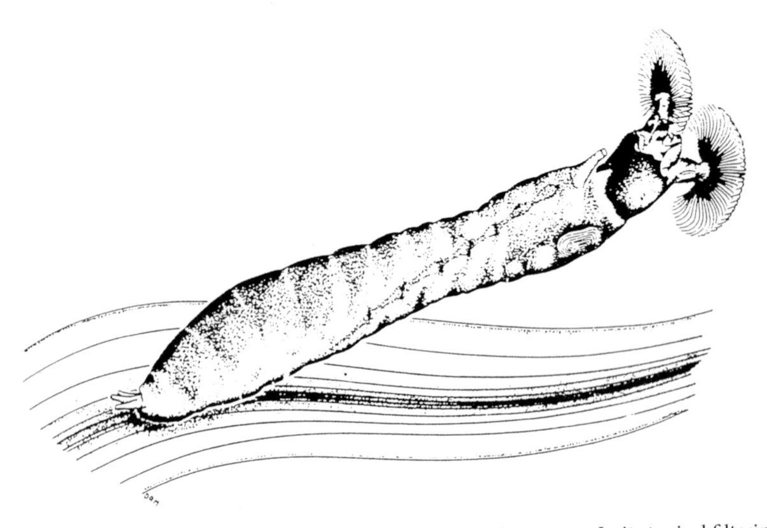

FIGURE 2.2 Adaptation of the larval black fly to life in swift currents. In its typical filtering stance the larvae, attached by hooks embedded in silk, is forced nearly horizontal by the current. Its body is slightly twisted, as can be seen by following the line of the ventral nerve cord. Filtering fans on the head capture particles as small as bacteria. (From Allan, 1995, with permission.)

fast, smooth currents, even on vertical faces. These larvae attach with circlets of outwardly directed hooks on the posterior proleg, which they embed in a mat of silk that they spin onto the stone surface. Should the animal become dislodged, an anchor line of silk allows the animal to regain the surface. Black fly larvae obtain their food from the water, which they filter with setae on fans attached to their heads, termed cephalic fans. Generally speaking, the more water that passes through the fans, the more food is filtered, and so the ability to live in fast currents provides real benefits.

Fishes exhibit a variety of adaptations to life in fast-flowing water. Fast-swimming fishes and fishes that swim in fast currents are streamlined and rounded in cross section, like a trout (Fig. 2.3a). They make a sharp contrast with the deep-bodied sunfish, found in slower rivers, pools, and lakes, and better suited for maneuvering than speed. Many stream-dwelling fishes live on the bottom, using its roughness for shelter, and exhibit a particular suite of adaptations. Many have a reduced or no swim bladder, having forsaken buoyancy. Bodies are flattened in the dorsal–ventral plane, eyes are dorsal, and pectoral fins often are enlarged and muscular to assist in holding position against current (Fig. 2.3b). Experiments in laboratory flumes show that as current velocity increases, such fishes may arch the back, as well as employ their muscular pectorals, to prevent slippage. Darters, sculpins, catfishes, and other groups are examples of fishes with such adaptations. Attachment suckers are found in a number of vertebrates, including the infamous lamprey, and amphibian tadpoles.

These examples illustrate that both fishes and invertebrates have plenty of ways to withstand current. Both also can be adept at avoiding current. Trout have been shown to favor feeding locations (where they "hold station" in the fish biologist's parlance) in slow-moving water adjacent to high current velocities. They expend little energy in holding station, while maximizing the number of food particles passing their vision. It once was thought that a zone of little current existing near a stone surface permitted an invertebrate with a flattened profile to avoid current forces. This now seems unlikely, but the flattened shape surely allows insects to move under and between the stones, and thereby avoid some current.

How much does current affect algae and microbes? Owing to their small size, these organisms pose even greater difficulties for experimental study. We know that floods can be devastating to attached mosses (rolling stones may land upside down, where mosses die for lack of sunlight; hence, "a rolling stone gathers no moss") and algae, which are scoured and sand-blasted from substrates during high-flow events. But flowing water provides a fresh supply of nutrients and removes harmful metabolic by-products, and so moderate current probably always is beneficial. Studies in laboratory flumes, and in some stable natural systems, document a continuing buildup of the algal mat, with a top layer of growing and dividing cells attached to underlayers of light and nutrient-starved senescent cells. As the mat grows thicker, its mass increases, the layers of dead cells become greater, and eventually the mat sloughs away. Somewhat faster currents

FIGURE 2.3 (a) A streamlined fish adapted for swimming in the water column against the current (trout). (b) A benthic fish adapted via its enlarged pectoral fins for holding position against the substrate (sculpin). (Photos by W. Roston.)

likely renew nutrients to greater depth within the mat. Yet higher currents will cause it to slough away. This pattern has been well documented in a desert stream in Arizona.

SUBSTRATE

The geologic parent material, transported, sorted, and deposited by the current, determines the inorganic substrate of a river. Wood, originating in the riparian and floodplain, transported, deposited, and perhaps partially buried, is the major organic substrate. All of these streambed components are potential obstructions that create heterogeneous currents and act as surfaces for algal and microbial growth. Look carefully at the bed of streams, or wade some distance and sense the bed through your feet. Differences in the size, shape, and roughness of stones are apparent to us; imagine what different habitat this constitutes for aquatic insects. The boulder that bruised your toe is a wonderful shelter for some fishes. Wood, from twigs to tree trunks, is abundant in many streams, although not where the riparian forest is naturally absent or has been cleared. A few small branches, together with some autumn-shed leaves and some sand, can create a nice little habitat for stoneflies and crane fly larvae. A downed tree and its root wad can create a deep pool in a large coastal river, which is likely to hold a pair of steelhead, and thus the interest of some angler.

Substrates of inorganic particles can be easily classified on the basis of particle size. A commonly used geological scale (Table 2.1) enables quantification of a sample of substrate (and informs us that small pebbles are larger than coarse gravel). Boulders are more a feature of mountain streams, where they combine with steep gradient to form step pools (Fig. 11.6). Cobbles and pebbles might typify a trout stream in the Rocky Mountains, and sand often dominates the bed of lowland rivers. However, northern Michigan trout streams (see description of Au Sable River in Chapter 6) are swift and sand-bottomed, because sand is the available material. Muddy and silty substrates are typical of very slow currents, and so they are most common in backwaters and side channels. Clays, in contrast, can form the

Table 2.1 The Classification of Mineral Substrates by Particle Size[a]

Size category	Particle diameter (range in mm)
Boulder	> 256
Large cobble	128–256
Small cobble	64–128
Large pebble	32–64
Small pebble	16–32
Coarse gravel	8–16
Medium gravel	4–8
Fine gravel	2–4
Very coarse sand	1–2
Coarse sand	0.5–1
Medium sand	0.25–0.5
Fine sand	0.125–0.25
Very fine sand	0.063–0.125
Silt	< 0.063

[a] Note that each size category represents a doubling of the next smaller size category.

bed of quite swift streams if the geology is favorable. This is because clays are adhesive, despite their small size. The result can be a slick and very smooth surface in rapid current, and anyone wading in a clay-bottom stream had best be wary of such conditions.

Pools and riffles provide further evidence that current influences bed material through daily flow as well as by cycles of scour and deposition. Finer materials accumulate in pools during periods of low flow, because current slows below the velocity at which fine particles settle. In riffles, flows are swift enough to prevent settling and to pick up ("entrain") most fines.

Flowing water always transports some inorganic material, from the clearest mountain stream to the famously sediment-rich Yellow River of China. Indeed, although something of a digression for this book, rivers are the great levelers of landscapes, "the gutters down which run the ruins of continents" as Luna Leopold put it. As described in Chapter 1, rivers scour their beds when in flood, increasing their sediment loads, then fill their beds as they subside; this is a never-ending cycle in rough equilibrium. Starve a river of its sediment load, and it will downcut in its erosive quest. This sometimes happens below dams, where the entering river deposits sediments within the reservoir, and then exits hungry for more. River beds may be downcut and become armored, almost like concrete, with significant loss of habitat complexity.

Organisms may display adaptations to particular substrates, and this is best seen for macroinvertebrates. Certain species are well known to be associated with moss, or sand, or with cobble–pebble mixtures or large boulders. Numerous studies have attempted to more finely gage substrate preferences down to the exactitude of Table 2.1. Some species show a definite affinity for, say, large cobbles in fast currents (a number of species of heptageniid mayflies come to mind), but often the associations are best represented as a statistical trend. Sand seems to be a poor substrate for many macroinvertebrates, although there are some specialists, such as burrowing mayflies, and a host of very small, mostly interstitial taxa, some of which are discussed in Chapter 17. Generally, however, diversity and productivity are reduced in sand substrates, relative to pebble–cobble substrates. Its instability makes sand a poor place for invertebrates to live and for their food supplies to be plentiful.

Any substrate in a stream can provide a surface on which algae will grow, and any roughness of the streambed can provide a depositional pocket in which organic matter and microbes can accumulate. Thus substrate is critical to the nourishment of stream food webs (Chapter 5). Stable surfaces generally support greater algal biomass. And roughness of the streambed creates heterogeneity in local conditions, thus enhancing diversity.

Finally, one must recognize the importance of wood as a substrate and as a determinant of channel structure. It is difficult to say how important wood was to different kinds of rivers in presettlement times. In many places today, removal of riparian forest and control of flooding have robbed rivers of their supplies of large woody debris. But many lines of evidence suggest that wood is of general importance. Wood and debris jams are plentiful in headwater streams of the northeast as well as old growth rivers of the Pacific Northwest, where hardly a pool can be found that isn't created by a tree that has fallen into the stream. In coastal rivers of the southeastern United States, sandy bottoms are relatively unproductive; invertebrate production in these systems is strongly associated with snags, and so are fishes. Historical records indicate that wood was much more plentiful in stream channels in presettlement times, when it likely created habitat, shaped channels, and served as a substrate for a variety of organisms.

TEMPERATURE

Stream temperatures are important for many reasons. Because few organisms of running waters are warm-blooded, the metabolic rate of virtually all organisms and overall system productivity are strongly influenced by water temperature. Physiological adaptations enable trout to be active and grow at relatively cool stream temperatures, while smallmouth bass perform optimally at somewhat higher temperatures. Mayflies in cold, northern climates often pass the winter as diapausing embryos and complete one generation per year, whereas in warmer, southern streams the egg stage is short, nymphs are active year-round, and several generations are likely to occur during the year. Climate warming predicted over the next century thus is likely to change the ecology of northerly streams as warm-adapted species replace cool-adapted species; system productivity is likely to increase as well.

There is reason for concern about loss of cool-adapted species under future climate warming. Some species will certainly shift their ranges northwards, but basin boundaries and east–west drainages may limit their opportunities. Species with very specific temperature requirements, termed *stenotherms*, may be out of luck. Cool groundwater springs often are home to cold-water stenotherms with relatively limited distribution, and these species are particularly vulnerable.

The annual range in temperature in most streams many of us will visit is between 0 and 25°C, although subtropical and tropical streams often reach 30°C and some desert streams reach 40°C. The upper range of water temperature at which organisms can survive depends upon their history and temperature adaptations. Cold-water fishes cannot long survive temperatures above 25°C. Most warm-water fishes have upper limits near 30°C, although a few desert fishes can withstand nearly 40°C. Specialized bacteria such as occur in the Yellowstone hot springs survive an amazing 75°C, which apparently is the record.

Water temperature is affected first by air temperature, and then by the mix of surface and groundwater, degree of shading, and stream size. Streams at more northerly latitudes and higher elevations are of course cooler because their air temperatures and solar radiation are less, compared to southern and lower elevation streams. The temperature of groundwater is usually very close to mean annual air temperature, and so spring brooks are very constant in temperature and strike us as cool in the summer but warm in the winter. As surface and shallow subsurface flows become a larger part

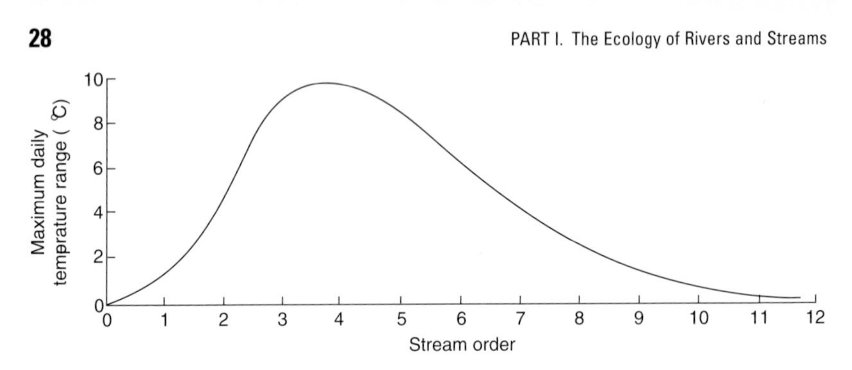

FIGURE 2.4 Maximum daily temperature range in relation to stream order in temperate streams. (From Allan, 1995, with permission.)

of the stream's hydrologic budget, it begins to take on the seasonal signature of mean monthly air temperature. Small streams warm by day due to solar heating of the streambed and the relatively small mass of water in the stream, and cool by night, creating a day–night temperature change that can be quite pronounced. Small streams high in the Rocky Mountains can warm to 15–18°C during the day, and cool to 4–6°C at night, due to the daytime radiation warming runoff from snowmelt. Water has much more thermal inertia than air, so it undergoes less violent "swings." And as the stream becomes a large river, the inertia of the mass of water buffers against such changes, so that day–night changes eventually become very small (Fig. 2.4). Lastly, shaded streams really are cooler, and streams denuded of their riparian vegetation really are warmer, because of the sun's radiant energy reaching the stream surface and streambed. A few kilometers of shade are sufficient to make a noticeable difference and determine whether trout can occupy that stream section. Fishes will select microhabitats based on very small temperature differences, presumably because of the physiological advantage of slightly lowering their metabolic costs of holding station in the current.

CHEMICAL FACTORS

The chemistry of a river is determined foremost by the geology of its catchment, but rain and human activity can provide significant inputs as well. There is a great deal of variation, especially among small, headwater streams, and many chemical variables one might consider. For simplicity we will describe dissolved oxygen, alkalinity, nutrients, and human contaminants.

Oxygen, a critical requirement of most organisms, usually is not a limiting factor. It is abundant in the atmosphere, dissolves readily in water, and becomes limiting only when water does not circulate between surface and deeper layers, and biological activity is very high. Small, turbulent streams typically are at saturation. Large, smoothly flowing rivers may experience some oxygen depletion near the bottom, if metabolic activity there is high. The saturation value decreases as water warms, and it is reduced at high elevations, where atmospheric partial pressure of oxygen is less than at low elevations. Very high levels of organic inputs, such as human or animal sewage, or waste from processing foods, can generate such high bacterial metabolism that all oxygen is used up. One famous example occurred in the river Thames in 1858, known as the "Year of the Great Stink," when parliament hung wet sheets over the open windows to try to cope with the smells. Wastewater treatment plants have indeed accomplished some good.

Alkalinity is a measure of the kinds and amounts of compounds that shift the pH of water into the alkaline, as opposed to acid, range. A pH of 7 is neutral, whereas lower values are acid and higher values are alkaline. Blackwater streams are naturally acid, due to decomposing plant matter, and streams can become acid when exposed to acid rain or the crushed rock created by some mining operations. Streams draining fertile soils often are slightly alkaline, and chalk streams can be strongly alkaline. Alkalinity is controlled primarily by the amount of bicarbonate and carbonate in water, which explains why chalk streams, so named for their limestone (calcium carbonate) rock, register high alkalinities. In general, more alkaline waters are more biologically productive, and more acidic waters are less productive. Liming lakes, to increase their productivity and neutralize their acidity, is common practice in some areas highly impacted by acid rain. It is occasionally practiced with rivers as well, although continuous flow makes it less practicable.

Nutrients are elements required for life by some organism, but especially by plants and microbes as they grow, producing the biomass that will support the food web. The list of elements and vitamins necessary to sustain life is quite long; usually we focus principally on nitrogen and phosphorus, well known for their capacity to makes lawns (and lakes) green.

Laboratory flume experiments demonstrate that changes in nutrient concentrations can affect the species composition of algae as well as total biomass produced. When light is adequate and current not too swift, long plumes of dense, filamentous algae signal the presence of excessive nutrients. Indeed, it seems probable that many of the streams and rivers carry sufficient nitrogen and phosphorus to stimulate substantial, even excessive, algal growth. Small streams draining relatively insoluble rock—probably in regions not well-suited for farming, either—are likely to have low concentrations of nutrients. But because algae in streams can be limited by any of a number of factors—shade, spates, shifting substrate, temperature, or grazing—we tend not to see eutrophication in streams and rivers to the extent we see it in lakes. But when a river enters a reservoir and currents slow, permitting sediments to settle and water to clarify, dense algal blooms inform us that the water contains a substantial nutrient load. The impact of those nutrients can be very serious because after an algal bloom comes death, sinking, and decay. The result can be an anoxic zone in the deeper waters caused by bacterial metabolism, and no different than what would occur if great quantities of sewage were dropped into a lake or sea. At this writing, the nutrient-laden waters of the Mississippi are having just this effect in the Gulf of Mexico, and a large area of the Gulf floor, once so productive of life and so critical to Louisiana shrimpers, is becoming a great dead zone.

Human activities are today a great contributor to the nutrient load of rivers. Phosphorus reaches streams mainly from sewage inputs and soil erosion. Waste water treatments plants have been effective at reducing this source of phosphorus inputs. Improved agricultural and "best management" practices have had some success in reducing soil erosion, which carries with it phosphorus attached to soil particles. This highlights the difference between point source pollution, such as sewage, which enters the river at a point where technology can be focused, and nonpoint source pollution, which enters from multiple, diffuse sources and so requires the combined efforts of a great many individuals to bring it under control. Today, nonpoint source pollution is the dominant contributor to the nutrient load of rivers.

Nitrogen also is an important nutrient, although for some time it received less attention than phosphorus. As a rough generalization the evidence suggests that phosphorus enrichment leads to eutrophication in lakes, whereas nitrogen enrichment has the corresponding influence on oceanic waters. In the early 1970s concern for a dying Lake Erie stimulated passage of the Clean Water Act and wastewater treatment that emphasized phosphorus removal. Presently, concern for oceanic dead zones is surfacing as a critical issue. Unfortunately, nitrogen will be difficult to control. Through the manufacture of fertilizer and burning of fossil fuels (which produce nitrogen oxides), humans now dominate the world nitrogen flux. In addition, there are very large atmospheric inputs, from fossil-fuel plants, car exhaust, and so on. Exchanges take place on a very large scale, driven by winds and the transport of large rivers. Those shrimpers in Louisiana are affected by farmers in Iowa, some of whose nitrogen fertilizer reaches the Gulf of Mexico, along with the rain-transported deposition of nitrogen oxides from all of our automobiles.

Chemical stressors constitute a final, unhappy category of abiotic factors in rivers. As with nutrients, some are nonpoint sources, while others are point sources. Acid rain has lowered the pH (increased the acidity) of many streams in western Europe, Scandinavia, the northeastern United States, and some regions of the Rocky Mountains. Partly a point source in origin (sulphur dioxide and nitrogen oxides from electricity-generating plants burning fossil fuel), partly nonpoint (nitrogen oxides from auto exhaust fumes), its delivery via the rain is diffuse. Some soils have enough alkalinity to prevent much effect, but when acid rain falls in catchments of hard igneous rock, the chemical signature of streamwater is acid, just like the rain. In the example of the River Duddon, mentioned previously, life in the stream is impoverished.

Mining is another source of acidity and toxic metals that together have poisoned many streams in Appalachia, the Rockies, and elsewhere. The crushed rock left over from extraction of the ore is easily weathered by rain. This generates acids and metals in solution that can be extremely toxic, virtually extinguishing life in even substantial segments of rivers. In

1997 there was much justified concern about a proposed gold mine just outside of Yellowstone. The mining itself raised concerns, but the biggest issue was the storage of waste rock or mine "tailings" in a location where failure of the containment mechanisms would have released contaminated water into Clark's Fork of the Yellowstone, a pristine trout stream. Mine developers proposed a dam as a barrier to release of water from the tailings pond, but, as one conservationist put it, that barrier would have to last "until the end of time" to ensure the river's safety.

Sad to say, one can continue with the list of effluents and chemical contaminants from poultry and livestock farms, agriculture, construction, and diverse industrial activities, both large and small. Our understanding of pollution has increased, stronger laws are on the books, and no one wants to live beside a polluted stream or river. The science is clear and a number of legal and policy tools are in place. However, the need for awareness and enforcement are as great as ever.

Recommended Reading

Allan, J. D. (1995). *Stream Ecology*. Kluwer Academic, Dordrecht, The Netherlands.

CHAPTER 3

Energy Resources

INTRODUCTION

Chapters 1 and 2 "set the stage" for our understanding of the biological communities found in rivers and streams. The various physical and chemical factors in a river act as limiting or controlling factors to biological activity because they influence which species can survive, and thrive, in a particular location. The amount and diversity of energy supplies are another important factor that determines which and how many organisms will be found in a particular stream. In this chapter, and elsewhere in this book, we will explore the various sources of energy for aquatic food webs—how it is produced, and the different ways it becomes available to consumer organisms.

An important first distinction is that between *autotrophs*, which produce their own energy from inorganic matter, and *heterotrophs*, which derive their energy from autotrophs. With a very few exceptions, such as certain deep-sea bacteria that use a process of *chemosynthesis* to derive energy from hydrogen sulphide, autotrophs are plants. Organic carbon compounds are formed out of carbon dioxide and other inorganic matter, capturing the energy of sunlight via the process of *photosynthesis*. We call this *primary production* because it creates new organic matter from inorganic precursors. All organisms unable to synthesize energy from inorganic matter, obtain energy by consuming the organic matter formed by primary producers. These are heterotrophs or consumers, and this includes animals, bacteria, fungi, and protozoans. When consumer organisms grow and reproduce, adding biomass to their populations, we call this *secondary production*. Virtually all life on earth derives its energy from the sun, via primary production. Autotrophs and heterotrophs use that energy to do metabolic work, and in the process convert the energy contained in organic carbon compounds back into inorganic matter.

In riverine food webs, all energy originates from primary production, but not necessarily from aquatic plants. In many instances, organic matter from terrestrial primary production enters the stream, is utilized first by microbes, and then, as a microbe-rich amalgamation, is consumed by other heterotrophs. The flow of energy through the food webs of stream communities is complex. In many types of rivers and streams, aquatic plants are important energy sources; in others, terrestrial plant production is very important and can be the dominant energy supply to running waters. The important categories of aquatic plants are the *algae*, the *mosses*, and the true vascular plants usually referred to as *macrophytes* because of their large size. Important terrestrial sources of energy include leaves, fruits, or other terrestrial plant materials which fall into or are blown into the stream. This nonliving organic matter is referred to as *detritus*, or *particulate organic matter* (POM). This same material also decomposes on the forest floor, enters the soil water, and eventually enters the stream as dissolved organic matter (DOM). Because aquatic plants also break down into POM and release DOM, it can be quite difficult to ascertain where this matter originated. We would like to know, however, whether the energy base of the riverine food web came primarily from aquatic primary production within the stream itself, or whether it was heavily subsidized by energy inputs from outside. Stream ecologists refer to energy produced within the stream channel as *autochthonous* production. Organic matter produced outside the stream which falls, blows or leaches into the stream channel is known as *allochthonous* production.

To better understand the complexities of aquatic food webs, we also need to define how energy moves from one food web level to another, or in other words how the organisms feed. Principal feeding or *trophic* roles are *herbivory* (plant feeders), *carnivory* (organisms which feed on other animals), and *detritivory* (detritus feeders). *Omnivory* refers to animals feeding on more than one level. A close study of Fig. 3.1 will help readers understand these terms and their relationships to each other.

The terms autotrophic and heterotrophic are also used by stream ecologists to describe the energy base of a stream as a whole. Ecologists use the production of oxygen by photosynthesis within a stream reach as an indicator of the amount of organic matter produced and the loss of oxygen by respiration as an indicator of the consumption of primary production by animals and microbes. Thus, if a given stream reach produces more oxygen

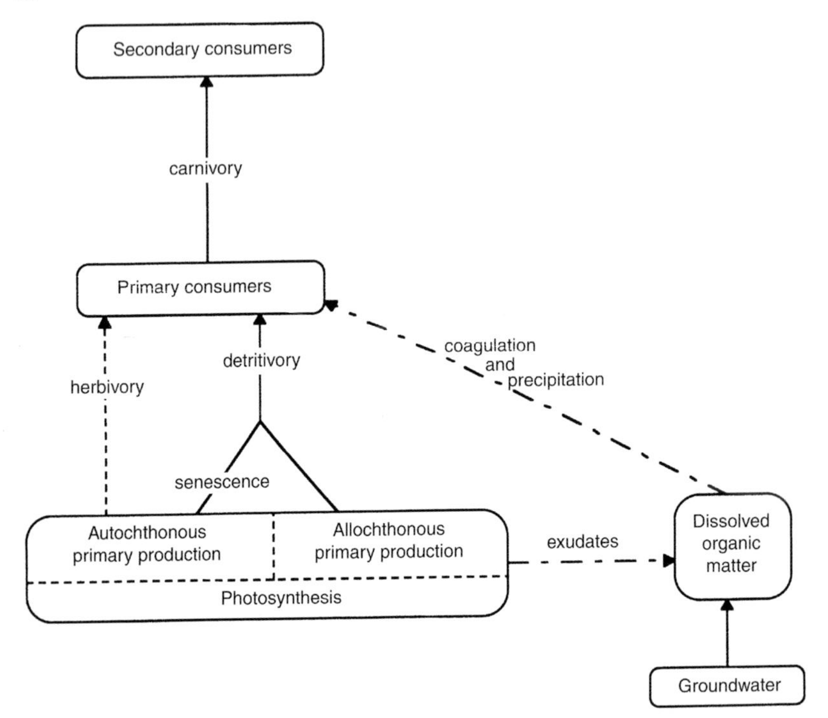

FIGURE 3.1 The sources of energy in streams (autochthonous and allochthonous primary production; DOM), energy pathways (dashed arrow = autotrophic; solid arrow = heterotrophic; dash-dot arrow = chemotrophic), and feeding processes (herbivory, carnivory, detritivory, omnivory) within stream food webs. Coagulation and precipitation are the processes by which DOM is converted to POM, and decay is the mechanism by which aquatic plant material is converted to POM after death.

than it consumes by respiration, it is autotrophic—it produces more energy than it uses. Conversely, a heterotrophic section of river consumes more oxygen than is produced within it, and must depend on an allochthonous source of energy to provide adequate energy for the stream community. Obviously, food webs in streams, or the pathways that energy follows as it is produced, utilized, and degraded to basic elements, can be simple or quite complex, with the latter being the more common.

INSTREAM (AUTOCHTHONOUS) ENERGY SOURCES

Now let's look in more detail at each of these energy sources. Plants living in rivers include the microscopic algae (although large colonies can be seen with the naked eye) and cyanobacteria (photosynthetic bacteria formerly known as blue-green algae; see Chapter 12) and larger plants or macrophytes. The latter includes the mosses and liverworts, and the true vascular plants, or angiosperms. The algae and cyanobacteria are the most important direct source of energy to heterotrophs, as most macrophytes are unpalatable, although they do contribute to instream production of DOM and POM.

In streams and rivers with sufficient nutrients (see Chapter 2), a suitable stable substrate, and adequate sunlight, algae will proliferate. They occur as single cells, colonies, or as long filaments. Three groups are particularly important and are described in more detail in Chapter 12. Green algae are common, generally palatable, and often conspicuous as long, green strands usually attached to solid objects, such as rocks, and trailing in the water current (Fig. 3.2). These are commonly, but mistakenly, called "moss" by many people. The filaments are usually bright green and can grow to several meters in length under favorable conditions. They usually take on a brownish color as the filaments mature, either from senescence, or from being covered with unicellular algae called diatoms. Diatoms are truly the "grasses of the water," numerous in both numbers and variety and the most important autochthonous energy source to stream food webs. They are unicellular, but may occur singly or in filaments or groups (Fig. 3.3). When you pick up a stone or other object from the streambed, it usually has a brownish, slippery coating on the sides exposed to sunlight. This coating is composed of billions of diatom cells and, incidentally, is what you usually slip on when you lose your footing while wading. Cyanobacteria can be unicellular or colonial, often secrete a mucilaginous coat, and have the ability to "fix" atmospheric nitrogen into other forms of nitrogen that can be utilized by other autotrophs. They probably are a less important energy source to stream food webs than are green algae and diatoms.

That slippery film on the surface of rocks and other hard substrates is referred to as a *biofilm*, and while algae are an important component, we

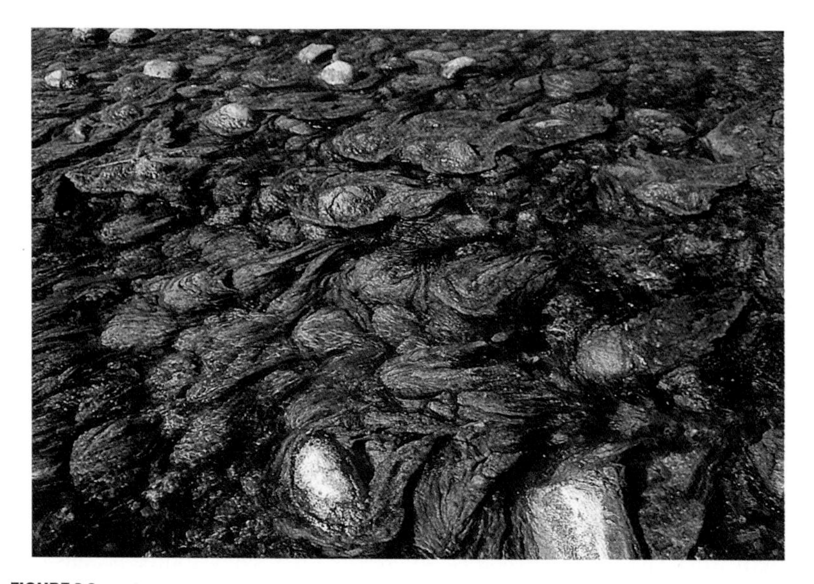

FIGURE 3.2 Filamentous green algae, San Juan River, New Mexico. (Photo by C. E. Cushing.)

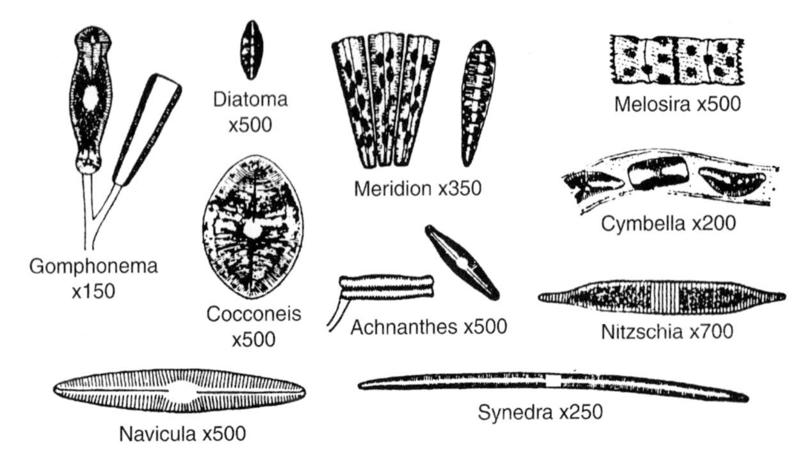

FIGURE 3.3 Common diatoms found in streams. (From Hynes, 1970, with permission.)

now know these biofilms are very complex microenvironments. Some algae, often those in the lower layers, die and their cell contents leak into the surrounding matrix, from which compounds diffuse away very slowly. This provides an environment in which heterotrophic bacteria can obtain energy from the breakdown of dead cells. In addition, actively photosynthesizing cells produce exudates of organic compounds that also nourish bacterial growth. The bacteria, in turn, convert these organic compounds back into inorganic compounds, which are needed by algae for continued photosynthesis. As Fig. 3.4 illustrates, both autotrophs and heterotrophs benefit from their close association within the biofilm. Any external source of DOM, perhaps from groundwater upwelling from the substrate, is potentially available to the heterotrophic bacteria, and so the biofilm can be seen to be an incredibly important region of energy production. Autotrophy often dominates within biofilms, but they will form under very low light or even in the dark. With the addition of microconsumers, including protozoans, nematodes, and tiny crustaceans, the biofilm becomes an entire ecosystem within itself.

Stream ecologists in North America traditionally have used the term *periphyton* ("peri" = around, "phyton" = plants) to describe this complex system, emphasizing the role of plants. In Europe, the German word *Aufwuchs* refers to the same entity. It is a vital community in the ecology of rivers and streams and an important food source for stream invertebrates.

In larger rivers, backwaters, and embayments, algae may be found suspended in the water column. These suspended algae are termed *phytoplankton* ("phyto" = plant, "plankton" = free floating), a term usually associated with the algae found in lakes and oceans. There do not appear to be any phytoplankton that are unique to rivers, and they are sparse or absent from fast-flowing streams, where the only likely source is cells dislodged from the stream bottom. Substantial amounts of suspended algae

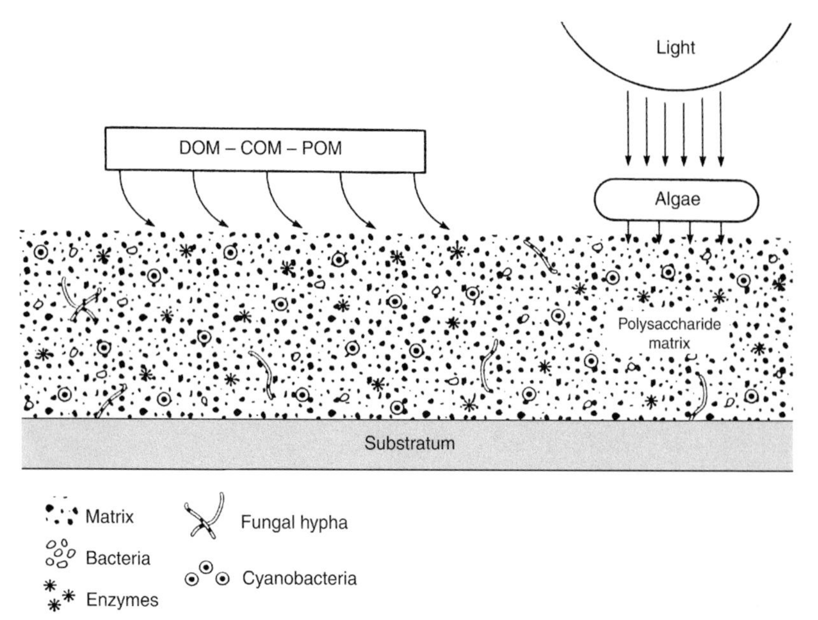

FIGURE 3.4 The biofilm found as a surface "slime" on stones and other submerged objects in streams. A polysaccharide matrix produced by the microbial community binds together bacteria, algae, and fungi, and is inhabited by protozoans and micrometazoans, which consume this material. Within the matrix, extracellular release and cell death result in enzymes and other molecular products that are retained due to reduced diffusion rates. (From Allan, 1995, with permission.)

may be found immediately downstream from reservoirs or lakes; however, they will not proliferate under these conditions and are usually lost to the stream within fairly short distances. They have been found to provide a rich source of energy to filter-feeding organisms (see Chapter 4), resulting in large concentrations of these animals for short distances below lakes or reservoirs.

Whenever current washes algal populations downriver more rapidly than they can reproduce, growth of phytoplankton populations is precluded. Hence slow currents and long river sections favor the greatest abundance of phytoplankton. Embayments and floodplain lakes, which can serve as reservoirs for phytoplankton, also can be important. Light and nutrients also can be limiting to river phytoplankton. In turbid, well-mixed rivers, adequate light for photosynthesis might penetrate much less than 1 m, yet if the river is 10-m deep, turbulence will ensure that the algal cell spends much of its time at light levels too low for photosynthesis. In general, the more lakelike the river, the more phytoplankton will develop, which is why impounded rivers often have the greatest algal blooms, sometimes reaching nuisance levels.

FIGURE 3.5 Moss in an Oregon stream. (Photo by J. M. Lyford.)

Macrophytes, primarily true mosses and angiosperms, comprise another autochthonous energy source. Mosses (Fig. 3.5) are attached plants having leaves, usually found in the colder, well-shaded areas of streams—essentially the headwaters—although they have the adaptive characteristics to grow in a variety of stream environments. These characteristics include the ability to grow in low light and low temperature conditions, a rapid rate of nutrient uptake, and a high resistance to being dislodged by spates—again, conditions found in headwater reaches of streams.

Common genera are *Fontinalis* and *Fisidens*. They grow attached to the rocks on the streambed (see Fig. 13.1) and also form thick coatings on rocks and logs along the stream banks. For some unknown reason, mosses are perhaps the least studied of the major constituents of stream ecosystems—at least from an ecological context. Mosses photosynthesize and produce organic matter and, in fact, may have higher rates of primary production that algae. However, because of their limited distribution within a given stream or river, they are of minor importance as an overall energy source throughout the entire continuum of the river.

As a stream increases in size and the current, at least in places, tends to decrease, silt will settle out and provide a suitable substrate for the rooting and growth of large water plants. Here you will find common genera of flowering plants, or angiosperms (Fig. 3.6), including *Potamogeton, Elodea, Ranunculus, Nuphar*, and others; in smaller springs and brooks watercress, *Nasturtium*, is common. Plants may be sparsely distributed or can form dense mats which clog watercourses, and they play various roles in the ecology of streams: an energy source (both before and after death), substrate for attachment of other organisms, cover from

FIGURE 3.6 Macrophytes growing in slow-flowing reach of the Madison River, Wyoming. (Photo by C. E. Cushing.)

predators, etc. Their role as an energy source in streams is probably more important after they die and decay, when their remains break down and become *fine particulate organic matter* (FPOM, < 1 mm in size). In addition, some insects with piercing mouthparts suck their juices and other, larger organisms eat the plants whole.

We will examine in more detail the role of each of the above autochthonous energy sources in a discussion of the entire food web, but first we should address the allochthonous energy sources—terrestrial plants and their products.

TERRESTRIAL (ALLOCHTHONOUS) ENERGY SOURCES

Until the mid-1970s, terrestrial plant production as a source of energy to stream ecosystems was largely unrecognized. Stream ecologists were quite near-sighted; we studied and described what we found in rivers and streams, but only what occurred from bank to bank. We essentially ignored the vital relationship between the stream and its catchment basin, until Noel Hynes, now retired from the University of Waterloo, published a paper in 1975 entitled "The Stream and Its Valley." When studies were made of energy inputs that included terrestrial as well as aquatic production, it became apparent that, at least in woodland streams, the former might dominate the latter.

All photosynthetically produced organic matter (primary production) from the terrestrial environment that reaches the river or stream becomes

FIGURE 3.7 Terrestrial organic matter input, White Clay Creek, Pennsylvania. (Photo by C. E. Cushing.)

a potential source of energy. Ecologists refer to leaves, fruit, twigs, logs, etc. (Fig. 3.7), as *coarse particulate organic matter* (CPOM), defined as any piece of organic matter larger than 1 mm in size. Stream ecologists actually break FPOM into several smaller size fractions, but for our discussion, this will suffice. Terrestrial DOM is any organic matter dissolved in water. It reaches the stream as canopy drip during rain, as surface flows, and via subsurface pathways.

So what happens to this CPOM once it reaches the water? Surely the vast majority of it cannot be used in its original size and form, so what processes occur to turn an essentially unpalatable chunk of dead organic matter into something that is nutritious to the organisms within the stream? Let's use leaves as an example because they are one of the main sources of CPOM. Figure 3.8 illustrates the sequence of processes that occur during leaf breakdown. First, as soon a leaf enters the water, soluble chemicals in the leaf are dissolved and removed from the leaf, contributing to the DOM pool. A leaf is a source of organic food, and so microbes, principally fungi, whose spores are present in the water column, and bacteria, colonize the remaining leaf material. A variety of fungi penetrate the wetted leaves with their *mycelia,* softening the leaf and essentially replacing the nonnutritious (to invertebrates) leaf matrix with a rich mass of nutritious fungi and bacteria palatable to the shredding invertebrates which can feed on altered CPOM (more about this in Chapter 4). One colleague suggests the analogy of peanut butter on crackers—the nutritious bacteria and fungi are the peanut butter, the leaf matrix is the cracker. Due to a combination of wetting, physical abrasion, microbial colonization,

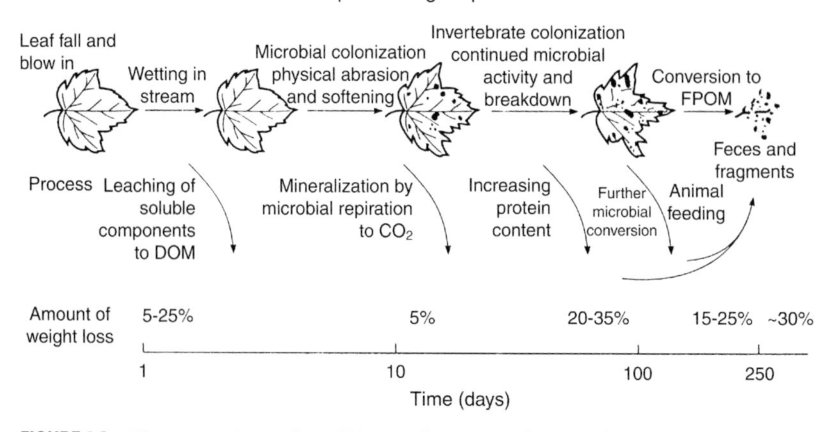

FIGURE 3.8 The processing or "conditioning" sequence for a medium-fast deciduous tree leaf in a temperate stream. First dissolved organic matter (DOM) is leached from the leaf. Microbial colonization follows, chiefly by fungi. Physical abrasion and leaf-shredding invertebrates help to fragment the leaf, and bacteria colonize these finer particles. As the leaf (coarse particulate organic matter, or CPOM) is converted to fine particles (FPOM), its fate becoming increasingly difficult to follow. The leaves of different plant groups break down at different rates. (From Allan, 1995, with permission.)

and shredding, CPOM is converted to FPOM. These fine particles are available for further colonization, especially by bacteria, because small particles present relatively more surface area, and in turn, can serve as food for animals that collect FPOM from the substrate and filter it from the water column. If one visits a woodland stream in the autumn, and observes the great quantities of leaves on the bottom and caught in debris jams, it is easy to see why so much attention has been paid to the fate of fallen leaves in streams.

Essentially the same process happens to the CPOM resulting from the death and decomposition of the mosses and macrophytes. Each is being converted by fungal and bacterial breakdown, ingestion, and fecal pellet production, or by physical maceration or abrasion, into smaller pieces of organic matter. Eventually, they reach that 1 mm dimension where they become part of the large FPOM pool found suspended in the water column or deposited among the stones on the bottom of the streams. FPOM, then, is another important energy source for stream invertebrates.

DISSOLVED ORGANIC MATTER

This source of energy for stream ecosystems does not conveniently fall into either autochthonous or allochthonous categories, because it originates both outside and within the stream. DOM, which is quantitatively the largest carbon pool in stream food webs, originates in the terrestrial environment when decomposing aquatic vegetation or exudates from living plants become dissolved in rain, surface, or subsurface water, and then enter the stream via canopy drip or groundwater. This DOM may be used directly by microbes, or it may even, under suitable conditions, form FPOM. Actively photosynthesizing aquatic plants also can be an important source of DOM. Studies of a Pennsylvania stream showed midday increases in water-column DOM, which corresponded to high rates of primary production by periphyton, and it evidently was exuded by living cells. When aquatic plants decompose, they also release DOM into the water.

Just as some plants are more nutritious than others, DOM varies in quality. That which enters the stream via soil water probably is the least nutritious (or most refractory, in the parlance of microbial ecologists). Soil microbes have had considerable opportunity to make use of the most available compounds, and what reaches the stream are definitely the leftovers. Canopy drip, on the other hand, is likely to contain higher quality DOM, and algal exudates also are of high quality, as indicated by microbial growth rates. This reinforces the idea that, for bacteria, living within the biofilm is beneficial, providing close association with algae, both photosynthesizing and decomposing.

Our understanding of DOM as an energy source is still quite limited. We know there is a lot of it, and that some forms result in higher bacterial growth rates than do others. Simple compounds, aminos acids, and sugars are some of the most available forms of DOM; humic and fulvic acids, so prevalent in blackwater streams, are among the most refractory.

A PRELUDE TO FOOD WEBS

We conclude with a brief review of what we have discussed, to set the stage for our discussion of feeding roles and food webs in the following chapter. Energy sources for stream food webs are many and diverse. Within the stream, various algae and macrophytes can be important primary producers, creating organic carbon compounds and releasing oxygen as a by-product. Plants can be an immediate source of food; when eaten by herbivores, they can break down to form particulate organic matter after death, and they can release DOM, either as exudate from living plants, or as leachate from decomposing plants. All of this is autochthonous production. Terrestrial plant production enters the stream as POM or DOM. The breakdown of CPOM into FPOM, and the abundance of the latter, suggest that FPOM may be a particularly important energy source to stream food webs. All of this is allochthonous production. Biofilms, layers that cover virtually every surface in streams, are a complex microenvironment in which algae often are the most evident member, but bacteria also can be very important. DOM, abundant but as yet poorly understood, originates inside and outside the stream, making it difficult to categorize as autochthonous or allochthonous.

These are the energy sources available to stream food webs. Efforts to understand the complexity of these food webs has proven interesting and challenging to generations of stream ecologists. Early work was based on direct observation of organisms feeding in streams or by evaluation of their stomach contents. Examination of the feeding apparatus has often led to a more refined understanding. Laboratory studies of feeding, including high-speed filming, have furthered our knowledge. Recent technological advances have provided stream ecologists with new tools to enable a more exacting investigation of food web relationships. Radioactive isotopes have enabled us to determine the uptake and assimilation rates of various elements and to trace them through the food web. Stable isotopes, in conjunction with their radioactive counterparts, can provide helpful clues to the origin of the tissue in a consumer's body mass. Stable isotopes are naturally occurring isotopes—for example, ^{13}C is the "heavy" isotope of carbon, ^{12}C is the "light" (normal) form. Stable isotope ratios for carbon, nitrogen, sulphur, and other elements can indicate whether algae or terrestrial plant matter is responsible for the growth of a particular insect.

We have now discussed the physical, chemical, and geomorphological aspects of streams and the energy sources available to the organisms inhabiting flowing waters. The following chapter will discuss the functional role of the biota within rivers and streams, emphasizing not only what they feed upon, but how they obtain their food.

Recommended Reading

Hynes, H. B. N. (1975). The stream and its valley. *Verhandlungen der Internationalen Vereinigung für theoretische und angewandte Limnologie* 19:1–15.

CHAPTER 4

Feeding Roles and Food Webs

INTRODUCTION

To live, grow, and reproduce, stream organisms must have sufficient food, in both quality and quantity. Moreover, food requirements change over the life cycle: the food requirements of an adult trout are quite different from those of newly hatched fry or of a small yearling. Adults consume larger invertebrates and small fish, while juveniles feed on tiny organisms such as midge larvae until they are larger enough to start taking larger food items. Some consumers are quite specialized, others less so; and most are adapted via their mouthparts or other feeding apparatus to capture some food items more readily than others.

Food webs include herbivores, carnivores, and omnivores, classified on the basis of the category of food ingested. Yet, as we learned in the previous chapter, the food sources in rivers are diverse, including algae and macrophytes, coarse (CPOM), fine (FPOM), and dissolved (DOM) organic matter, and the complex biofilm. We need a classification of the feeding roles of aquatic organisms that addresses this complexity. In addition, such a system should take into account that consumers differ in how they acquire their food. For example, two aquatic insect species might each contain FPOM in their guts, but one might gather small particles deposited on the substrate, the other might filter particles out of the water column. Likewise, several fish species might each consume insects, but one might capture its prey from the substrate, another might eat bottom-dwelling insects when they are dislodged and drifting in the water column, and yet another might feed on surface insects, many of which would be terrestrial. Ecologists use the terms "functional group" to refer to invertebrates that feed in the same way (e.g., filter-feed on FPOM) and "guild" to refer to fishes that feed in the same way (e.g., water-column invertivores).

FEEDING ROLES OF INVERTEBRATES

The functional role of organisms in rivers and streams is largely based on how they obtain and utilize their food resources. The broad categories within food webs, mentioned in Chapter 3, are *herbivores*, organisms which feed on plants (i.e., primary production); *carnivores*, organisms which feed on other organisms (= predators); and *omnivores*, organisms which feed on both plants and animals. A fourth category is usually included—*detritivores*, organisms which feed on organic detritus (small bits of organic matter). This is textbook biology and most people are familiar with these terms. Classifying organisms, in this example aquatic insects, into these categories describes how they are linked into a particular food web; an example of such a simplified food web is shown in Fig. 4.1. The insects are the herbivores and detritivores in this example, feeding directly on algae or on both algae and macrophytes after they have been converted to detritus, and the trout are the carnivores, feeding on insects.

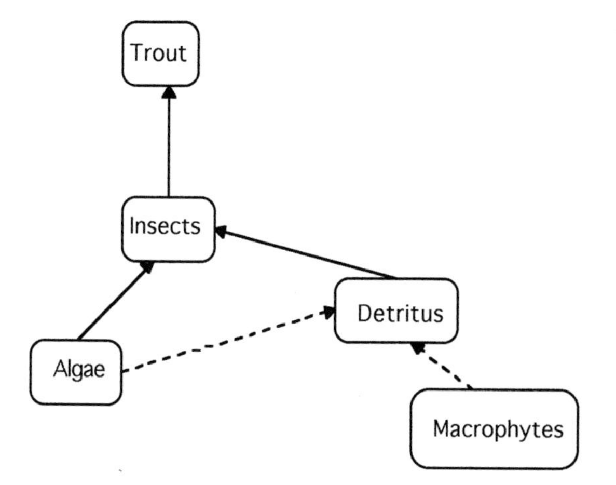

FIGURE 4.1 Diagram of food web based on type of food. Solid arrows indicate feeding pathways; dashed arrows indicate decomposition pathways.

But the great majority of aquatic insects are omnivores, based on examination of their gut contents and watching them feed, and so this classification is not terribly helpful. We need to know more than just what these consumers eat—we need to know how they obtain their food and how diet might change with prey availability, location, season, and so on. This will provide a more complete picture of the true pathways of energy flow through the food web.

Stream ecologists use *functional feeding groups*, a classification scheme developed by Ken Cummins to augment the herbivore–carnivore–omnivore–detritivore system. The major categories are *shredders, grazers* (sometimes called *scrapers*), *collectors* (both *gatherers* and *filterers*), and *predators*. Other categories have been added since inception of this scheme to address specialist organisms, and we will describe these below. This system was mainly developed with insects in mind because they are by far the most numerous organisms present in most streams and have the greatest diversity of feeding methods. Other organisms, such as birds, mammals, and amphibians, are less diversified in their feeding habits, and most function almost exclusively as predators. We will have more to say concerning their feeding habits in later chapters. We concentrate here on the insects in order to explore functional feeding groups in streams, and how this provides us with a more complete understanding of food webs.

Shredders (Fig. 4.2) are an important functional group in many streams, and particularly in small streams of deciduous forests or streams which receive considerable inputs of CPOM. Common shredders include some stoneflies such as *Pteronarcys californica*, the famous salmon fly of the Madison and Deschutes rivers and other streams, many dipteran larvae such as those of the family Tipulidae (the crane flies or "leather jackets"), and most of the caddisflies in the family Limnephilidae that construct cases of organic material (Fig. 4.5). The eastern caddis species *Pycnopsyche lepida*, however, shows a role reversal during its immature life. Early life stages (instars) have an organic matter case and are shredders; later instars construct sand grain cases and become grazers that exploit the spring periphyton bloom on rocks.

Not surprisingly, shredders are commonly found where there are large accumulations of CPOM—in forested headwater streams and in larger streams where CPOM accumulates against obstructions, in pools, and in other depositional zones. For the CPOM to be used by shredders, the location must remain fairly well oxygenated (also a requirement of the fungi that colonize, soften, and nutritionally enrich the leaf [Fig. 3.8]). This means that obstructions that trap CPOM in the current are excellent places to find shredders as long as the material stays in place. It also means that usually only the surface layers of CPOM, such as leaf litter, are used in pools and backwaters, because internal layers may lack oxygen.

The mouthparts of shredders are adapted for maceration of the CPOM particles, which they tear and shred while feeding. Their feeding results in the initiation of the conversion of CPOM to FPOM by physically breaking up the CPOM and by production of FPOM in the form of fecal pellets. Shredders obtain energy from the leaf itself and from the microbes, primarily fungi, which colonize it. Returning to the peanut butter (microbes) and

FIGURE 4.2 Typical shredders: (a) cranefly larvae (photo by R. N. Newell), (b) stonefly nymph, *Pteronarcys* (photo by R. W. Merritt).

cracker (leaf matrix) analogy introduced in Chapter 3, ecologists have debated how important each is to the shredder's nutrition. The current view is that both are important, but leaves without microbes are by themselves a very poor food source.

Another functional group is the grazers (Fig. 4.3). They are most likely to be found where light reaches the stream bottom, promoting algal

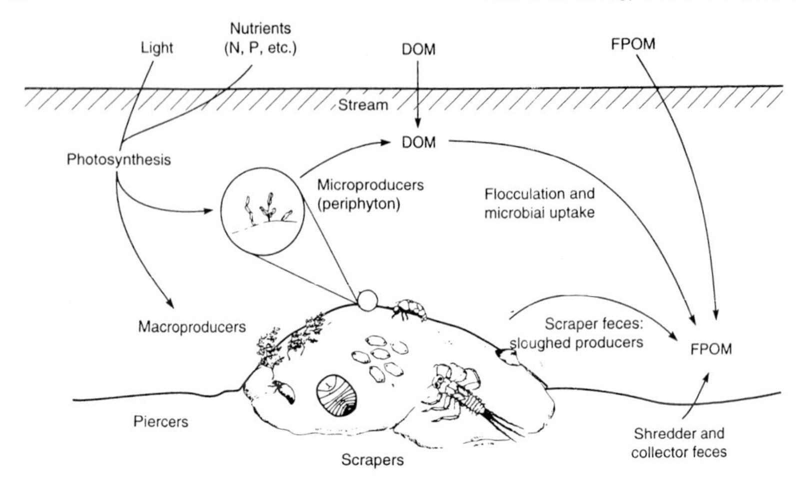

FIGURE 4.3 The grazer: periphyton and piercer: macrophyte linkages in which aquatic plants are the primary energy source. (From Allan, 1995, with permission.)

growth, because this is their main food source. These organisms have mouthparts adapted to scraping the film of algae (Fig. 4.4), or periphyton, growing on the surfaces of rocks and other large objects; thus, they literally scrape off, or graze, this food source, analogous to cattle grazing in a pasture. Some grazers, including some caddis larvae, have mouth parts adapted to scraping or rasping diatoms that lie very tightly adjacent to

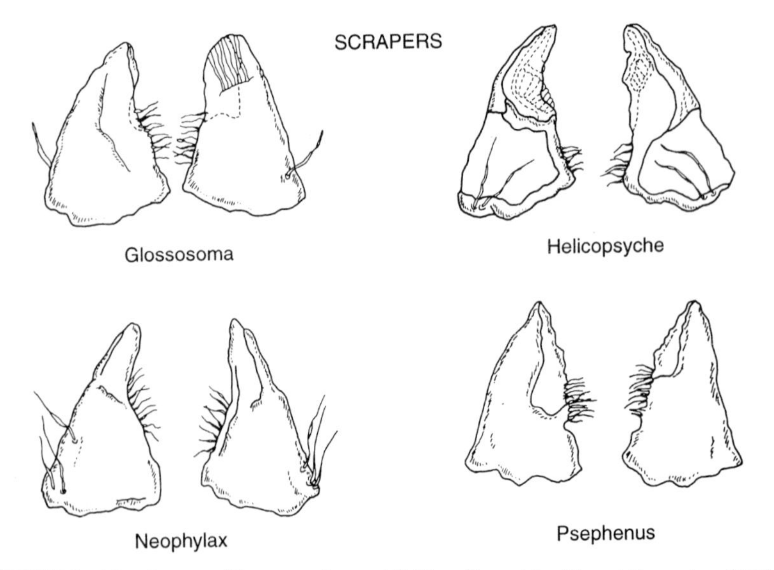

FIGURE 4.4 Mouthparts of four scraping caddisflies. (From Merritt and Cummins, 1996.)

FIGURE 4.5 Grazing caddisfly *Dicosmoecus* on stone, Yellowjacket Creek; Washington. (Photo by C. E. Cushing.)

stone or stick surfaces. Others, including many mayflies, are more adept at "browsing" the overstory and more loosely attached algae. This reminds us that functional groups are useful broad classifications that can be more finely dissected to reveal further differences in feeding strategy. Grazers, too, produce copious amounts of FPOM through the production of fecal pellets and dislodgment of algal cells during their feeding.

Several familiar organisms are scrapers. The genus *Dicosmoecus* (Fig. 4.5), the October Caddis of fly fishers, is a stone-cased grazer (the general rule is that caddis larvae with cases made from organic matter are shredders and those with cases made from inorganic matter are scrapers). However, like many large grazers, they may incidentally ingest some larger items such as small midge larvae and CPOM fragments while scraping the algal film. Many common mayflies are grazers, including many members of the family Heptageniidae; this includes such genera as *Epeorus, Stenonema*, and *Rhithrogena*. Another common grazer is the caddisfly genus *Glossosoma*, a small caddis which carries around a small, tortoise shell–shaped case made of coarse sand grains cemented together. On a stone, the case is about the size and shape of an aspirin tablet; like all caddis grazers, it moves over the surface of stones shearing off the algae while carrying its case along.

The largest functional group is the collectors (Fig. 4.6). This large group is further divided into filtering-collectors and gathering-collectors, and the names essentially describe how they obtain their food. Both groups feed almost exclusively on FPOM. Filtering-collectors obtain their food, as the name implies, by filtering FPOM from the water, and although there are a

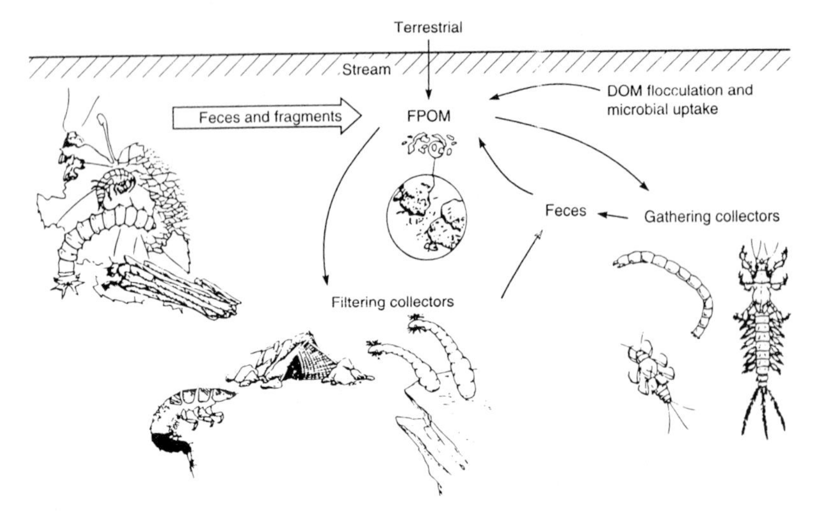

FIGURE 4.6 The collector; FPOM linkage illustrating both gathering- and filtering-collectors, in which FPOM colonized primarily by bacteria is the energy source. (From Allan, 1995, with permission.)

number of refinements and modifications, the two main methods are by filtering with nets and with specially adapted body parts. Net-weaving filterers are mostly caddisflies (Fig. 4.7), which construct some kind of net in conjunction with their nonportable case or fixed retreat. The larva retreats into the case and lets the current carry FPOM through the net, where it is caught onto the fine meshes of the net. Periodically, the insect will either feed on the entrapped particles or may even devour the net plus its catch, FPOM and all, and then spin a new net. Probably the most common net-spinners are members of the family Hydropsychidae, one of the most ubiquitous families of caddisflies in the world. They can be found from the smallest headwater streams to the largest rivers as long as suitable solid substrate and a current are present. The species may change in relation to such factors as temperature and FPOM size, and the catch nets of different species appear adapted to function at different current velocities. In the downstream direction, the change in species usually includes decreases in body size and in net mesh size as CPOM decreases and FPOM increases. It should also be remembered that these filtering nets may also entrap microscopic animals; however, feeding on these does not make the filter-feeders predators in this classification system. How the organism acquires its food takes precedence over the occasional ingestion of animal prey.

Other filtering-collectors include organisms that have developed specialized body parts to enable them to filter FPOM from the current. Three common insects which accomplish this in different ways are the caddisfly *Brachycentrus*, the larvae of the black fly (order Diptera), and the mayfly *Isonychia*. *Brachycentrus* is common in streams. Most people who have looked at stream organisms are familiar with their small, four-sided cases

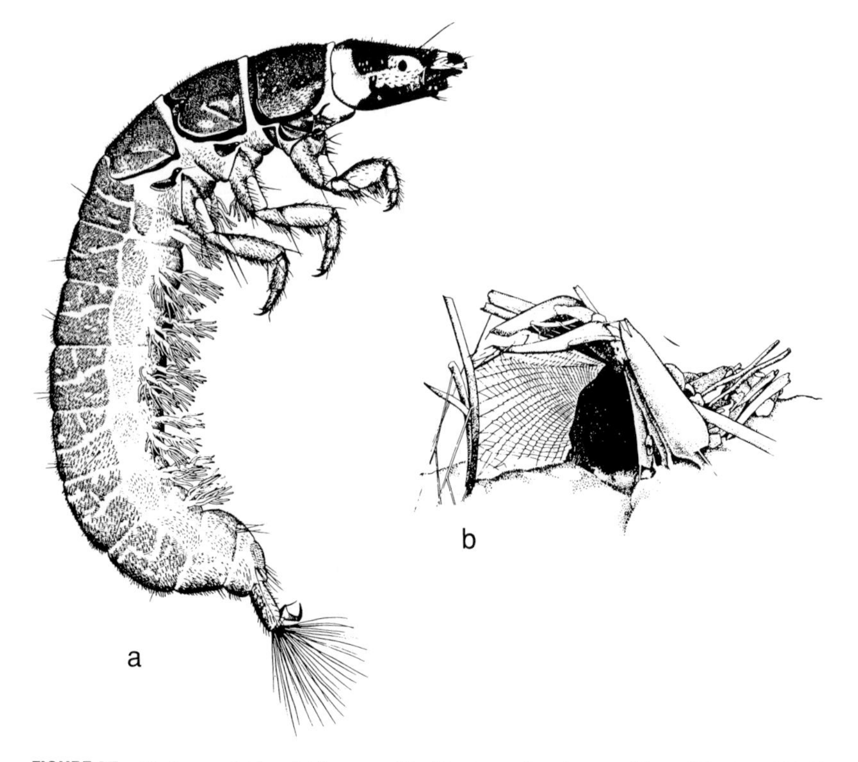

FIGURE 4.7 Hydropsychidae: (a) larvae, ×15, (b) retreat showing net (From Wiggins, 1996.)

built like slender, tapering log cabins, each case being 2–3 cm long and usually attached to aquatic plants that trail in the current or to submerged stones, logs, or branches. The larva cements one edge of the four sides at the mouth of the case to a solid substratum with the opening of the case facing into the current (Fig. 4.8). The larva has many fine hairs on the middle and hind legs and, with the abdomen inside the case, it extends its legs out like a tackler about to sack a quarterback. But in *Brachycentrus*, it is the hairy fringes on the legs which do the filtering, removing FPOM from the current until the larvae has enough to comb the material from its legs with its mouthparts. The blackfly larva accomplishes its filter-feeding differently. Each larva constructs a small silk pad on a rock or stick and anchors its abdomen to this pad with a series of anal hooks encircling the rear of the abdomen. One set of its mouthparts, the labia, are modified to resemble fanlike structures (Fig. 2.3) fitted with fine hairs and coated with sticky mucous, which the larva extends into the current and with which it filters the FPOM. When the filtering fans are full, it collapses them and stuffs them into its mouth, then combs the particles off as the fans are withdrawn for use again. The mayfly *Isonychia* has heavy fringes on the front legs, forming a basket held under the head. The nymph stands up on

FIGURE 4.8 Drawing showing feeding position of *Brachycentrus* larvae.

the middle and back legs allowing water to pass through the basket and under the body. When the basket is full, the nymph raises the front legs and removes the FPOM with its mouth parts.

Gathering-collectors, as their name implies, obtain their food, largely FPOM, by simply gathering it from wherever they can find it—under rocks, in deposition zones, on the surface of stones. FPOM accumulates in many places on the streambed wherever the current slackens enough to permit it to settle from the water column and accumulate in these deposition zones. Mayflies of the genus *Baetis* are a common example of this functional group. Most gathering-collectors have rather generalized structures and many make-do simply by scurrying around the stream bottom picking up particles wherever they find them, or "ingesting" their way through the sediments in earthwormlike fashion.

The final functional group is the predators, animals that eat other animals. They occur throughout the stream community and have many different adaptations to enable them to pursue and capture prey. Most stoneflies are predators, but only one family of caddisflies, the Rhyacophylidae, are active predators. Interestingly enough, this is also the only family of caddisflies that does not construct cases—their free-moving lifestyle may enable them to better pursue and capture prey. Other common predators are all the species of dragonflies and damselflies of the order Odonata, and the well-known hellgrammites (order Megaloptera).

Ecologists have also given names to some other insects which do not readily fit into the functional feeding groups described above. These additional categories include *miners* for some larvae that feed on detritus buried in fine sediments, *piercers* for insects with sucking mouthparts that feed on plant fluids, and *gougers* for larvae that burrow into large woody debris while feeding on the fungal and bacterial colonies that develop on their surfaces.

We should say a few words here about the functional role of fishes, although more will be said about this later. There are fish species that fit each of the feeding types; some suckers are herbivorous, grazing periphyton from rock surfaces; carp are omnivorous; and trout are predaceous. Fishes have also been classified into feeding *guilds*, depending on where they feed in the water column. Thus, there are *top-feeders, midwater-feeders*, and *bottom-feeders*; some obviously fit into more than one of these categories.

Now that we've learned what the different functional groups are, let's look at another food-web diagram which uses these functional groups rather than the groupings shown in Fig. 4.1. Figure 4.9 shows a more complex and illustrative food web which provides much more information about the pathways of food and energy. It allows an ecologist to have a much better grasp of the various niches being occupied in the stream, and provides a much fuller picture of the ecosystem being studied.

Chapters 1, 2, and 3 have provided us with information on the basic chemical, physical, and biological variables of stream ecosystems. This chapter introduced information on the functional role of the organisms. In the following chapter, we will put all of this information together and describe how these factors interact to produce the communities of organisms that we find in rivers and streams.

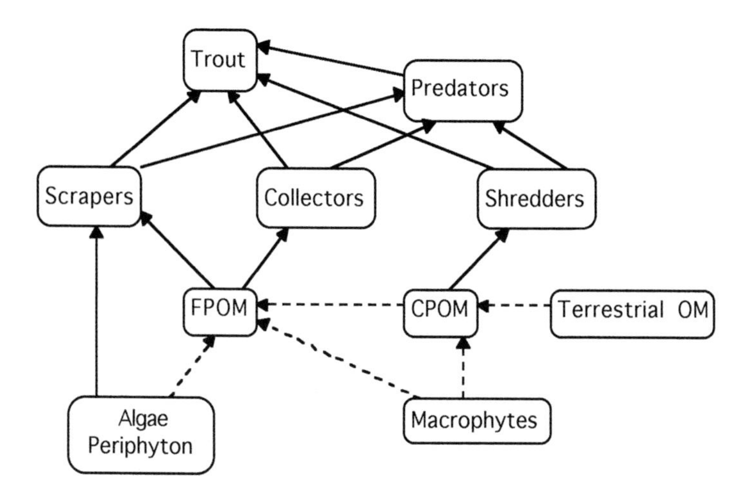

FIGURE 4.9 Diagram of food web based on functional feeding groups. Solid arrows indicate feeding pathways; dashed arrows indicate decomposition pathways.

Recommended Reading

Allan, J. D. (1999). *Stream Ecology*. Kluwer Academic. Dordrecht, The Netherlands.

Cummins, K. W. (1973). Trophic relations of aquatic insects. *Annual Review of Entomology* 18:183–206.

Cummins, K. W. and Klug, M. J. (1979). Feeding ecology of stream invertebrates. *Annual Review of Ecology and Systematics* 10:147–172.

CHAPTER 5

Ecology: The Structure and Function of Riverine Ecosystems

INTRODUCTION

Building on the material of previous chapters, we now wish to show how riverine ecosystems reflect the integration of the physical, chemical, and biological factors discussed in previous chapters. Consider the cross section of a typical stream (Fig. 5.1). We can see the interrelationships among the energy sources produced within the stream as well as terrestrially derived primary production, and how these energy sources are utilized by different functional groups of organisms within the stream. This is a generalization of what is going on at a particular place in a stream, but it captures the essential processes and players. How might these relationships change along an entire stream continuum from its headwaters to its mouth? Some clues might have caught your attention in previous chapters, particularly when thinking about different sources of energy. Leaves and CPOM might be of greatest importance in forested headwater streams, whereas algae

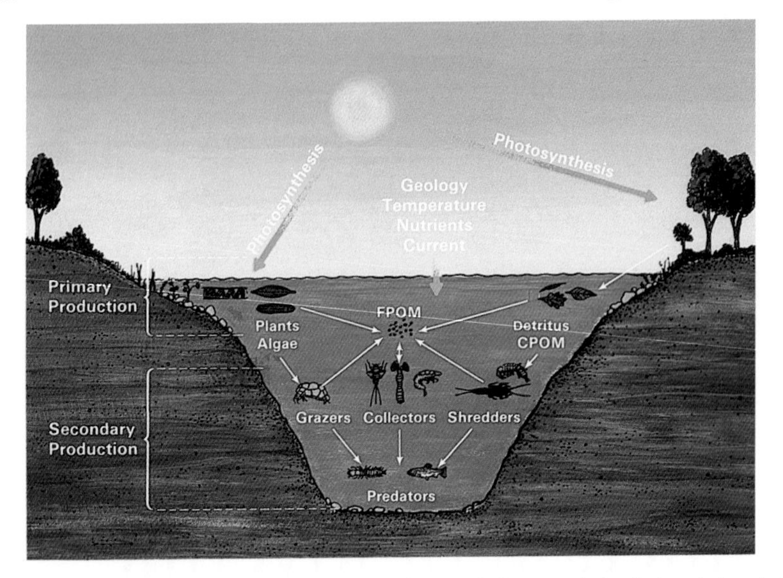

FIGURE 5.1 Cross section of a typical stream showing influence of physical and chemical variables and interrelationships among energy sources and functional feeding groups.

might grow best where ample sunlight reaches the streambed. In 1980, Robin Vannote, Wayne Minshall, Kenneth Cummins, James Sedell, and Colbert Cushing published a seminal paper in stream ecology, in which they outlined how a river changes along its length, in physical dimensions, but especially in its ecology, due to changing energy inputs, which in turn cause changes in the community of organisms. Known as the River Continuum Concept (RCC), this model generates many useful predictions about patterns that can be seen in any geographical region or biome. Of course, no model is applicable to every situation, but the generalities of this model have been supported by considerable research.

THE RIVER CONTINUUM CONCEPT: A MODEL

As our model we will use a hypothetical stream ecosystem located in a deciduous forest, typical of those found in the eastern part of the United States. After we have described this model, we will discuss how it would change under differing geological, geographical, or other conditions. Since its formulation and first testing in the late 1970s, the RCC has since been refined, adjusted, and modified, but these details need not concern us here. The basic model will serve us well in our discussion.

Our hypothetical river, located in a deciduous forest drainage basin has its headwaters flowing through a heavily shaded forest; it then flows into more open country, and eventually becomes a large, deep, heavily silted river (Fig. 5.2). We will break the river into three general regions for

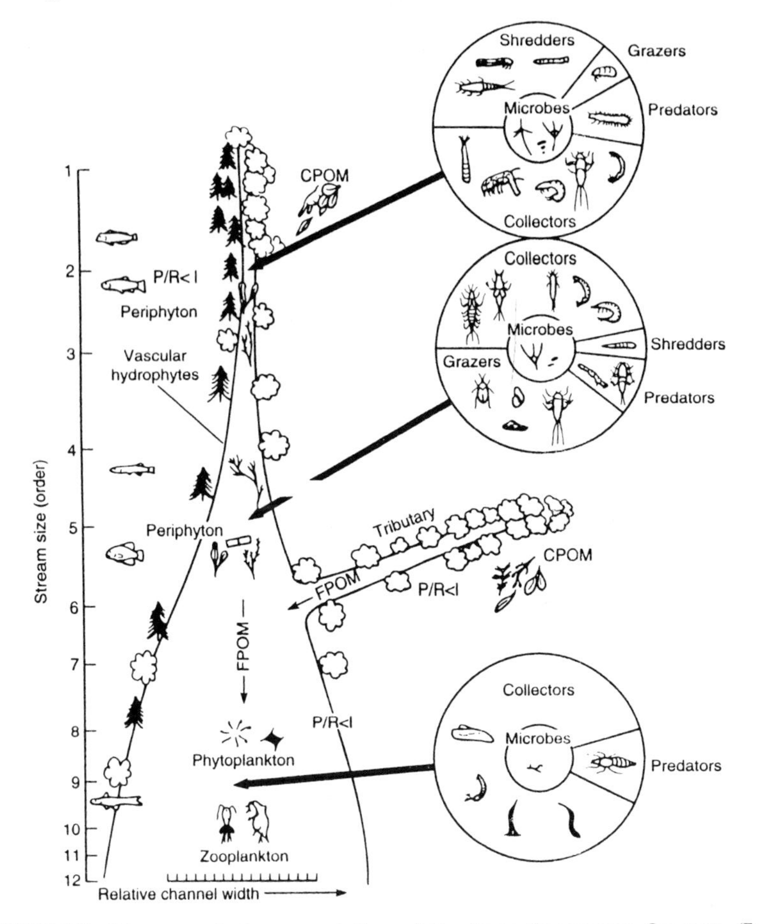

FIGURE 5.2 Diagrammatical representation of the River Continuum Concept. (From Allan, 1995, with permission.)

discussion: the headwaters (orders 1–3), the mid-reaches (orders 4–7), and the lower reaches (orders 8 and above).

In the headwaters, the stream is narrow and generally well shaded by the riparian canopy (Fig. 5.3). The stream bottom may be rocky or sandy, or a combination of the two, depending on the geological characteristics of the drainage basin. Because of the riparian canopy, insufficient light reaches the streambed to promote much algal growth, and the current is too fast, substratum often unsuitable, and nutrients usually too low to allow growth of macrophytes. Mosses are the dominant aquatic primary producers. Because riparian vegetation is abundant and the stream channel is narrow, considerable CPOM from the terrestrial environment enters the stream. This may be in the form of leaves, especially during autumn leaf fall, or twigs and branches which fall or reach the streambed by gravity

FIGURE 5.3 Photo of a shaded headwater stream, Ledyard Spring Branch, Pennsylvania. (Photo by C. E. Cushing.)

from the usually steep hillsides found in this part of the stream. If we compare the amount of primary production (measured as amount of dissolved oxygen produced by the instream primary producers) with the *respiration* (measured as removal of dissolved oxygen by the aquatic community), we will find that respiration exceeds primary production. Instream primary producers are not meeting the demands of the stream organisms; thus, energy from outside the stream is necessary for the continued existence of the stream community. This energy comes from the CPOM from the terrestrial environment. Ecologists term this a *heterotrophic* stream reach, meaning, roughly, that it doesn't produce enough of its own food. If the diversity of the functional groups of insects present is examined as either numbers or weight, we would expect shredders to compose about 35% of the total and collectors about 45%; these are the most numerous of the functional groups present. Grazers are essentially absent, making up only about 5% of the population, and predators compose the remaining 15%. This is because the huge supply of CPOM can support a large shredder population, and the FPOM generated by shredder feeding, mechanical breakdown, and other sources supports the large population of collectors. Because low light and nutrient levels do not support an abundance of algae, grazer populations are low. The percentage of predators present is fairly constant throughout the river continuum; however, different groups of organisms perform this function in the different reaches. This reach of the continuum exports a large amount of FPOM to the midreaches. Fish species present in these reaches include various minnows, trout, sculpins, and other typical fishes tolerant of seasonal and daily cold temperature regimes.

FIGURE 5.4 Photo of mid-reach of a stream, Salmon River, Idaho. (Photo by C. E. Cushing.)

As we move downstream to the mid-reaches, the streambed has widened, the bottom is well lit by direct sunlight (Fig. 5.4), temperatures have warmed, and nutrient concentrations have increased, all leading to a proliferation of algae—filamentous greens and/or diatoms on the bottom. The stream bottom is usually composed of rubble, rocks, and pebbles, with sand and silt accumulations where the current slackens. Rooted macrophytes will occur in protected places where sediments have accumulated and current speed is reduced. At the same time, CPOM inputs have decreased on an areal basis, because the stream is wider and due to a decrease in extent of riparian canopy. Primary production by algae and macrophytes in this reach would exceed respiration by the instream community; hence, this is an *autotrophic* stream reach, meaning that the stream produces more energy than is needed to support itself. Here the functional group proportions would show that the collectors are about as numerous as they were in the headwaters, composing about 50% of the population, although they might include different species than found in the headwaters, but perform the same function. High FPOM export from the headwater extends into the mid-reaches, and, along with FPOM generated within this reach, provides the rich food source for these collectors. However, there is a major change in the proportion of the shredders and grazers. In this reach, shredders are reduced to about 5% of the population while grazers have increased to about 30%. This is a reflection of the low input of CPOM to this reach and the proliferation of algae on the stream bottom. Predators again make up about 15% of the population. This section of the continuum also exports a large amount of FPOM to the lower reaches of the river. Indeed, because this reach produces more food than it respires, the excess production is exported downstream. Fishes present in the mid-reaches are typical of those species that can tolerate wider fluctuations in daily and seasonal temperatures, although there is considerable overlap with some of the headwater species. Trout are usually present, along with suckers and many minnows. This section of the continuum is highly productive, and it is the only autotrophic section in the continuum. This is the reason that many of the most famous trout streams in the country are found in the mid-reaches of streams.

In the large, usually slow-flowing and deeper lower reaches of our hypothetical river, several changes occur (Fig. 5.5). The increased turbidity of the water prevents sunlight from supporting algal growth on the bottom; this is also affected by the fine-grained, shifting nature of the bottom. Nutrients are in high concentrations. Because algae cannot grow on the stream bottom, instream primary production now takes place within the water column where suspended algae (*phytoplankton*) may flourish, and by macrophytes, which are abundant along the margins in the slow-flowing water. The phytoplankton come from a number of

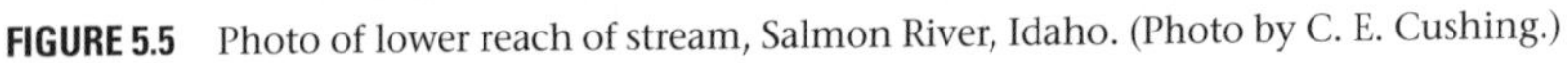

FIGURE 5.5 Photo of lower reach of stream, Salmon River, Idaho. (Photo by C. E. Cushing.)

sources—from lakes along the river's course and from detached periphyton which can remain viable while floating—and they become abundant if the current is slow and the water reasonably clear. Indeed, the presence of phytoplankton and rooted plants hints at the lakelike characteristics of many high-order rivers. Terrestrial input of CPOM is small, sometimes negligible. The water column contains large amounts of suspended FPOM transported from the mid-reaches. Respiration of this material and of senescent phytoplankton exceeds the primary production of the small algae crop and macrophytes, thus making this a heterotrophic reach. The benthic community is largely made up of collectors, both filterers and gatherers, which comprise about 85% of the community population. Shredders and grazers are generally absent because of the lack of CPOM and periphyton. Grazing snails may be found feeding on the surface layers of algae which coats plant stems or other places. Most of the collectors present are sediment dwellers, such as mollusks or dipteran larvae, which construct burrows or tubes in the mud and scavenge or filter the suspended FPOM. Again, predators are present and compose about 15% of the population, but of course, are different species. Fishes present in the lower-reaches are those found in environments where temperatures fluctuate widely. Species such as suckers, carp, and chubs are typical.

This, then, is the word story of the predictable patterns found in a hypothetical pristine stream continuum from its headwaters to its mouth. The RCC has also been tested in streams around the world and, for the most part, its principles have been supported. Problems, however, occur when comparing its predictions in streams in differing geographical or geological conditions and in streams where changes occur because of man-made alterations in the pristine pattern. For instance, many New Zealand streams are short in length and have a tendency to have the CPOM flushed from them by floods. Hence, there is a paucity of shredders in many of their streams.

Changes in the patterns predicted by the RCC occur from both natural and human causes, and we call these *reset mechanisms*. For instance, when a small tributary typical of a headwater reach enters a larger, mid-size or large stream, the insect community immediately below the confluence contains functional groups characteristic of the tributary, rather than that of the receiving stream. This is because the conditions in the receiving stream have been "reset" to those higher in the continuum. The insect community in the receiving stream reverts to expected proportions a short distance below the tributary inlet. The same thing happens when the riparian canopy in headwater reaches is removed; sunlight reaches the stream, periphyton grows, grazer organisms are found, and autotrophic conditions prevail. Thus, the stream pattern has been reset to the patterns expected in the mid-reaches. This condition also occurs in many western streams (Fig. 11.4) that have less dense riparian canopies in their headwaters than those of eastern deciduous forests where the RCC model was formulated.

Human actions also induce "resetting" of the predicted biological patterns. Removal of riparian vegetation, logging, damming, etc., interrupt the pristine conditions in a stream continuum. Damming drastically resets the downstream continuum pattern, and the changes are related to where on the continuum the dam is located, whether it has a surface (warm) or bottom (cold) release of water, and several other factors.

OTHER ECOSYSTEM CONCEPTS

Other aspects of the river ecosystem are important to mention and influence their ecology. The *hyporheic* community consists of those organisms occurring in the interstitial spaces between the rocks and stones (*hyporheic zone*) making up the bottom of the stream. Obviously, they are usually fairly minute, although the larvae of several stoneflies and midges have been found in this habitat far below the streambed and many meters laterally from the existing stream shoreline. More about the inhabitants of the hyporheic zone will be discussed in Chapter 17.

It should also be kept in mind that stream habitats and communities exist in differing scales, from those existing on a single grain of sand or rock to those characteristic of the entire stream reach. The study of these habitats is called *patch dynamics* (Townsend, 1989), and it is important to consider these when describing the animal communities and their interactions. Another important relationship in some river and stream ecosystems is the interaction and exchanges that occur between the stream and its floodplain. This has been termed the *flood-pulse concept* (Junk *et al.*, 1989) and describes the exchange of nutrients, organisms, and organic material that occurs when a stream or river floods and then recedes. Aspects of the flood-pulse concept are discussed in Chapter 8 in relation to the exchange of nutrients and other materials between the floodplain and main channel of blackwater rivers. The *nutrient spiraling concept* (Webster and Patten, 1979) explains how the normal cycle of nutrients in aquatic ecosystems is stretched into a spiral by the stream's current, allowing for the calculation of spiraling distance as a measure of a stream's ability to recycle nutrients.

Also remember that the dynamic exchange of organisms and energy in riverine ecosystems occurs in several directions—up- and downstream, laterally and vertically within the hyporheos, and with time. Upstream movement occurs by active movement, either within the stream itself or above it in the case of flying insects. Downstream movement is either by active movement or by drifting of organisms dislodged from the stream bottom. Lateral movement occurs by adult insects dispersing into the terrestrial environment, and both lateral and vertical movements take place within the hyporheos and between the hyporheos and the water column. The fourth dimension, time, describes the seasonal changes that occur among the populations as they grow, mature, and die.

Chapter 1 describes how to determine stream order and Chapter 14 presents information on how to identify the major groups of stream insects and their functional groups. Using this information and the description of the three major regions along the river continuum described in this chapter, it would be an interesting exercise to choose a reach along a favorite stream and see how it fits into this model. Visit a stream and look at such things as the width of the stream and shading by the riparian vegetation. Pick up some rocks and see if they have a "slimy" feel; if so, it

FIGURE 5.6 Collecting screen. (From McCafferty, 1981.)

means they are coated with diatoms, indicating that sufficient sunlight reaches the bottom to allow algae growth. If the rocks are slippery, you should be able to find some insects (stone-cased caddisflies or mayflies) on top of the rocks where they are grazing the algae. If shading extends across most of the streambed, the rocks will probably feel rough, indicating little algae growth. Pick up some rocks or submerged sticks and see if you can identify any of the insects you might find. If you are in a headwaters reach, look for accumulations of leaves behind or under rocks; you should find some shredders in these. You may want to construct a hand-screen using two dowels and some window screen to aid in collecting the insects; a simple one is shown in Fig. 5.6.

This chapter completes the first part of our book and describes the basic structure and function of stream ecosystems—their ecological nature. In the following part, we will present descriptions of several kinds of rivers and streams that occur in different geographical regions.

PART II

Types of Rivers

Part I explained how rivers and streams function as ecosystems. Part II, containing 6 chapters, examines several types of rivers characteristic of different regions and unique settings, presenting information on their ecological characteristics and how they fit into the predictions of the River Continuum Concept.

Chapter 6 describes a favorite type of river to many—the trout stream. Trout streams are both widely distributed and extremely popular—for more than just the trout. Many avid fly fishers may read no further, but it is our hope that they will continue reading to more fully appreciate how their favorite streams compare with others. Chapter 7 singles out two major large rivers of the West, the Colorado and Columbia rivers, and describes how these once magnificent free-flowing rivers have been altered by human actions and the ecological implications of these changes. The slow-flowing, blackwater rivers of the southeastern United States are described in Chapter 8, and the warm rivers of the Midwest are discussed in Chapter 9. Chapter 10 emphasizes the desert region of the southwestern United States, where some of the most fascinating and unique stream ecosystems are found. In a region where flowing waters can be few and far between, some truly unusual ecosystems are present. Chapter 11 concludes Part II with short descriptions of some special flowing water ecosystems, such as hot water springs, alpine streams, tropical streams, and others.

Part II describes the ecological characteristics of some characteristic rivers and streams found in the United States, and, together with Part I, presents a comprehensive examination of flowing-water ecosystems. Part III will describe the major groups of plants and animals found in or closely associated with rivers and streams.

CHAPTER 6

Trout Streams

INTRODUCTION

The names come easily to mind—the Madison, the Battenkill, the Henry's Fork, the Au Sable, the Gunnison, Silver Creek—and all of the many more classic trout streams throughout the country. Yet look at the range of geographical locations in which they occur: from deciduous eastern forests to coniferous forests to sagebrush steppe-desert regions. Obviously, they must have something in common if all support robust trout populations. What are the common conditions that produce a productive and viable trout habitat? In this chapter we will examine several habitat factors and see what must be present to produce a good trout stream. We will also compare some classic trout streams and briefly discuss some problems and solutions facing trout streams.

CHARACTERISTICS OF TROUT STREAMS

Temperature

Trout streams are generally cool- to cold-water streams, and most trout species prefer cold water. Brown trout, however, generally tolerate warmer temperatures than other trout, and they can be found at lower elevations or in warmer stream sections. Preferred temperatures range from 10 to 16°C. As water temperatures approach about 21°C, trout are less able to compete with other fish species for food and other resources. Lethal temperatures for trout range from 23 to 26°C, depending on the species. Each life stage and function (e.g., swimming) has optimal temperatures and these may vary with species, age, and ambient thermal regime where trout occur.

The influence of fluctuating water temperatures on growth of trout in good trout streams is somewhat conflicting. Experimental studies of adult brown trout have shown that fishes raised under fluctuating temperature regimes had faster growth, but whether this holds true for juveniles or other species is unknown. Conversely, the relatively constant temperatures found in tailwater fisheries below dams (e.g., the Green River below Flaming Gorge Dam and the San Juan River below Navajo Dam) have been cited as the reason for rapid growth of trout in these places and the development of blue ribbon trout fisheries; the fishes are acclimated to the ambient temperature and are able to feed and grow essentially all year long.

Cover

Cover is vital to trout survival; thus, good trout streams contain an adequate amount and diversity of cover to provide shelter from predators and strong currents. This cover can be quite diverse and includes riparian vegetation, boulders in the streambed, overhanging banks, logs, root wads with undercut areas, and even shade from overhanging objects. Trout are territorial; thus, here we are talking about aspects of their habitat, that is, the relatively small area that they occupy in terms of the larger stream, and this varies with both age and species. Thus, adequate cover must be present throughout the stream or river, not just here and there.

One of the most important and characteristic features of good trout streams is the alternating riffle—pool sequence, which is especially well developed in gravel-bed streams (Chapter 1). This configuration allows trout both the luxury of having productive riffles in which to feed and the deeper pools in which to rest and take cover. Trout have been found to hold a "station" behind a rock or other obstruction characterized by a sharp, neighborhood transition in current velocity. This permits the fish to expend little energy in holding its location and in a short foray to readily capture drifting prey.

Food

Trout frequently are drift feeders, taking dislodged insects floating downstream in the water column or on the surface of the water. They will also actively pick insects from the stream bottom. Thus, a good trout stream supports an abundant population of macroinvertebrates (insects, arthropods, and crustaceans) capable of supporting the trout biomass in the stream. Terrestrial insects are also important food items in many situations, and, of course, larger trout feed extensively on small fishes—minnows, small trout, or other species. Below we address more fully the relationships between the productivity of macroinvertebrates in a stream and the trout biomass it supports.

Spawning Area

Successful, reproducing trout populations require streams or rivers which provide suitable spawning areas for the fishes to construct their nests (redds) and lay their eggs. These are areas with extensive gravel-bottom areas which are largely free from excess amounts of silt so that the water can flow freely through the eggs in the redd once they are buried by the fishes. The water must be high in dissolved oxygen to adequately aerate the developing eggs; thus spawning areas are usually in cooler reaches because cold water holds more dissolved oxygen.

Stream Bottom

Typical successful trout habitat in streams is composed of a combination of large rocks, rubble, and smaller amounts of sand and gravel. This, of course, doesn't mean that successful trout fisheries are not found in streams with predominately sand and silt bottoms. However, trout inhabiting these areas must have access to stream sections with gravel bottoms to construct their redds and lay their eggs. The typical rocky substratum provides many factors which are conducive to viable trout populations. These include cover from predators, resting places from the current, abundant niches for a variety of invertebrate food items, and the aeration resulting from the water tumbling over and among the rocks enhancing the absorption of dissolved oxygen by the water.

Dissolved Oxygen

Trout require ample dissolved oxygen, and, except in polluted rivers, this condition usually is met. The amount of oxygen that can dissolve in water (its saturation value) is greater at cooler temperatures and vice versa, and streamwater is readily aerated by its falling over rocks and flowing through rapids—any movement that increases its contact with the air. Oxygen is also contributed by photosynthesis from aquatic plants and algae. Mechanisms that remove dissolved oxygen from the water include respiration by aquatic organisms and decay of dead organic matter within the water. Lack of oxygen is likely to be a problem for a trout stream only when untreated human or animal wastes promote excessive decay, or when silt and other fine materials limit water flow within the stream bottom, smothering redds.

Let's look at some classic trout streams: the Beaverkill in the East, the Au Sable River in the Midwest, and the Madison River in the West.

Although their geographical settings differ, each possesses the habitat characteristics described above—or they wouldn't be great trout streams. So, rather than repeating the habitat characteristics of each stream, we will describe each stream's setting and background, ecology, and environmental problems, and we will examine how its physical and biological characteristics compare with our basic RCC model.

THE BEAVERKILL

The Beaverkill (Fig. 6.1) can truly be called the "cradle of fly-fishing in America." Its history is intimately related to the development of fly-fishing in terms of flies and tackle, the presence of some of the most famous fly fishers, and the history of trout management. The Beaverkill flows for about 120 km through Delaware and Sullivan counties in lower New York State. It originates in a steep, rocky ravine at an elevation of about 880 m and initially flows through forests consisting of beech, birch, maple, and hemlock. Its name derives from the numerous beaver dams in the headwater reaches; "kill" in Dutch means "river"—hence the Beaverkill.

Setting and History

The valley of the Beaverkill is steep and narrow, and mountain slopes come to the stream edge with little development of floodplains. The bed of the stream consists of sand, gravel, rubble, and boulders. As it flows through the forest, it receives water from many sources—springs, tributaries, and

FIGURE 6.1 The Beaverkill River, New York. (Photo by E. VanPut.)

side valleys. The first major tributary is Alder Creek, 19 km below its source, and the second is Willowemoc Creek, which essentially doubles the size of the Beaverkill. Large, deep pools and riffles are found throughout the length of the Upper Beaverkill before Willowemoc Creek joins it. Gradient lessens perceptibly over the last 25 km, and the Big or Lower Beaverkill also becomes much wider—up to 60 m wide after receiving the East Branch of the Delaware River. Water temperatures are cool in the upper forested reaches, but the river warms considerably once it becomes wide enough for sunlight to reach the surface. Trout populations throughout the river depend upon the upper reaches in terms of spawning areas, cool water, and permanent flow.

Paleo-Indians were the first inhabitants of the region. The Lenni-Lenape (Delawares to the English and Loups to the French) called the Beaverkill the Whelenaughwemack, and it was shad, not trout, that drew their attention to the river. Fur traders entered the region around 1610 and were followed soon after by other colonizers. Logging was the primary industry originally pursued in the watershed of the Beaverkill and the logs were transported to the sawmills by floating them downriver in huge rafts. As logging efforts subsided, a number of tanning factories were located along the river, dumping noxious wastes into the stream. Later, "acid factories," industries which produced chemicals from wood products, were located along the Beaverkill. The region of the Beaverkill today is sparsely populated; fewer residents reside in some areas today than previously. Reforestation of the landscape protects the watershed, and lands earlier cleared for agriculture and denuded by the wood-chemical industry have now become second-growth woodlands and mature hardwood forests.

Ecology

The history of fishing in the Beaverkill is as storied as that of any stream in the United States. Initially, the only trout present was the native eastern brook trout (*Salvelinus fontinalis*). The fishes were, and still are, small but plentiful; in fact, early catches were reported as total pounds of fish rather than size. Initial fishing pressure was, of course, low, but the fame of the stream grew and fishing pressure increased as the first railroads reached the area in the late 1860s, followed soon after by roads. Trout numbers began to plunge as there was no stocking or planting of trout at this time. The first stocking of the Beaverkill with brook trout took place in 1876, followed by rainbow trout (*Oncorhynchus mykiss*) in 1880 and brown trout (*Salmo trutta*), in the lower river, in 1887. The U.S. Fish and Wildlife Service reports that the Beaverkill was one of the first streams in the United States to receive brown trout imported from Europe. These fishes were well established within 10 years. Stocking during the 1930s consisted of fry and fingerling trout, and after 1935, emphasis shifted to yearlings and adult trout. Today, most stocking is a combination of yearlings (20 cm) and 2-year-old (30–40 cm) fishes.

The development of local fly patterns by Beaverkill anglers began around 1900. The rich mayfly hatches, which attracted early fly fishermen, gave rise to one of the most popular fly patterns still in use today, the

Hendrickson, which was tied to simulate *Ephemerella invaria*, a gathering-collector. The main hatch of *E. invaria* usually occurs in the last week of April or first week in May. Other famous fly patterns developed in this region include the Parmachene Belle, Petrie's Green Egg Sac, and the Fan-Wing Royal Coachman; the first and last were essentially attractor patterns, and the Petrie was an early caddis imitation. Another widely used fly pattern developed on the Beaverkill, the Quill Gordon, was tied to imitate the adult of the collecting mayfly *Epeorus pleuralis*. The Beaverkill contains many famous pools along its fishable length, which, although they experience tremendous fishing pressure, still manage to produce excellent trout catches, including several trophy-sized trout each year.

In its headwaters, the Beaverkill is a narrow, well-shaded stream approximately 1- to 2-m wide. Its bottom is covered with boulders and rocks. No collections of macroinvertebrates have been done in the upper 20 km of the stream, but given its other characteristics—little light reaches the streambed and inputs of CPOM are high—it is likely that shredders and collectors predominate, as predicted by the RCC.

In the middle reaches, from approximately river kilometer (Rkm, measured upstream from the mouth) 23–43, the river broadens to 15–30 m; rock and rubble make up the stream bottom, and summer water temperatures reach 22°C. Dissolved oxygen is sufficient to support cold-water organisms—invertebrates and trout. The riparian canopy has decreased to about 5–10%, indicating that sufficient sunlight reaches the streambed to support algal growth on the stream bottom. As would be expected, scrapers and collectors, both gatherers and filterers, dominate the macroinvertebrate populations in this reach of the stream. The important mayflies include the genera *Leucrocuta, Drunella, Paraleptophlebia, Acentrella, Epeorus,* and *Baetis*. *Brachycentrus, Hydropsyche,* and *Cheumatopsyche* are the most numerous caddisflies found here. Several midge larvae (e.g., *Polypedilum*) are also well represented.

The lower reaches of the Beaverkill, Rkm 0–23, are not significantly different from the middle reaches in terms of ecological structure. Here the river is a bit wider, 40–50 m, but all other physical and chemical characteristics are the same. This also holds true for the functional feeding group abundances; the Lower Beaverkill still shows a community dominated by scrapers and collectors. It is interesting to note, however, that one shredding caddisfly, *Lepidostoma*, is more numerous in the lower reaches than it was in the mid-reaches. The numbers in both reaches were not high, so no particular significance can be placed on this.

Threats to the River

Although the human impacts to the river documented above—logging, rafting, tanneries, and chemical inputs—have been eliminated or curtailed, all is not serene along the river. Today, open dumps and landfills are found along the river, although significant efforts have been made to eliminate or cover these. Where they occur, they are not large, but they are unsightly and detract from the experience of fishing one of the country's premier streams.

Highway construction along the river was once a contentious issue when a new four-lane expressway was proposed to run alongside part of the river. Much opposition was voiced by those who loved the Beaverkill and thought that construction of the expressway and its associated bridges would do lasting harm to the river. The expressway was built, and the river has not been "ruined," but fishing is now accompanied by the drone of increased traffic and semitrailer trucks rushing by.

The continued pressure on the trout population has resulted in the usual litany of changes—posting, leasing of private reaches, obtainment of access easements. Getting local fisherman to adopt catch-and-release regulations was not an easy sell, although the success of this approach in both the Beaverkill and Willowemoc has resulted in significant expansion of the initial small reaches devoted to this approach to the inclusion of several miles of stream where all fish must be returned unharmed. Fishing has noticeably improved in these areas.

THE AU SABLE RIVER

The Au Sable River (Fig. 6.2) is justifiably famous among anglers, and especially fly fishermen, as one of America's premier trout streams. Located in the northeastern region of Michigan's Lower Peninsula, the main stream of the Au Sable originates between Grayling and Gaylord, at an elevation of 441 m. Joined by the north and south branches, the river drains in an easterly direction, entering Lake Huron at Oscoda, at approximately 183 m ele-

FIGURE 6.2 The Au Sable River near Mio, Michigan. Canoe traffic is heavy on this scenic river, sometimes leading to conflicts with anglers. (Photo by J. Abdella.)

vation. This lovely, swift, and cool river is amazingly resilient. Its fishes and its landscape have a long history of abuse. The river tolerates a high level of recreational use, from both anglers and canoeists, and along many reaches the banks are lined with homes. Yet somehow the Au Sable River maintains its beauty and its appeal, as well as its productive trout fishery, making it one of the most enjoyed rivers in the midwestern United States.

Setting and History

The Au Sable originates in gently rolling terrain of low moraines and glacial outwash plains that characterize much of the north-central Lower Peninsula of Michigan. Early French explorers named it the "Riviere aux Sables"—River of Sand—an apt name that the river retains in modified form. Riffles with coarser materials can be found, and some sections of the Au Sable have gravel bottoms, but Michigan's sand-bottom rivers typically include many long runs and few riffles; pools occur mainly at bends and undercut banks. The geology of this region is a principal reason that the Au Sable has remained a productive trout stream despite so much abuse. The soils are highly permeable, allowing rainwater to percolate rather than cause surface runoff, and so rivers are fed mainly by cold, clear groundwater. Discharge measurements (see Fig. 9.1) reveal that the Au Sable and other rivers of this region have very constant flow. The river will rise after a hard rain, and appear somewhat turbid, but much less than one might expect.

Of course, the Au Sable and its sister rivers of northern Michigan were not trout streams initially. The grayling (*Thymallus arcticus*), a prolific but vulnerable fish, was the native salmonid in these waters. Possibly a recognizable subspecies, and certainly a distinct stock, the native grayling was extinct by about 1900, succumbing to the multiple insults of over-harvest, habitat damage, and the influence of introduced salmonids. Brook, brown, and rainbow trout, introduced into the Au Sable in the late 1800s, are now the quarry of the many anglers who visit the Au Sable.

Ecology

The Au Sable River probably is the best brown trout water east of the Rockies. In the flies-only, no-kill section reverently called the "Holy Water," insect hatches are plentiful, wading is easy, and the stream-bred brown trout are plentiful and include a goodly number of "lunkers." Especially during the hatch of the misnamed "Michigan caddis" (*Hexagenia limbata*), described shortly, anglers catch some very big browns. Brook trout, although mostly small, also are abundant in the Au Sable and readily take flies, and so they provide steadier sport than the sophisticated big browns. Rainbow are present, but in small numbers. Such is the popularity of the Au Sable among anglers that one can find accounts in many books about trout fishing. The 1994 book *The Angler's Guide to Twelve Classic Trout Streams in Michigan* by Gerth E. Hendrickson is an excellent source.

Although to anglers it is first and foremost a trout stream, the Au Sable supports some 40–50 fish species and prior to 1880 had no trout. An extensive survey in the 1920s turned up 45 fish species. When the Michigan Department of Natural Resources repeated that survey in 1972, mimicking as closely as possible the techniques used and sites fished, they found 44 fish species, but 8 had disappeared and 7 newcomers had arrived. Besides trout, common fishes include sculpins, shiners, white and other suckers, several dace, and Johnny darters. The minnow family is especially well represented. Wherever impoundments occur, lake-dwelling species flourish and often maintain a presence in stream environments due to their success in the man-made environment. Of course, all of this fish diversity helps to make a trophy brown trout fishery. At approximately 30-cm-length, brown trout increasingly prey on other fish. Minnows and small brookies nourish the trophy browns that lurk in deep pools and in the dreams of avid fly fishers. Although the Au Sable is a coldwater stream, summer temperatures exceed 21°C and trout are less successful in headwater sections below lakes, in river sections below impoundments, and in some low-lying river sections receiving little groundwater input.

No aquatic insect features as prominently in Au Sable fishing lore as the curiously misnamed "Michigan caddis," *Hexagenia limbata*. In reality a burrowing mayfly (family Ephemeridae), with tusked mandibles, flanged legs, and prominent gills, the *Hexagenia* hatch in late June–early July is the high point of the season. Emergence begins near dusk and continues until about midnight, sometimes so dense that the rustle of insect wings is audible. Even the large brown trout lose their normal wariness and actively feed at the surface. In the fading light anglers must cast to the sound of "slurps" close to brush and debris jams—not an experience for the novice. *Hexagenia* hatches are greater in the sand and silt sections above and below the Holy Water, where the substrate is mainly gravel, but *Hexagenia* swarms move some distance.

Many other mayflies of the Au Sable and like rivers are of interest to trout and the angler. *Tricorythodes* (family Caenidae), known as the tiny white wing black mayfly to anglers, is a small but abundant mayfly whose nymphs occupy areas of silt and debris. Emergence occurs early in the morning and late in the season. The family Ephemerellidae includes numerous species of *Ephemerella*, which are especially abundant in gravel reaches and rapids. Owing to differences in life cycles, one or another species can be found hatching between April and November. Subimagos are imitated by various flies known as blue-winged olive duns. *E. subvaria* is an important early season species, imitated by the Dark Hendrickson pattern. *Siphlonurus* (family Siphlonuridae) is a large mayfly known to anglers as the Gray Drake. Its emergence in June-July is another high point for Au Sable fly fishers, but a nuisance to motorists, whose windshields and radiators they sometimes clog. The small, streamlined *Baetis* (family Baetidae) mayflies include many species; they are abundant, and their hatches occur from spring until autumn. Subimagos may emerge at almost any time of day, although cloudy mornings seem popular. When trout are feeding during a *Baetis* hatch, the angler has little choice but to fish with long, fine leaders and # 20 flies. While this is not an exhaustive survey of Au Sable mayflies, we will close with one last, important family—the Heptageniidae. *Stenonema, Heptagenia, Epeorus,* and *Rhithrogena* are four common genera that share a flattened dorsal–ventral aspect, a large head with prominent eyes, and a preference for large cobbles in swift water. Mating flights are common at dusk, although some species will swarm on cloudy afternoons. The Light and Dark Cahill and the Quill Gordon are famous dry flies used to imitate the imagos of these mayflies.

On the Au Sable, numerous caddis (Order Trichoptera) hatches occur between April and November. Fly fishers can imitate caddis larvae, some of which dwell in cases, as well as the pupa and adult. Many well-known dry flies imitate adult caddis—emerging, ovipositing, and spent. As is true of mayflies, some caddis females oviposit on the water surface; others enter the water to oviposit on the bottom, and so there are many variations to imitate.

Brachycentrus (family Brachycentridae) is a filtering-collector which gives rise to a large spring hatch on the Au Sable; the Little Black Caddis is a good imitation. Several species of caddis produce large daytime hatches in autumn. These include the filtering-collector family Hydropsychidae (*Hydropsyche* and relatives), simulated by the Cinnamon Sedge and Little Olive Sedge; the grazers *Glossosoma* (family Glossosomatidae) and *Neophylax* (family Uenoidae), imitated by the Little Black Short-Horned Sedge and Small Dot Wing Sedge, respectively; and the predaceous *Rhyacophila* (family Rhyacophilidae), mimicked by the Green Sedge. The peeping caddis (e.g., *Neophylax*) describes the caddis larvae that dwell in cases of mineral or plant material. The head and the thorax with its three pairs of legs protrude from the case, allowing the caddis to forage and move about.

All of the important genera of insects found in the Au Sable are either grazers or collectors (both filtering and gathering) with the exception of *Rhyacophila*, which is a predator. This indicates that there is a plentiful supply of both periphyton and FPOM in the Au Sable and suggests that the mainstem river has the characteristics of a mid-order reach as described by the RCC.

Threats to the River

Logging activities had devastating effects on the water quality and stream habitat of the Au Sable a century ago. Today the watershed, and its threats, are different. Reforestation throughout the twentieth century has produced an uplands of mixed hardwoods and pine, while cedar, spruce, fir, willow, and alder dominate in wetter areas and along river banks. The human population is surging, and the popularity of second and retirement homes fuels much development. Some 67 dams, some constructed long ago as mill ponds, others in this century for power-generation, and ranging in height from < 1 m to > 10 m, impound river sections, modify habitat, and raise water temperatures in many sections of the Au Sable and its tributaries. Wastewater discharge from urban areas has largely been curtailed, but septic fields of second and retirement homes are likely to cause nutrient enrichment. At present, fortunately, the Au Sable's water quality ranks among the best of Michigan's Lower Peninsula rivers. Sprawl, and soil erosion at roads, bridges, and streamside access, are now principal dangers to the health of this river. Recreationists risk "loving the river to death."

Fortunately, state and federal laws provide protection to the Au Sable. A 37-km segment of the main stream from Mio to Alcona Pond was designated under the National Wild & Scenic River Act in 1984. The original plan called for a total of 145 km, but this quantity was reduced to assuage private landowners, who objected to federal designation. Then, in 1987, a total of 560 km received designation under the State Natural Rivers Program, including 157 km of main stem. However, with approximately 60% of the watershed in private ownership, user conflicts are inevitable.

The Au Sable River has taken some of the worst abuse to which a river can be subjected. Yet, protected by its glacial history that ensures abundant, stable, and cool flows, it remains beautiful, clean, and productive. Paradoxically its greatest strength also creates its greatest risk. Highly valued by the many who enjoy the river, its users create a powerful constituency for wise management. Because human pressures can only increase further, we find ourselves with a familiar dilemma. We must somehow summon the will to protect, and if need be, limit, the inevitable impact of our own actions upon the places we revere the most.

THE MADISON RIVER

The Madison River (Fig. 6.3) flows through a variety of settings during its approximately 200 km journey from its origin at the junction of the Gibbon and Firehole rivers in Yellowstone National Park, at approximately 2090 m elevation, to its terminus at about 1400 m. It is one of the most heavily fished trout rivers in the western United States. These two streams

FIGURE 6.3 The Madison River above Ennis, Montana. (Photo by C. E. Cushing.)

also provide one other important asset of the Madison—a rich supply of calcium bicarbonate derived from the Yellowstone rocks, which essentially turns the Madison into a large chalk stream.

Setting and History

The river was named by the explorers Lewis and Clark in 1805 after then Secretary of State James Madison, during their epic journey to the Pacific Coast. Salish (Flathead) Indians were the original users of the area, later driven from the region by the Shoshoni, who in turn were chased out by the Blackfeet. These last inhabitants made life miserable for the first white people to venture into the area, the fur trappers, for it wasn't trout, but beaver that brought the first visitors to the region.

From its origin, the Madison flows as an open, fairly gently flowing stream through the remainder of its journey through the Park and into Hebgen Lake, a man-made impoundment. Hebgen Lake reduces the temperature of the water, by allowing dissipation of heat from the thermally heated river, to levels almost ideal for trout production. From Hebgen Lake, the Madison flows a short distance into Quake Lake, a relatively small lake formed by the damming of the river by an earthquake-generated rock slide in 1959. Quick earth-moving efforts established a new outlet for the river through the rock debris, and the Madison downstream experienced only a short time of dewatering. From Quake Lake to the town of Ennis, the Madison River flows about 80 km through what has been called "one long riffle." This is the most popular and heavily fished reach of the river, not only because of its excellent trout populations, but also for its easy fishing conditions throughout this section. Near the town of Ennis, the Madison divides into several braids before entering Ennis Lake, another man-made impoundment formed originally for electrical generation, but now little used. This lake is silting in, and summer temperatures reach dangerous levels for trout populations. Below Ennis Lake, the Madison River flows about another 50 km through Beartrap Canyon to the Three Forks area, where it joins with the Jefferson and Gallatin rivers to form the headwaters of the Missouri River. Major tributaries are absent; the South Fork of the Madison enters Hebgen Lake and the West Fork of the Madison enters the river from the west about 15 km downstream from Quake Lake. The Madison River flows through a combination of western coniferous forest and sagebrush steppe-desert during most of this journey. Hay meadows are abundant; most of the adjoining land is occupied by cattle ranches.

Ecology

The Madison River is one of the classic, productive trout streams occurring in the western United States. The water flows over rocks and rubble-sized stones, providing abundant habitat for macroinvertebrates and resting and foraging areas for fish. Water temperatures, except in Ennis Lake and the sections of the river below it, remain cold enough for good fish growth.

Historically, the Madison's trout populations were large—both in numbers and size. Catch limits prior to about 1950 were as high as 25 fishes per day, and anglers could expect to land 1.5–2.5 kg fish daily. As its fame grew and fishing pressure increased, limits had to be reduced. Today, catch-and-release fishing is encouraged and, in fact, required in certain portions of the river. Another devastating blow to the Madison fishery, particularly the rainbow trout, was the introduction of whirling disease (see following discussion) into the river. This has resulted in the near complete loss of some age-classes of rainbow trout.

The original native trout of the Madison River was the cutthroat trout (*Oncorhyncus clarkii*); in all likelihood, the trout that Meriwether Lewis illustrated in his journals was a cutthroat. Subsequent planting of nonnative rainbow, brook, and brown trout, together with increased fishing pressure, has resulted in the virtual elimination of the native cutthroat. The introduced trout, of course, were more popular with anglers, biologists, and fish hatcherymen at that time. Rainbows, easier to raise in numbers and weight, and popular with anglers, appeared to be the answer for management. Browns, although not as easy to raise as rainbows, provided trophy fishing because of their ability to better resist angling pressure than rainbows and grow to "trophy" size. Perhaps one of the most significant changes in the trout fishery of the Madison River came as a result of studies in the early 1970s of the impact of rainbow stocking on the resident trout population. Recommendations for the elimination of all stocking of rainbow hatchery trout were put into place in 1974 when biologists discovered that the river could maintain better, healthier, and equally abundant trout populations by allowing natural reproduction to sustain the river. All stream stocking throughout Montana was stopped in 1976, and other states have begun to follow these practices.

The Madison River proper does not have the 1st–3rd order reaches, because these are contained in the Gibbon and Firehole portions of this continuum. The Madison River itself, therefore, originates as about a 4th-order stream at the confluence and receives no tributaries that increase its order.

The Madison supports an extremely rich macroinvertebrate population dominated by mayflies, stoneflies, and caddisflies, and the reach of the Madison from Quake Lake to the town of Ennis supports an abundant trout population, though less than before the whirling disease epidemic. It also has a world-class population of fisherman in pursuit of the fish! This quest reaches its apex during the annual hatch of the "salmonfly" (*Pteronarcys californica*), a large, shredding stonefly (Fig. 6.4), which occurs from mid-June to the end of July. This large (5–7 cm) insect emerges in huge numbers; it is a weak flier and often falls into the water along the shoreline where the lunker trout slurp them up in a true feeding frenzy. The hatch proceeds upstream as water temperatures rise, and the pursuing anglers proceed apace. Note that this species is a shredder—yet here it is occuring in the mid-reaches where CPOM input would be expected to be minimal. The explanation is that the willow and alder riparian vegetation increases downstream along the shoreline of the Madison from Quake Lake to Ennis, thus providing sufficient CPOM to support the population of shredders.

Although this is the most famous insect pursued by anglers on the Madison, many other popular species contribute to huge hatches of insects familiar to fly fisherman. Common mayflies (order Ephemeroptera)

FIGURE 6.4 The "salmonfly" *Pteronarcys*, which is actually a stonefly. (Photo by H. Daly.)

include *Baetis, Rhithrogena, Ephemerella, Drunella*, and the popular "trico," *Tricorythodes*. Although the "salmonfly" *Pteronarcys* is the most prominent stonefly (order Plecoptera) in the river, several other genera occur including *Pteronarcella, Skwalla, Isoperla*, and *Sweltsa*. Caddisflies (order Trichoptera) rival the mayflies in diversity in the Madison River; it is no wonder that many common fly patterns used by fisherman are imitations of adult caddisflies. There are representatives of all of the functional feeding groups—shredders, grazers, collectors (both gathering and filtering), and predators. Six different species of net-constructing filtering-collectors in the family Hydropsychidae alone are present. Anglers thus have a wide variety of fly patterns to choose from in attempting to duplicate this smorgasbord of trout food; popular patterns include tricos, Adams, Blue-Wing Olives, Elk Hair Caddis and Goddard Caddis, and the Wulff attractor patterns, to name just a few. But when the *Pteronarcys* hatch comes on, preferred imitations of this large, awkward stonefly include Sofa Pillows, Bird's Stonefly, Fluttering Stones, and other large fluffy patterns. Then, of course, many of the popular nymph patterns are also successful—Hare's Ear, Prince, Bitch Creek, and others. Hatches of these insects extend through the summer and fall, providing anglers something to duplicate during most of the season.

Well, this is fine for the angler, but what can we learn of the ecology of the Madison River, including its headwater streams? The lower orders of this ecosystem, which are incorporated in the Gibbon and Firehole rivers, are fairly open and so considerable sunlight reaches the streambed. Studies have shown that the section of the Madison River within Yellowstone National Park is autotrophic, and invertebrate populations in the Gibbon

River are dominated by collectors and grazers as would be expected because lack of shade favors periphyton growth. The openness of the Gibbon, Firehole, and upper Madison River is characteristic of many western streams and is a significant departure from the basic RCC model. Now, if you determine what functional feeding groups are represented by the dominant hatches of insects in the Madison, you will find that the vast majority are collectors (both gathering and filtering) and grazers. This tells us that the Madison downstream from Quake Lake compares favorably with predictions of the RCC model for a mid-reach stream with little shade. Obviously, algae are present to support the grazer population (as our model predicts), and FPOM must be present in large amounts to feed the huge diversity of mayfly and caddisfly collectors.

Threats to the River

Several problems, other than those discussed later in this chapter that are pertinent to all trout streams, are of special concern in the west. These include overgrazing, logging and mining impacts, urbanization, and the scarcity of water.

For years, cattleman have exercised their rights to unrestricted grazing over much of the western public lands, often with little regard for either the condition of the range or where livestock was allowed to graze. Overgrazing of vast areas has resulted in high runoff, heavy siltation of streams, and subsequent destruction of aquatic organisms. Unrestricted access to stream channels by livestock has resulted in overgrazed and trampled stream banks where cattle have watered. Streambank erosion is heavy, riparian vegetation is destroyed, and stream ecosystems are degraded. Fencing of streamside banks to allow cattle access only at selected places has proven to be an effective tool in the restoration of streamside riparian areas. The Madison Valley suffered its share of damage from overgrazing, but the ranchers in the area have taken a progressive look at past excesses, and new practices are resulting in improvements throughout the watershed.

Logging has created severe impacts to rivers and streams as a result not only of unwise timber cutting schemes, but also from the related activities of road building, log transportation, and waste disposal. Cutting on steep hillsides and clear-cutting have resulting in severe impacts to streams and rivers because of the propensity of these lands to have high rates of silt runoff into streams. Road building has been especially detrimental because most logging roads are built along streams for ease of access; subsequent roads then cut along steep hillsides. One can easily see how heavy loads of silt reach the streams during periods of high runoff. These silt loads smother algae and macroinvertebrates, destroy salmonid spawing beds by filling the gravel beds with silt, and increase water temperatures by destroying riparian vegetation and opening up the waters to direct sunlight. Inadequate programs for waste disposal result in streams being choked with lumbering slash, thus destroying natural habitat.

Mining impacts are largely concerned with the introduction of waste products into streams where they may either kill organisms outright or allow certain chemical elements, such as heavy metals, to accumulate in the food web to the point where they pose potential or real harm to humans. Mine drainage typically increases the acidity of receiving streams; if severe enough, no life exists in the streams. One waste product of concern in gold mining regions of the west is cyanide, which is used in the extraction process. Processing of ore results in large ponds of heavily cyanide-laced water; failure of dikes or other restraining structures could result in huge outwashings of this highly toxic material into nearby streams. Obviously, this could not only wipe out all life in the stream, but pose serious threats to human use of the streamwater. Mining may also introduce silt and other products into streams and rivers, which can adversely impact the aquatic ecosystem. This can have a double impact on macroinvertebrates; it can directly smother and kill them if dense enough, or it can cover the periphyton, thereby eliminating a major food source. One other serious impact of mining on streams occurs in places where the overburden is disposed of into small, headwater valleys. In these cases, entire stream reaches are buried and eliminated from existence.

Only one brief story needs to be mentioned in relation to the urbanization and encroachment upon rivers and streams by people. Robert Redford's film version of Norman Maclean's novella *A River Runs through It* could not be filmed on the Blackfoot River, the original river in the story, because it was too highly built up by homes; an alternate, less-developed river had to be filmed instead.

RIVER PRODUCTIVITY AND TROUT BIOMASS

The concept and measurement of productivity in streams, whether of plants, invertebrates, or fishes, has a scientific basis, both in terms of how and what is measured and the units of measurement. It is fairly obvious to anybody who spends much time in different streams and rivers what a productive stream is. It has rich populations of plants, insects, and fish. But how is this measured and expressed? Basically, the organisms of interest are collected from a known area, then counted, weighed, or analyzed for energy units (calories); electroshocking is a common method for fish collections (Fig. 6.5). Repeated measurements over time provide a rate, or time, unit. These data are then expressed as an amount in a given area produced over a given amount of time. For fisheries, these units are usually pounds per acre per year (1 lb/acre = 0.9 kg/ha).

Many measurements have been made of trout production in various streams, and it has been found that production, on the average, is higher in hardwater limestone streams than in softwater freestone streams. Limestone streams with excellent fish populations have production rates of 110–220 kg/ha/year, with some measurements up to 340 kg/ha/year being reported. Contrast this to much lower production values of 6–70 kg/ha/year in freestone streams, even though these can provide excellent fishing opportunities. The reason for these differences are easily explained. Limestone streams generally drain geographical regions with highly soluble rocks that provide high concentrations of nutrients, thus ensuring rich growths of algae and, in turn, invertebrates upon which the trout feed.

FIGURE 6.5 Electroshocking fish, Mack Creek, Oregon. (Photo by C.E. Cushing.)

Freestone streams, on the other hand, drain igneous rocks that are insoluble, thus resulting in more infertile water in streams draining these regions. Remember that food webs are generally based on the "food pyramid" model; each succeeding food level depends on a larger base—few fishes depend on many invertebrates which depend on even more algae. The rule of thumb often cited is that it takes 10 times more energy to support each succeeding level, that is, it would take 100 kilocalories of invertebrates to support 10 kilocalories of trout.

In early studies of trout production, a situation which came to be known as Allen's Paradox resulted from a study by K. Radway Allen on a population of brown trout in a stream in New Zealand. Essentially, he found that each time he sampled the stream, the energy content of the invertebrate population in the stream was insufficient to support the amount of trout present; yet, there they were. Several reasons have been suggested for this discrepancy—terrestrial invertebrate drift, small fishes, and other food resources not included in the analysis—which could make up the difference for the inadequate invertebrates population. A second explanation arises from the fact that the amounts of invertebrates and fishes present were reported as standing crops, that is, the amount of energy of invertebrates and fishes per unit area at each sampling time. What was not considered was the growth and development time of the two populations; the time factor was ignored. Thus, the energy provided by the invertebrates to the trout was underestimated because it was continually reproducing and replacing what was eaten, whereas the trout population remained relatively static. To this day, aspects of Allen's Paradox continue to be addressed in stream studies of fish and invertebrate production. Some scientists believe that trout feeding on invertebrates is overestimated; others believe that invertebrate production is underestimated—all of which provides exciting avenues of research for stream ecologists. (In 1994, Tom Waters presented much of the above information on trout and river productivity in an article published in *The Quill*.)

ENVIRONMENTAL PROBLEMS FACING TROUT STREAMS

Trout streams throughout the country face a wide variety of detrimental factors: dams, diversions, pollution, loss of habitat to encroachment, and logging impacts are some we have mentioned in our description of the Beaverkill, Au Sable, and Madison rivers. Here, we want to address two significant impacts that will affect trout streams for some time to come: whirling disease and stocking policies.

Whirling disease is an infection of trout and salmon caused by a protozoan parasite, *Myxobolus cerebralis*. The parasite has a two-host life cycle, tubifex (*Tubifex tubifex*) worms that live in sediments and ingest the spores released by dying, infected fishes when they decay, and the trout themselves. The spores, which can survive for up to 30 years in wet or dry sediments, undergo development in the worm's intestine and multiply rapidly, changing into the form of the disease that infects trout. When released by the worm, the water-borne spores infect susceptible fishes by attaching to their bodies, or they are ingested if trout eat tubifex worms.

Young fishes are most at risk when heavily infected with the parasite because the cartilage of young fishes is not hardened, allowing the parasite to cause deformities (Fig. 6.6). Infected fishes may display a distinctive rapid whirling, thus the name. In instances of high infectivity rates, the disease is usually fatal to young trout. Fishes that survive carry the spores throughout their lifetime.

Whirling disease has been identified in all western states, except Arizona, and several states in the northeast. It has devastated several popular fisheries, notably the rainbow trout in the Madison River and upper Colorado River drainage. Conversely, California has identified the parasite in many streams; yet, they have not noticed declines in any trout populations. Strangely enough, it appears that rainbow trout are more susceptible than other species. Eradicating this parasite is going to be difficult, if not impossible. Many trout hatcheries are severely infected by the parasite; eradicating them from the rearing ponds and the water supply will cost millions of dollars per hatchery. Keeping them clean will add to these costs. Controlling the disease in nature likely will be even more difficult. The tubifex worm is widespread in nature; the spores spread easily in mud on waders, boats, and other equipment; and little is known of the relationships between the parasite and the various salmonids. These facts all point to a long and difficult research effort to eradicate or control this parasite.

Not all anglers are looking for the same fishing experience. Some want to take home a lot of fishes, others prefer catch-and-release fishing, some are satisfied with a few big fishes, others disagree as to whether catching native fishes is better than hatchery-reared exotics. Since anglers, through

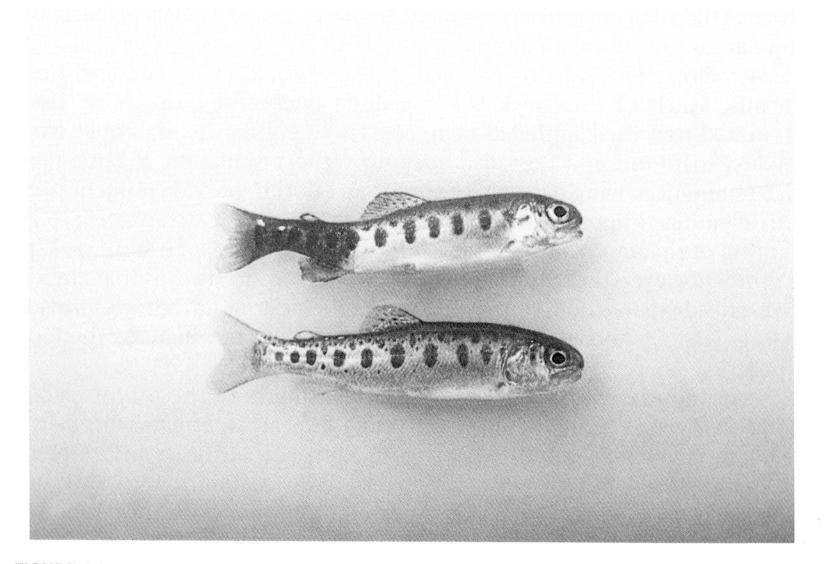

FIGURE 6.6 Young trout with whirling disease (top) and normal fingerling (bottom). (Photo by G. Schisler.)

their license fees, provide a significant portion of the fishery manager's budget, it requires the wisdom of Solomon to please everybody. To make this worse, most locations are experiencing more and more anglers fishing in fewer bodies of water as populations explode and development eradicates more and more fishing habitat.

One of the biggest controversies in trout management is the conflict between those people who promote catch-and-release fishing, and those people who promote minimum restrictions as to how to fish with as much stocking as feasible. Catch-and-release fishing necessitates the development and designation of "blue ribbon" trout waters having naturally reproducing trout populations, in other words streams having the characteristics mentioned in the first part of this chapter. The latter type of fishing can be developed in streams having adequate habitat characteristics except for spawning areas; in other words, they have essentially all of the habitat characteristics necessary to allow trout to live, but not to reproduce. Fishes caught by anglers must be replenished with hatchery-reared fish because natural reproduction is not available.

SOLUTIONS

Much effort is being expended by individuals, fishing organizations, professionals, and others to help solve the problems encountered with protecting and enhancing the trout streams that we have. A large literature has been devoted to this, but we want to highlight some of these here;

more exhaustive treatments of each of these can be found in both the popular and scientific literature.

A major effort is being devoted to the restoration of damaged trout streams. Much of this work is being done under the auspices of Trout Unlimited and their affiliated chapters. These efforts usually entail work devoted to restoring a degraded stream by the installation of structures, such as boulders, wing dams, logs, etc., in an effort to provide resident trout with conditions approximating those found in natural streams (Fig. 6.7). Planting of riparian plants, such as willows, is also popular. It must be kept in mind, however, that these efforts are best devoted to streams that already have an adequate food supply for trout; the restoration efforts provide resting, feeding, and hiding places for trout, but do not increase the food supply.

Education efforts devoted to changing attitudes of the fishing public are badly needed. Convincing the inveterate meat fisherman who insists on taking home his limit on every trip to change his attitudes and embrace catch-and-release fishing, or at least vastly reduced limits, is a hard, but necessary sell. The public must also be taught that they must take an active role in helping their local conservation departments protect the good stream habitat remaining and encourage the wise management of these rivers and streams.

There is another fresh wave of thinking wafting through both lay and professional people in terms of stream reclamation. This is the thought of dam removal. As more and more dams are coming up for relicensing, increasing thought is being given to the question, "Is this dam still necessary?" Many older dams constructed early in the twentieth century have

FIGURE 6.7 Example of stream restoration efforts. (Photo by Trout Unlimited.)

outlived their purpose, either economically or structurally. Already, funds have been allocated by Congress to remove some of these dams, and the salmonid populations are often the major benefactor of these actions.

Clearly, the preservation, protection, and enhancement of trout streams is a priority among conservationists today; indeed, healthy trout streams are virtually synonymous with what we mentally picture when we visualize a stream the way we would like to see it exist. In the following chapter, we will examine what happened to two classic western salmonid streams, the Colorado and Columbia rivers, when they were severely abused by man.

Recommended Reading

Hendrickson, G. E. (1994). *The Angler's Guide to Twelve Classic Trout Streams in Michigan.* University of Michigan Press, Ann Arbor.

Leonard, J. W. and Leonard, F. A. (1962). Mayflies of Michigan trout streams. Cranbrook Institute of Science.

Richards, C. and Braendle, B. (1997). Caddis super hatches. Frank Amato, Portland, Oregon.

Van Put, E. (1996). *The Beaverkill.* Lyons & Burford.

Waters, T. F. (1994). Productivity of streams—can we increase it? *The FFF Quill,* Spring 1994, pp. 14–15.

CHAPTER 7

Large Rivers of the West

INTRODUCTION

Several large rivers have headwaters in the western United States, including the Columbia, Colorado, Missouri, Yellowstone, Snake, Arkansas, and Rio Grande rivers. In this chapter we will emphasize the Colorado and Columbia rivers, both well studied and important rivers. Physically, they are similar in length (Colorado: 2320 km, Columbia: 1936 km) and drainage area (635×10^3 km^2 and 670×10^3 km^2, respectively). But the mean annual discharge of the Columbia, at 7960 m^3/s, is more than ten times greater than that of the Colorado, at 640 m^3/s. The Colorado is an arid-land river, with a low discharge to drainage area ratio. Issues of water supply dominate its management. The Columbia's ample discharge has been harnessed to provide so much hydroelectric power that the region enjoys the cheapest power in the nation, and it can market its surplus to California and elsewhere. The economic benefits of its dams dominate its management. But it is the ecology of these rivers which is of interest to us here, so let's have a look at what we know of the ecological characteristics of these two major river systems.

The ecological history of these two rivers can be summed up in a single word—abuse. This abuse largely has been due to the construction of large hydroelectric and storage dams to produce cheap and abundant electrical energy, to supply irrigation water to crops throughout the western United States, and to control floods. This has happened, but at a cost which people only now are truly beginning to appreciate. From the 1930s to recent years, the Corps of Engineers and Bureau of Reclamation have held fast to their philosophy that every drop of water that flows to the sea unused is wasted. Fortunately, as public awareness of the ecological harm caused by dams began to grow, the missions of both the Corps and the Bureau evolved from dam-building to other pursuits, including recreation.

In this chapter, we also will examine how these short-sighted policies have impacted the ecological characteristics of the Colorado and Columbia rivers.

THE COLORADO RIVER

Fig. 7.1 shows a map of the Colorado River Basin and the location of the major dams. The Colorado River Basin is divided at Lee's Ferry (Fig. 7.2) into the Upper Basin (Wyoming, Colorado, and Utah) and the Lower Basin (Arizona, California, Nevada, and New Mexico). The historic origin of the Colorado River (then called the Grand River) was the outlet of Grand Lake, Colorado. Current maps have redesignated the former North Fork of the Colorado as the Colorado River (Fig. 7.3), so that it now originates in Rocky Mountain National Park. From there, the river travels 2320 km to its outlet in the Gulf of California. Along the way, it passes through seven mainstem dams, the two largest being Glen Canyon Dam, at the divide between the upper and lower basins, and Hoover Dam in the Lower Basin. Additional dams (over 100) are present on most major tributaries; one dam each on the Green, Gunnison, and San Juan rivers, and four in the Gila River system. Dam development is not as extensive on the Colorado River, especially the upper part, as it is on the Columbia River; however, a few large dams have severely impacted the ecology of this river system.

Preimpoundment Ecology

Most of the free-flowing reaches of the upper Colorado River remain relatively pristine, with a rich and diverse fauna and flora of fishes, macroinvertebrates, and algae. These streams have their share of perturbations that accompany increasing encroachment of people in these developing regions, but many retain the ecological characteristics of healthy trout streams. Little is known of the preimpoundment ecology of the lower Colorado River Basin—that region most severely impacted by the large dams. The Green River below the Yampa River and the Colorado River from central Colorado to Lake Powell are the only reaches where the present flowing water environment resembles historic conditions.

FIGURE 7.1 Map of Colorado River Basin, showing demarcation of Upper and Lower Basins (dotted line) and location of major dams. (From the U.S. Department of the Interior Bureau of Reclamation, with permission.)

FIGURE 7.2 Photograph of the lower Colorado River at Lee's Ferry. (Photo by C. E. Cushing.)

FIGURE 7.3 Photograph of headwater reach of the Colorado River. This reach was formerly known as the North Fork of the Colorado River, but current maps show the headwaters at a site approximately 28-km north of where this picture was taken. (Photo by C. E. Cushing.)

Postimpoundment Ecology

In general, the aquatic habitat of the postimpounded lower Colorado River is characterized by more constant thermal regimes (though waters immediately below dams usually are cooler), higher salinities, lower turbidity, and reduced scouring by bed sediments resulting in a more profuse benthic algal (periphyton) population. Macroinvertebrate fauna and fish populations remain diverse, but are composed of species different from those found prior to extensive impoundment. The altered water temperatures now found in the river and reduction in the transport of organic material have resulted in significant changes in the availability of macroinvertebrate food resources for fishes in the river. In the Upper Basin, high densities of macroinvertebrates are found on the gravel bars and these are used extensively by the fishes. Benthic macroinvertebrate populations in the Grand Canyon region are productive, but composed of few species, consisting mostly of chironomids, simuliids, oligochaetes, and an introduced amphipod.

The distribution of fish species within the Grand Canyon has changed significantly over the past several decades due largely to the altered temperature and sediment discharges, but it is also related to changes in spawning habitat conditions. Early fish populations were highly endemic; 78% of the 32 fish species found in the Colorado River Basin in 1895 were endemic (see Chapter 10), but several of the native species that were once restricted to the Grand Canyon have been eliminated. Cold-water conditions have adversely affected reproduction of several native fish species; however, some continue to spawn in available tributaries where suitable spawning conditions still occur. Competition and predation by nonnative fish species have increased, resulting in further declines of native fish species.

Ecological interactions within the Grand Canyon are interesting. Whereas the native fish populations evolved in an environment characterized by highly variable discharges, large annual temperature fluctuations, high turbidity, high inputs of organic matter, and basin-wide opportunity for migration, the present environment is much less variable. Water temperatures do not vary seasonally, the water is clearer, and the annual silt loads no longer course through the system. Much of the water quality of the present Canyon water is determined by processes occurring in Lake Powell.

The changes have affected not only the fish populations, but also the macroinvertebrates, which, in turn, is the food supply for the fishes. The large amounts of organic debris formerly present also provided a suitable substrate for insects. Most surprising, over 65% of the aquatic plant and invertebrate biomass present in the Grand Canyon is produced upstream from Lee's Ferry! The riparian community has also changed dramatically. It was once characterized by perennial vegetation such as mesquite, catclaw, native willows, and others; it is presently being replaced by rapid expansion of saltcedar (*Tamarix* spp.).

Comparison of historic and present hydrographs and temperature conditions illustrate how severely dams have altered the river (Fig. 7.4). These altered flow and water temperatures have significantly impacted, both positively and negatively, the ecology of the river below the dams. On

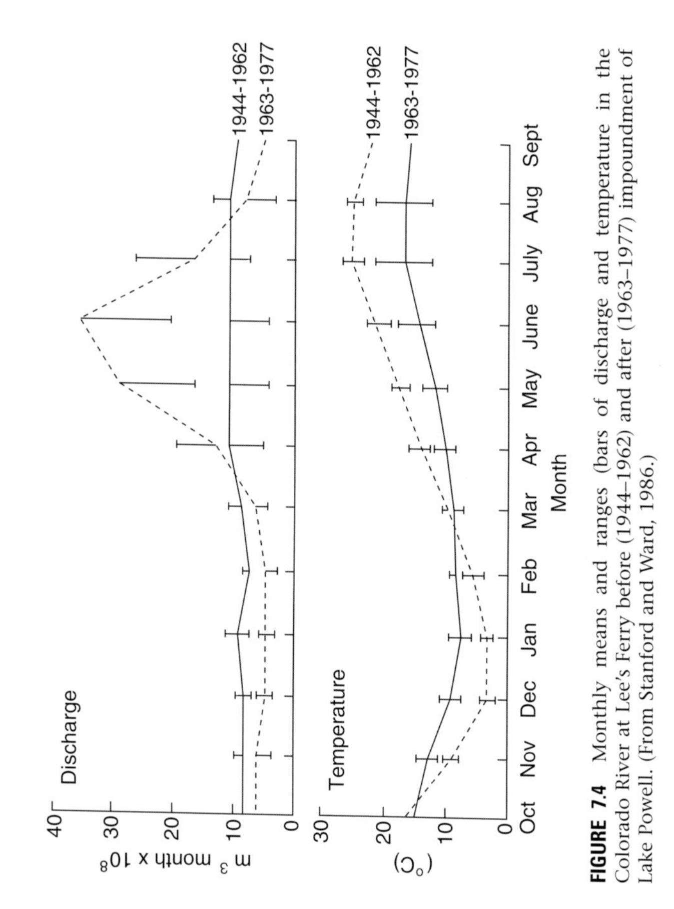

FIGURE 7.4 Monthly means and ranges (bars of discharge and temperature in the Colorado River at Lee's Ferry before (1944–1962) and after (1963–1977) impoundment of Lake Powell. (From Stanford and Ward, 1986.)

the positive side, the constant, cool, release of water below Flaming Gorge Dam on the Green River has resulted in an extremely productive reach of river supporting rich algal and invertebrate populations, and, in turn, an abundant trout population. On the negative side, the constant flow release has deprived the lower reaches, especially within the Grand Canyon, of the seasonal flood peak which was necessary to suspend and redeposit sediments as well as inputs of sediments from upriver. This resulted in a decrease in the number of large sandbars and establishment of extensive tamarisk communities. Indeed, the ecology of the Colorado River below Glen Canyon Dam was historically related to the seasonal wax and wane of flow and the suspended sediment load carried by the river. Alternating deposition and erosion of sandbars by seasonal high flows carrying high suspended loads of sand and silt resulted in the characteristic sandbars used by river-floating camping parties for decades. Also, the release from Glen Canyon of virtually silt-free water has resulted in the Colorado River changing from a silt-laden, turbid stream to a clear-flowing one below Glen Canyon Dam. This would, on the surface, seem to be a positive impact, but the silt load performed a valuable service. It should be noted, however, that all input of sediment to the Colorado River below Glen Canyon Dam has not disappeared. Several undammed tributaries, such as the Little Colorado River and the Paria River still contribute significant sediment loads during periods of high flow.

One of the saddest aspects of the ecology of the present Colorado River is the fact that it no longer reaches the sea in most years; less than 1% of its virgin flow now reaches its mouth. The Colorado River sustains more consumptive water use than any other river in the United States. Water withdrawal has resulted in its terminus now being defined by an extensive area of wetlands in the vicinity of its mouth; boats are unable to travel upstream from the sea. Once the Colorado River reaches the point where it has passed through all of its impoundments and has received the Gila River and its two overused tributaries, the Verde and the Salt (see Chapter 10), one-third of the flow is then diverted to California. Some of this withdrawal irrigates the Imperial Valley; the rest goes to Los Angeles and San Diego to allow their citizens to live in lands never meant to support the populations existing there.

The history of politics, intrigue, and outright lawlessness that allowed this huge water transfer to occur is beyond the scope of this book. The interested reader is encouraged to read the excellent book *Cadillac Desert* by Marc Reisner, perhaps the best book in print addressing the water problems facing the American West. Yet it would be remiss of us if we did not say something about the impact of this diversion upon the ecology of the affected aquatic ecosystems. The increased salinities in the lower Colorado River, resulting both from concentration of dissolved chemicals by evapotranspiration from the large reservoirs (Lake Mead and Lake Powell) and from extensive irrigation withdrawal and return, have made the water unusable for drinking or irrigation needs. This is not true just in the Lower Basin, but also pertains to that water diverted to California and even more to that water which eventually reaches the Mexican border. Obviously, water with extremely high salinity concentrations will have adverse impacts on biological communities found within any stream channels carrying this water. Diversity will be restricted to those organisms able to tolerate high salinities; all other organisms will disappear. Further, those saline tolerant organisms will be impacted by the artificial flow regimes associated with man-made diversion schemes.

The overall effects on fish populations have resulted in a reduction of the native fish populations to the point that several endemic species face the possibility of extinction. Conservation agencies have initiated an extensive program directed toward protecting and enhancing populations of those species at risk.

There are four native fish species of major concern; they are the Colorado pikeminnow (*Ptychocheilus lucius*), the bonytail (*Gila elegans*), the razorback sucker (*Xyrauchen texanus*), and the humpback chub (*Gila cypha*). The highest numbers of these fishes, except for the Colorado pikeminnow, are found in the Lower Basin. Unfortunately, most of the natural habitat needed to sustain these species is found in the Upper Basin. The various basin states have initiated a variety of recovery programs intended to save these native species. The programs include transportation of fishes within the system, stocking of hatchery-raised fish, especially in the Upper Basin, and engineering alterations such as construction of fish ladders.

But probably the most important program being instituted within the Colorado River system to aid in the recovery of native fish species is the attempt to restore the annual hydrograph to something approximating natural river-flow conditions. Many hydroelectric dams are operated to provide electric power when demand is greatest—usually during daylight hours. Thus flow fluctuations occur within 24-h intervals in a river which historically experienced a hydrograph which had fairly uniform daily flows interrupted by a high spring runoff period. Experiments have been conducted to artificially reproduce historic hydrographs to examine the impact on bedload sediments, formation and erosion of sandbars, and impact on native fish habitat. Results of these experimental manipulations are still being analyzed, but early observations revealed that the accumulated sandbars were eroded, large pieces of organic matter were again transported downstream, and the rich periphyton populations established below Glen Canyon Dam were not eliminated by the high flows and scouring.

It was early recognized that there would not be enough water flowing in the Colorado River to meet the demands of the growing populations served by water withdrawal from the system, especially in California. To adjudicate distribution of the water, the Colorado River Pact of 1922 divided the Colorado River Basin into an Upper and Lower Basin, and water was apportioned equally to the two basins based on estimated annual flow at Lee's Ferry, the dividing point between the two basins, of about 18.5×10^9 cubic meters; the Pact also included a contingency of 1.2×10^9 cubic meters for the satisfaction of Mexican claims. This estimate has subsequently been shown to be based on a time interval when discharge was higher than it has averaged through the remainder of this century.

Subsequent agreements have impacted water manipulations and the ecology of the Colorado River system; these are described in Table 7.1.

Table 7.1 Agreements Governing Allocations of Colorado River Water

Agreement	Purpose
1928 Boulder Canyon Project Act	Required California to limit its use to 4.4 million acre-feet of the lower basin's allotment.
1948 Mexican Water Treat	Agreed to provide Mexico with 1.5 million acre-feet per year.
1948 Upper Basin Compact	Upper Basin states agreed to allocate among themselves the water reserved by the 1922 Compact.
1963–1964 Supreme Court decision in *Arizona v. California*	Complicated series of rules, the result of which was that Arizona received the water right it needed to seek federal funding of the Central Arizona Project (CAP).
1968 Colorado River Project Act	Authorized CAP and partially reversed *Arizona v. California*.
1972 and 1977 Clean Water Acts	Salinity level reduction became a national commitment.
1973 Endangered Species Act	Calls for consideration of all dam operations as they relate to potential or real impacts on listed species.

Major challenges face all agencies concerned with water flow of the Colorado River, not the least of which is deciding upon an appropriate strategy for restoration of the Colorado River below Glen Canyon Dam. Five management goals have been suggested by some scientists to address potential restoration of the Colorado River ecosystem. These include (1) traditional river management, (2) managing the river as a naturalized ecosystem, (3) rehabilitating it as a simulated natural ecosystem, (4) rehabilitating it as a substantially restored ecosystem, and (5) reestablishing a fully restored ecosystem. Indeed, serious questions arise as to whether any major restoration efforts should be made in light of the complex and interrelated impacts to different parts of the ecosystem that result from the different options available.

THE COLUMBIA RIVER

The Columbia River and its drainage basin, together with the location of mainstem dams and those on its major tributary, the Snake River, are shown in Fig. 7.5. Before getting into our story of the Columbia, we should describe some of the characteristics of the Snake River itself because it is both large in size and large in importance to our ecological story.

The Snake River originates in Wyoming and flows 1690 km through Wyoming, Idaho, Oregon (along the eastern border), and Washington to its confluence with the Columbia River. The waters collected in its mountainous headwaters drop some 2,900 m. Its waters, once they get to the irrigated regions in Idaho, are some of the most contested waters in terms of usage of any in the West. Tim Palmer, in his 1991 book *The Snake River*,

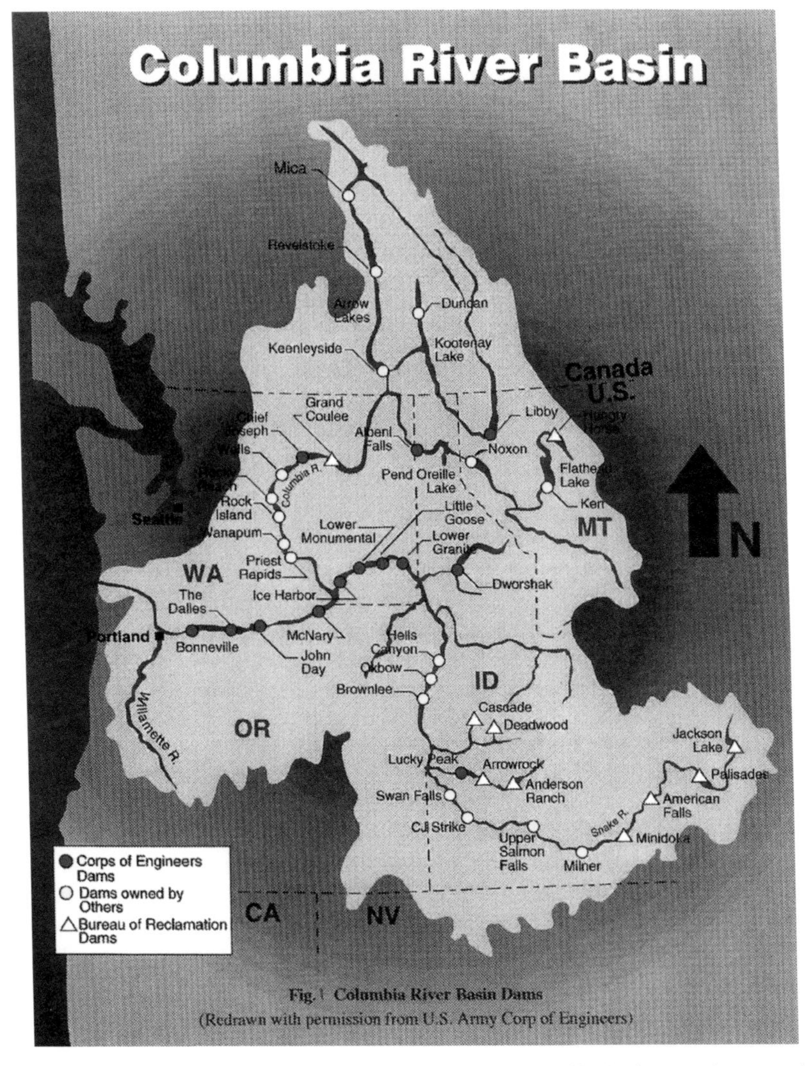

FIGURE 7.5 Map of Columbia River Basin showing location (dots) of major dams on the Columbia and Snake rivers. (From the U.S. Army Corps of Engineers, with permission.)

tells the truly remarkable story of this great river—its history, political shenanigans, philosophies of the various users—and leaves one to wonder if truly rational people have been making the decisions governing appropriation of the river's waters. One of the great stories is that the Snake River is not really one river, but it is two! How so? The river from its source to Milner Dam, halfway through Idaho gradually diminishes as it is appropriated by various users until, believe it or not, it simply stops at Milner Dam. No water flows in the reach immediately below the dam. Further downstream, seepage and return flows finally begin to rebuild the rivers flow and the second Snake River begins. Enlightened irrigators and other users have started returning more water above Milner Dam; perhaps in time it will be a single river again.

The Snake River is, or was, of vital importance to the salmonid runs of the Pacific Northwest. Historically, the river carried huge runs to spawn in the river and the headwaters of its main tributaries. The four mainstem dams on the lower Snake River in Washington State have essentially eliminated natural spawners from returning to the Clearwater and Salmon rivers. The formerly huge run of kokanee salmon to their native spawning location in Redfish Lake (the lake got its name from these fishes) in the headwaters of the Salmon River dwindled to a single returning fish in the early 1990s! Intensive restoration efforts utilizing hatcheries have since increased this number. Hell's Canyon Dam, Oxbow Dam, and Brownlee Dam in the Snake River Canyon have done the same for spawning runs in the upper tributaries in southern Idaho; Hell's Canyon Dam does not have fish passage facilities. Efforts to replace some of these runs with hatchery-produced fishes, such as the federal hatchery below Dworshak Dam on the North Fork of the Clearwater River, have been largely unsuccessful. Serious consideration is presently being given to breaching Ice Harbor Dam, Lower Monumental Dam, Little Goose Dam, and Lower Granite Dam on the lower Snake River to return river flows to near-normal conditions. The listing of several runs as threatened and/or endangered under the Endangered Species Act has prompted this consideration as perhaps the most reasonable way to restore the historic runs of salmonids. Certainly, hatcheries and barging of smolts downstream have not been the answer, so despite the considerable economic tribulations certain to impact irrigation, transportation, and electrical production interests, breaching may occur.

The story of the Columbia River ecosystem—what it once was and what it is now—is intimately tied to this extensive array of dams and their impact on the natural resources, especially the historically abundant salmonid runs. The Columbia River originates in Columbia Lake, British Columbia (Fig. 7.6); from there it flows 1936 km (740 km in Canada) and drops 780 meters in its journey to the Pacific Ocean. It drains a basin of about 670,810 square kilometers of which 15% is in Canada; the majority of the area is in the states of Washington, Oregon, and Idaho, with smaller areas in Wyoming and Utah. Other major tributaries are the Spokane, Okanogon, Wenatchee, Yakima, Cowlitz, and Lewis rivers in Washington State, and the Umatilla, John Day, Deschutes, and Willamette rivers in Oregon.

Preimpoundment Ecology

Little is known of the existing ecological conditions of the Columbia River in terms of its aquatic communities prior to closure of the first dam, Bonneville, in 1938. Original flows ranged from 1980–18,690 cubic meters per second from May to July and around 1980 cubic meters per second from September through March. One study in the mid–Columbia River

FIGURE 7.6 Headwaters of the Columbia River just below where it exits from Columbia Lake in Alberta, Canada. (Photo by C. E. Cushing.)

basin in the vicinity of River Kilometer 478–644, showed an extensive and diverse macroinvertebrate fauna typical of large, free-flowing rivers. An abundance of caddisflies, stoneflies, mayflies, and other taxa characteristic of high quality waters was present. Also found was a phytoplankton community representative of clean-water conditions, and the water quality conditions were judged to be of very good quality in general. Both cold- and warm-water fish species were present.

Postimpoundment Ecology

The organisms most closely identified with the Columbia River are, of course, the anadromous fish runs consisting of steelhead trout and several species and races of salmon. The postimpoundment change in the ecological character of the Columbia River has severely impacted these fish runs, and it is the aspect most worthy of discussion concerning the present ecological conditions. Impacts to organisms other than salmonids reflect the overall change of the aquatic environment from a free-flowing riverine system to a series of lakelike pools connected by short flowing stretches. Aquatic organisms in the reservoirs are largely composed of animals capable of living in soft sediments—oligochaetes, midge larvae, and mollusks.

The construction of 11 dams on the main stem of the Columbia River within the continental United States, two more in Canada, plus over 100 additional dams on its tributaries, has greatly altered a mighty, free-flowing river system that once supported over 200 anadromous stocks and returned 7–30 million adult salmon and steelhead to the river annually. Today the

FIGURE 7.7 The Hanford Reach of the Columbia River, the last free-flowing stretch of this once mighty, but now tamed, river. (Photo by C. E. Cushing.)

system is characterized by a series of reservoirs, which produced about one million returning fish in 1997, less than 10% of historic levels. With the exception of short flowing stretches between the spill of one dam and the head of the next downstream reservoir, the only free-flowing stretch of the river remaining in the United States is an 80-km reach within the Hanford Nuclear Reservation. This stretch, named the Hanford Reach (Fig. 7.7), lies between Priest Rapids Dam and the head of Lake Sacajawea, the reservoir behind McNary Dam. It is subject to daily water level fluctuations as a result of power generation at Priest Rapids Dam, but still retains many characteristics of the original river and is the last place where chinook salmon and steelhead can spawn in the main stem of the Columbia River. Thus, what was once a flowing river has been transformed, except for the Hanford Reach, into a series of pools having ecological characteristics more similar to lakes than to rivers. Trichoptera, mainly filter-feeding members of the family Hydropsychidae, still occur in the Hanford Reach, but the once abundant mayfly and stonefly populations are virtually gone.

The dams exact a toll on anadromous fishes traveling in both directions; not all fishes are successful in negotiating the ladders provided for upstream migrants, and the downstream migrating smolts are subject to many hardships, including lack of spring flows to flush them to the sea, passage through turbines, increased predator pressure in pools downstream of the dams, and so on. A number of technological fixes have been tried to enhance both fish numbers within the system and their passage around the dams. Fish hatcheries have long been the choice of fish managers to enhance existing populations, but their value has come under increasing scrutiny in recent years. Although some change is being seen, their use will probably continue for years to come. Bypass facilities, either fish ladders for upstream migrants or passageways around turbines for downstream migrants, have also been engineered but certainly have not been the lodestone for the alleviation of the extremely high mortality, especially of downstream migrating smolts. Attempts have been made to modify the spill gates at dams to prevent smolts passing over the dams at high water flows from being plunged into deep pools where they encounter nitrogen supersaturation and may succumb to its lethal effects. Screening of turbine intakes and irrigation diversions have had mixed success. One of the most controversial programs in place today is the collection of downstream migrating smolts at upstream dams, transferring them to either barges or trucks, and then transporting them downstream to be released below Bonneville Dam (Fig. 7.8). Fishery biologists and agencies have long argued the merits of this program; the bottom line is that the downward decline of the anadromous fish populations continues. Listing of some races, and consideration of others, under the Endangered Species Act has resulted in a flurry of management plans, new studies, etc., aimed at the long-term goal of restoring the historic salmon runs, or at least increasing them significantly. This has resulted in a complex argument among agencies, irrigators, power companies, environmentalists, commercial interests, and Indian tribes over how best to run the Columbia River system.

Other factors enter into any plan that might successfully restore salmonid runs to this river system. Probably the most important is that of providing suitable spawning area for the fishes once they have successfully

FIGURE 7.8 The lower Columbia River below Bonneville Dam, the lowermost dam. (Photo by C. E. Cushing.)

passed the dams to reach native spawning habitat. Unfortunately, except for the extreme, but increasingly advocated, suggestion of breaching the four lower Snake River dams to restore more natural river conditions, none of the plans address what has been most significantly lost—original habitat for successful spawning and rearing of smolts together with proper ecological conditions for their downstream return to the sea. If successful runs can become reestablished, the adults still must find adequate, pristine spawning areas to successfully spawn in, and it is in the headwater reaches where this occurs that man is still inflicting some of the worst ecological havoc—timber cutting, increased road access, and other activities which degrade the spawning streams by introduction of silt, removal of riparian vegetation, and other perturbations which destroy spawning habitat.

A sound, scientifically based recovery program has been promulgated for the recovery of the salmonid stocks by a group of scientists called the Independent Scientific Group of the Northwest Power Planning Council. Based upon this group's extensive review of the current salmonid management scheme in the Northwest, they suggested nine recommendations based on the existing (implied) conceptual foundation and the science related to salmonid restoration. These recommendations are quite lengthy and the interested reader can find them in ISG (1999). Successful implementation of their plan will need to be not only scientifically sound, but socially and economically acceptable. The current solution emphasizing hatchery production to maximize the number of smolts with augmenting flows and barge transportation to move fishes downstream past the dams has only shown a continuing downward spiral of returning numbers. New approaches are necessary if true recovery of these fish populations is to be realized, but it will be extremely difficult to get agreement by all concerned users of the river.

All in all, the Colorado and Columbia rivers represent severely impacted riverine ecosystems. The Upper Basin of the Colorado remains fairly typical of pristine streams of this region; however, the Colorado River in the Lower Basin and the entire Columbia River within the continental United States are mere ghosts of their former ecosystems. These significant, and largely irreversible, impacts were initiated during a time when there was little appreciation for the value of unperturbed wilderness or pristine environments.

The Colorado and Columbia rivers are truly "gems" in terms of their uniqueness. They have played significant roles in the history and development of the western United States, from the historic descent of the Colorado River by John Wesley Powell and the early fur trade days of the Columbia River by the North West Company under Alexander Mackenzie. Today's ecological awareness has shown just how high the cost to the environment and its resources was, and the enlightened populace has essentially said, "Stop." Indeed, there is even serious consideration being given to restoring the lower Snake River to its free-flowing state by breaching the lower four dams—all in the interest of restoring the historic salmonid populations. This is only one step in the restoration of the total habitat needed for successful restoration of these fish runs, but it does indicate that thinking has changed significantly since the days of building dams wherever the river was running free.

Recommended Reading

Anon. (1998). Angling for answers: The quest to save the Colorado's endangered fish. Colorado River Project, River Report, pp. 5–9. Summer 1998.

Independent Scientific Group (ISG). (1999). Return to the River: Scientific issues in the restoration of salmonid fishes in the Columbia River. *Fisheries* 24:10–19.

Long, M. E. (1997). The Grand Managed Canyon. *National Geographic Society*, July 1997, pp. 117–135.

Palmer, T. (1991). *The Snake River: Window to the West*. Island Press, Washington D.C.

Reisner, M. (1993). *Cadillac Desert*. Penguin Books, New York.

Schmidt, J. C., Webb, R. H. Valdez, R. A. Marzolf, G. R. and Stevens, L. E. (1998). Science and values in river restoration in the Grand Canyon. *BioScience* 48:735–747.

Sibley, G. (n.d.). Building and rebuilding the River. *Canon Journal*, pp. 4–27. Water Education Foundation. Layperson's guide to The Colorado River.

CHAPTER 8

Diverse Rivers of the Southeast

INTRODUCTION

The southeastern region of the United States contains a wide diversity of streams. These range from high-gradient, clear-flowing headwater streams which exhibit ecological characteristics in common with similar streams found elsewhere in the country, to low-gradient, blackwater streams found on the Coastal Plain region of several southeastern states. Picture a large, slow-flowing, tea-colored river bordered by extensive forests and swampland and you have a typical blackwater river. It is these riverine systems which we will emphasize in this chapter because of their significant water quality, and physical and ecological differences from streams of similar size found elsewhere in the country. Truly, these unique streams are worthy of separate attention.

THE OGEECHEE RIVER—A BLACKWATER RIVER

Much is known about the ecology of blackwater rivers. The Satilla River, the Edisto River, Colliers Creek, the Ogeechee River, and others have been extensively studied. We are going to emphasize the ecological characteristics of the Ogeechee River because it has been studied in detail; we also will include pertinent information from other streams. All of the information that we will present in the following synthesis comes from research conducted by Art Benke (Univ. of Alabama), Judy Meyer and Bruce Wallace (Univ. of Georgia), Len Smock (Virginia Commonwealth Univ.), and their colleagues who have published extensively on these unique riverine systems.

The 6th-order Ogeechee River (Fig. 8.1), in Georgia, flows for about 400 km from its source to discharge into the Atlantic on the southeastern coast of the state. It drains 13,500 km^2 of land forested by typical blackwater stream floodplain trees: bald cypress (*Taxodium distichum*), Ogeechee lime (*Nyssa ogeechee*), sweetgum (*Liquidamber styraciflua*), water oak (*Quercus nigra*), swamp black gum (*Nyssa biflora*), water tupelo (*Nyssa aquatica*), and willow (*Salix* spp.). Because the floodplain is extensive, little human development has occurred alongside these rivers. The Ogeechee River, as with other blackwater rivers, differs from similar size rivers in other parts of the country in that it has no large impoundments for hydropower. The low gradient is not conducive to the generation of power.

As with most rivers, the present Ogeechee River differs in its ecological conditions from those present several hundred years ago. Historically, the Ogeechee had much greater standing stocks of woody debris, and beaver

FIGURE 8.1 The Ogeechee River. (Photo by R. T. Edwards.)

created extensive habitat modification. Much of the woody debris has been removed to improve navigation and the beaver have been essentially eliminated. River habitat almost certainly is less complex as a consequence.

Physical Characteristics

The Ogeechee River drops only 200 m along its entire length, resulting in a mean gradient of 0.05%. The bottom is sandy, again characteristic of low-gradient, blackwater rivers. The morphology of the Ogeechee River channel has been severely modified by the removal of woody debris snags. This practice ended in the 1950s, but seriously deprived the river of a major habitat site for benthic macroinvertebrates living on their surfaces (see later discussion). Water temperatures fluctuate between 8° and 16°C in winter and between 24° and 30°C in summer.

Water Quality

As with similar rivers, water of the Ogeechee is heavily colored by high concentrations of dissolved organic matter (DOM) leached from the terrestrial environment. Dissolved oxygen concentrations are low, particularly in summer, due largely to the high respiration rates of the large amounts of entrained particulate and dissolved organic matter. Record lows during the 1986 drought reached 3.4 mg/L (41% saturation). Both alkalinity and pH are low, with mean values of 23 mg $CaCO_3$/L and 6.5, respectively. Concentrations of the nutrients nitrate and phosphate are also low, averaging 0.10 and 0.05 mg/L, respectively. Suspended sediment load content is low.

Ecological Characteristics

The broad floodplain of the Ogeechee River is inundated for several months each year and annually floods during the winter, with important ecological consequences. First, flooding results in considerable exchange of water between the main channel and waters on the floodplain. Over 96% of the DOM in the main channel is derived from floodwaters draining back into it. Second, flooding provides for exchange of POM and invertebrates between the floodplain and the river. The extensive amounts of CPOM that fall from the trees growing in the floodplain are converted to FPOM and entrained into the main channel as the floodwaters recede. Figure 8.2 illustrates the floodplain—channel interactions. The fact that these broad floodplains actually increase in size and importance as the river flows downstream means that the contribution of FPOM, derived from CPOM, to the main channel actually increases downstream. This is exactly the opposite of the pattern predicted by the RCC, in which the inputs and importance of CPOM are highest in the small, headwater reaches and of lesser importance in 4th- to 6th-order reaches. FPOM is important in these reaches, but it is largely derived from instream sources; thus, CPOM plays a major role in energy inputs in the larger reaches of blackwater rivers. This is supported by the fact that 47–64% of the macroinvertebrate production

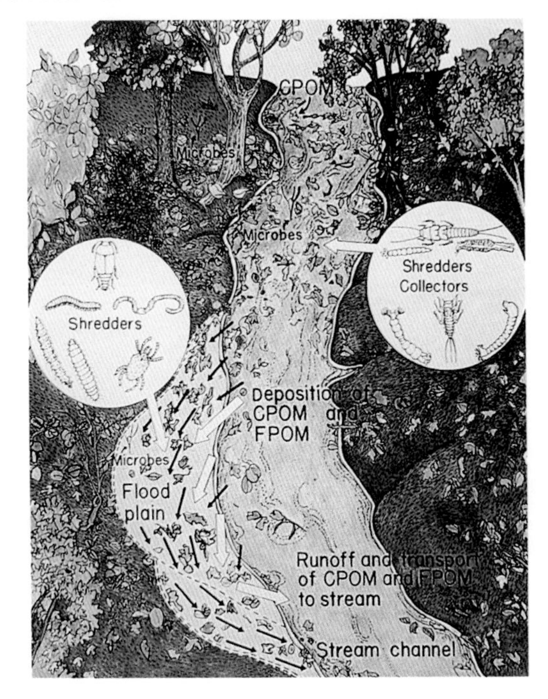

FIGURE 8.2 A conceptual model of the ecological processes and interactions between the floodplain and channel. Black arrows indicate deposition of CPOM and FPOM onto the floodplain from the stream during flooding, and white arrows indicate movements of CPOM and FPOM from floodplain to stream. (From Merritt and Lawson, 1992, with permission.)

in a smaller 2nd-order blackwater stream, Cedar Creek in South Carolina, was dependent upon FPOM, rather than CPOM.

The sandy bottoms of these rivers would suggest that benthic invertebrate communities would be small and the streams unproductive. This is not the case. The sandy bottoms contain oligochaetes, several groups of dipteran larvae, and mollusks, and the production of these invertebrates is similar to that in other sandy bottom streams. This community relies on FPOM within the substrate and extends up to 20-cm deep. The full role of the hyporheic zone of blackwater rivers remains to be studied.

The site of highest macroinvertebrate production in the Ogeechee River by far, and in other blackwater streams, is on the characteristic, extensive amounts of woody debris found in these streams. This is referred to as "snag habitat," and although much reduced from historic levels, it is still significant. These solid substrates provide a site for extensive populations of filtering-collectors (caddisflies and black flies) and gathering-collectors (mayflies). The snags themselves are a source of food as they disintegrate. The annual production rates of the snag-dwelling invertebrates are among the highest of any streams in the country, as well

FIGURE 8.3 Black Creek, a tributary of the Ogeechee River showing the characteristic sandy bottom and snags. (Photo by R. T. Edwards.)

as being the highest of any of the other microhabitats in blackwater streams (e.g., sandy bottom or hyporheic zone). The diverse mayfly population is attributed to several factors, including a continuous, rich food supply, high drift rates, and short generation times. These insects, in turn, are important in the development of the fisheries in these rivers. It has been estimated that up to 60% of the food of sunfish, an important sport fish in these rivers, is derived from snag-inhabiting invertebrates. Figure 8.3 is a photograph of Black Creek, a tributary of the Ogeechee River, at low water and shows the characteristic sandy bottom and snags typical of these blackwater streams.

There are both macro- and microfilterers inhabiting the snags. The macrofilterers derive most of their energy from drifting animal material, whereas the microfilterers and gathering-collectors rely mainly on amorphous detritus. Overall, the basic energy source for the invertebrates in the Ogeechee River is closely related to the inputs from the allochthonous FPOM from the floodplain.

In a study of the Satilla River, another blackwater stream in southeast Georgia similar to the Ogeechee River, it was found that 72–82% of the drifting invertebrates were those inhabiting the snag habitats; the rest were from the sandy-bottom benthos. Drift was most active at night, especially in the summer, and the daily drift patterns of some taxa, such as black flies, were related to discharge volume.

The net movement of macroinvertebrates between the floodplain and main channel in small 1st-order blackwater streams is minimal in terms of annual mean density, biomass, or production rate. Movement occurs by both crawling and drifting, with the latter correlated with seasonal discharge. The floodplains also are important in the life history of many fish species.

Microbial organisms also play important roles in these streams, via the "microbial loop" (described in Chapter 17). Bacterial populations feeding on dissolved organic carbon and particulate detritus become food for protozoans, other microorganisms, and eventually for higher trophic levels, such as filter-feeding black flies, gathering-collector mayflies, and sediment dwelling chironomids and oligochaetes. Suspended bacterial populations are quite high, accounting for 31% of the biomass transported in the water column. The main source of bacteria appears to be from the floodplain swamps rather than production within the river channel. Fungi also appear to be an important food source for higher organisms in blackwater streams, although much remains to be learned concerning fungal productivity in riverine ecosystems.

Because of the high respiration rates resulting from the high inputs of particulate and dissolved organic matter, and the relatively low production of dissolved oxygen by instream primary producers, these streams are basically heterotrophic in nature (i.e., they use more energy than they produce). The degree of heterotrophy also increases downstream, again unlike the pattern predicted by the RCC where autotrophy increases downstream, at least in the small- to mid-sized streams that we are discussing here. At the same time, this high load of transported POM is the prime reason for the high productivity of the snag-dwelling filter-feeding and gathering-collectors mentioned above.

Figure 8.4 depicts the food webs present in a moderate-sized, 6th-order blackwater stream. This figure presents conditions during low-flow; thus the food webs of the floodplain are not shown. Note also that there are three separate food webs: a food web in the water, a food web on stable substrates (snags), and a food web on unstable substrates (sandy bottom). The caption more fully explains the complexity of these webs and how they are interconnected.

BIODIVERSITY AND CONSERVATION OF ALABAMA RIVERS

Another interesting aspect of rivers in the southeastern United States is their rich endemic fauna and the problems with preserving the biodiversity of these flowing waters. We'll talk first about some background related to preservation of biodiversity, then present aspects of the characteristics of these rivers and streams, and conclude with some statistics that put the preservation aspects of the fauna into perspective.

Although many people associate the extinction of species with terrestrial ecosystems and rain forests, freshwater species within the United States actually are more at risk than are terrestrial species (Chapter 22). Biological inventories provide a scientific basis for recognizing the regions where biological diversity is greatest, as well as the regions where it is most seriously imperiled. An extensive analysis by the Natural Heritage Network identified 327 watersheds as critical for the conservation of fish and mussel

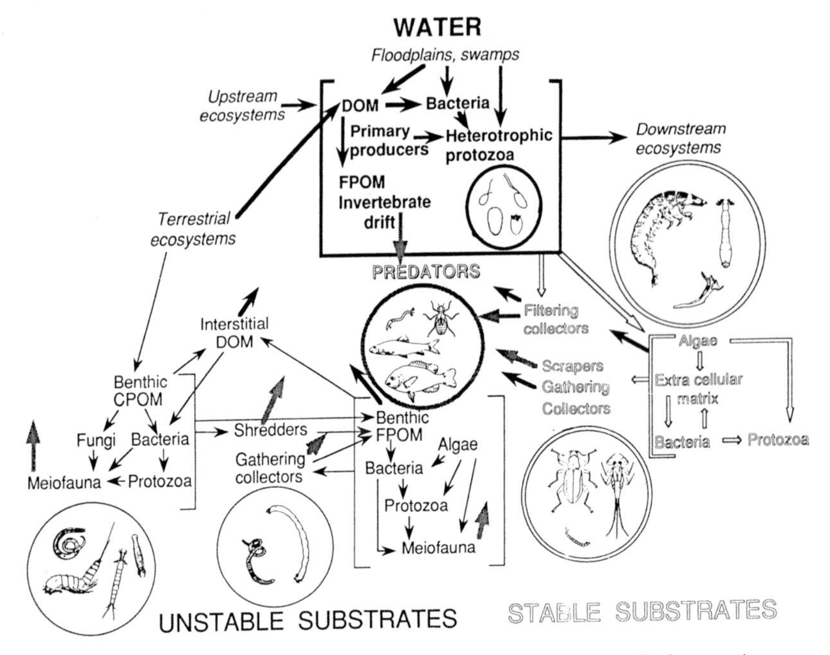

FIGURE 8.4 A conceptual model of the food web of a southeastern blackwater river as a mosaic of food webs characteristic of geomorphically defined habitats linked by flowing water. Sketches indicate consumers representative of each food web in a southeastern blackwater river. Each habitat is denoted with a different typeface and arrow: water (bold), unstable substrates (medium), and stable substrates (outline). Gray arrows denote consumption by predators, and bold arrows indicate movement into the water. Outside sources of organic matter are indicated in italics. (From Meyer, 1990), *BioScience* 40:48, Fig. 3. © 1990 American Institute of Biological Sciences.)

species. Most of these "hot spots" of aquatic biodiversity occur in the Southeast; two regions so identified are the Tennessee-Cumberland River basins and the Mobile River basin, all flowing through Alabama. They include 35% of all the vulnerable fish and mussel species; 70% of these at-risk species occur nowhere else in the world.

Let's look more closely now at the unique biodiversity of rivers and streams in Alabama. Much of this uniqueness can be attributed to the diverse nature of the rivers and streams and the variety of geological areas that they drain. The Mobile River, the major river draining the state, draws water from rivers crossing four geographic provinces, the Appalachian Plateau, the Valley and Ridge, the Piedmont, and the Coastal Plain (Fig. 8.5). Rivers draining smaller basins on the Coastal Plain and originating in Alabama include the Escambria-Conecuh River, the Yellow River, the Choctawhatchee River, and part of the Apalachacola River; these all flow into the Gulf of Mexico. The Tennessee River flows through the northern part of the state, but originates in Tennessee. It should not be surprising, then, to find such a high diversity of flora and fauna in

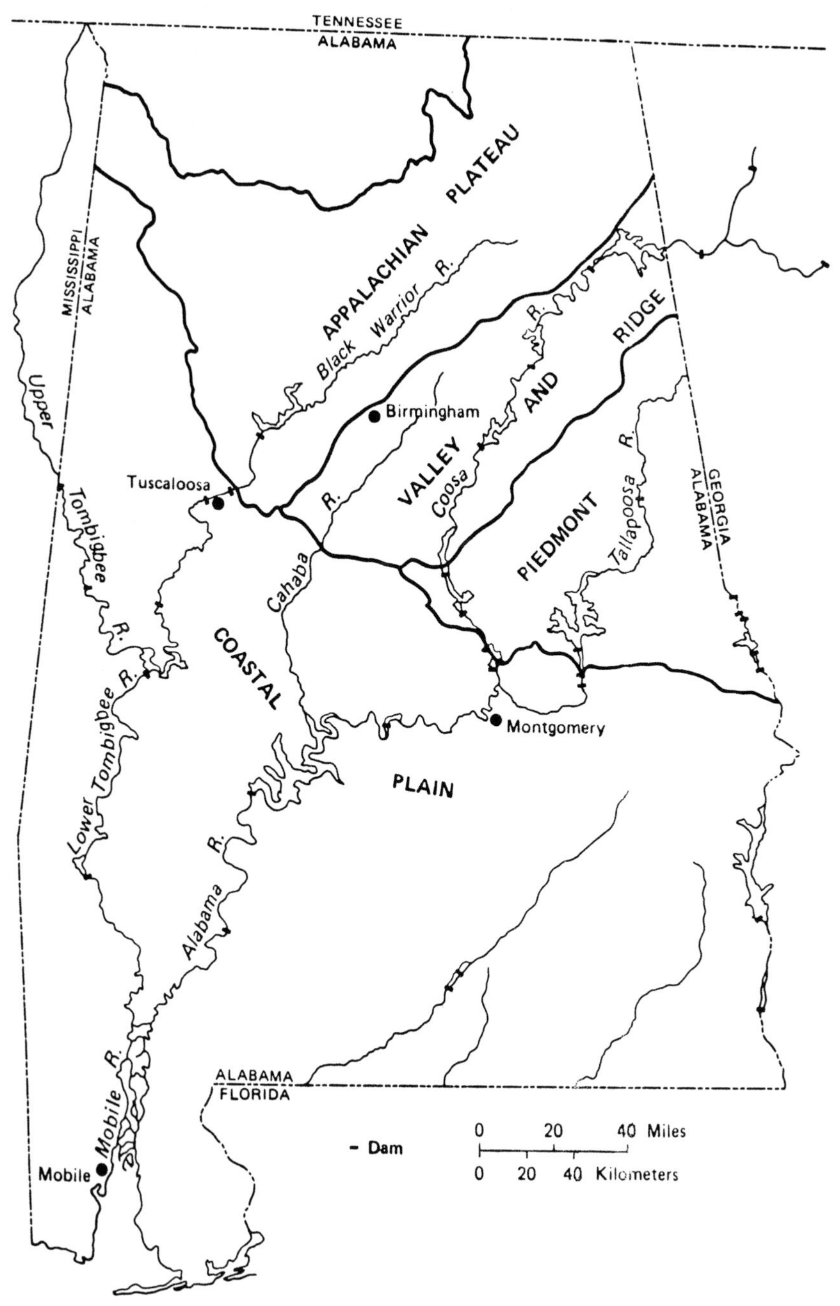

FIGURE 8.5 Map of Alabama showing the major rivers and the major physiographic provinces and subbasins of the Mobile River drainage. (From Ward *et al.*, 1992, with permission.)

rivers and streams draining such a wide variety of geological and topographical regions. Indeed, the underlying geology has resulted in rivers with widely varying chemical and physical characteristics, again leading to a diverse flora and fauna. Another factor leading to the diversity of organisms found in these riverine systems is that the area was never subjected to Pleistocene glaciation, a catastrophic disturbance leading to the extinction and redistribution of many plants and animals.

This provides an overall description of the rivers and streams of Alabama and the basic underlying factors contributing to the richness of the flora and fauna of these systems. Now let's examine some of the details of this diversity. Bear with us, this will entail a lot of numbers and percentages, but careful reading will reveal the severity of the problem facing the fauna of these systems. We will present information on four groups of animals: native freshwater fishes, native freshwater gill-breathing snails, native mussels, and native freshwater turtles. We will refer to them as fishes, snails, mussels and turtles. Alabama contains 38, 43, 60, and 52%, respectively, of the total North American fauna of these four groups, obviously significant numbers for a single state. Some 41, 77, 34, and 22%, respectively, of the fishes, snails, mussels, and turtles, occurring in Alabama and an adjacent state are endemic to that region, meaning they occur nowhere else. This is a remarkable degree of endemism, and emphasizes the points made earlier in regard to the high diversity of riverine habitats in the region and their history.

What is being done to conserve this remarkable biodiversity? Unfortunately, the answer is, very little. Again, for our four respective groups, 10, 65, 69, and 43% of the fishes, snails, mussels, and turtles are considered either extinct, endangered, threatened, or of special concern by professional biologists. This indicates the potential for considerable problems, especially for the snails, mussels, and turtles. Only 40% of the fishes, 1% of the snails, 32% of the mussels, and 20% of the turtles are formally listed as either threatened or endangered under provisions of the Endangered Species Act of 1973. Further, no critical habitat has been protected. Rather than listing more statistics, suffice it to say that conservation efforts, both in Alabama and elsewhere in the country, will help preserve many genetically and taxonomically distinct species.

Many factors contribute to the loss of aquatic biota in Alabama. Habitat destruction from dam construction, channel modification, and other physical impacts on rivers and streams is perhaps the most important cause. Water-quality degradation from siltation, point and nonpoint source pollution, and other sources also contributes to the decline of diversity. Turtle populations have suffered from overcollecting by the pet trade as well as severe habitat loss from a variety of factors. One interesting aspect of species loss, not usually described, is that resulting from ecological interactions, whereby impacts on one species result in a cascading impact on other species. For instance, larval mussels, known as *glochidia*, must attach to fishes as parasites for a few weeks before dropping to the bottom to mature (Chapter 15). If the fishes are eliminated, the mussles cannot reproduce. This is only one example of the complex interactions existing in ecological food webs; direct food web links are another.

THE KISSIMMEE RIVER: LIFE, DECLINE, AND RECOVERY

Background

The Kissimmee River is a part of the huge Everglades fluvial system. It originates from Lake Kissimmee, which receives water from the 4229 km^2 upper basin. The Kissimmee River itself drains the 1200 km^2 lower basin and then flows into Lake Okeechobee in central Florida (Fig. 8.6). Thus the headwaters of the Kissimmee River are part of the huge Kissimmee–Okeechobee–Everglades fluvial ecosystem that drains essentially all of southern Florida. The Kissimmee River was 166 km in length prior to alteration, with a floodplain ranging from 1.5- to 3.2-km wide. The Kissimmee River ecosystem consisted of a mosaic of wetland plant communties that supported a diverse assemblage of waterfowl, wading birds, fish, and other wildlife.

The Everglades is one of the largest marshes in the world, some 1.3 million hectares in extent. The area is dominated by grasses, particularly saw grass, *Mariscus jamaicensis*. It is aptly known as the "river of grass," not only because of the extensive area of grass but also because it has a moderately strong to weak current and is relatively clear as compared to most marshy or swampy waters. Pliocene and Pleistocene limestone underlies the Everglades, most of which has been covered by up to 5 m of peat in places. The entire basin slopes slightly to the south, carrying the overflow of water from Lake Okeechobee to the south and southwest. How much does this flow amount to? Much water is drained for cultivation, and extensive channel alterations have affected the flow, but in the early 1960s, the overflow from Lake Okeechobee and the Kissimmee River basin ranged from 925×10^3 to 2713×10^6 cubic meters of water annually. Add to this the rainfall on the Everglades proper amounting to about 125-cm per year, and you can realize that there is a lot of water flowing through the Everglades ecosystem.

Channelization and the Everglade's Decline

In the early 1960s, the Corps of Engineers began to straighten and dredge the Kissimmee River for flood-control purposes. From 1962 through 1971, the Corps transformed what was once a complex, braided channel with an extensive floodplain wetland into a canal 90-km in length, 10-m deep, and 100-m wide, with a series of five waterflow control structures and five impoundments. In the process 12,000–14,000 ha of wetlands were lost. Channelization not only impacted the morphology of the river, but also had severe impacts on water-level fluctuations and the vegetation and animal communities. Habitat loss reduced populations of wading birds, waterfowl, and fishes. Migratory waterfowl populations decreased by as much as 90%. Nutrient cycling was affected in Lake Okeechobee.

It was almost immediately recognized that this was an ecological disaster, and in 1976, the Florida legislature passed the Kissimmee River Restoration Act to restore much of the river to its original configuration. This was only 5 years after completion of the channelization—talk about

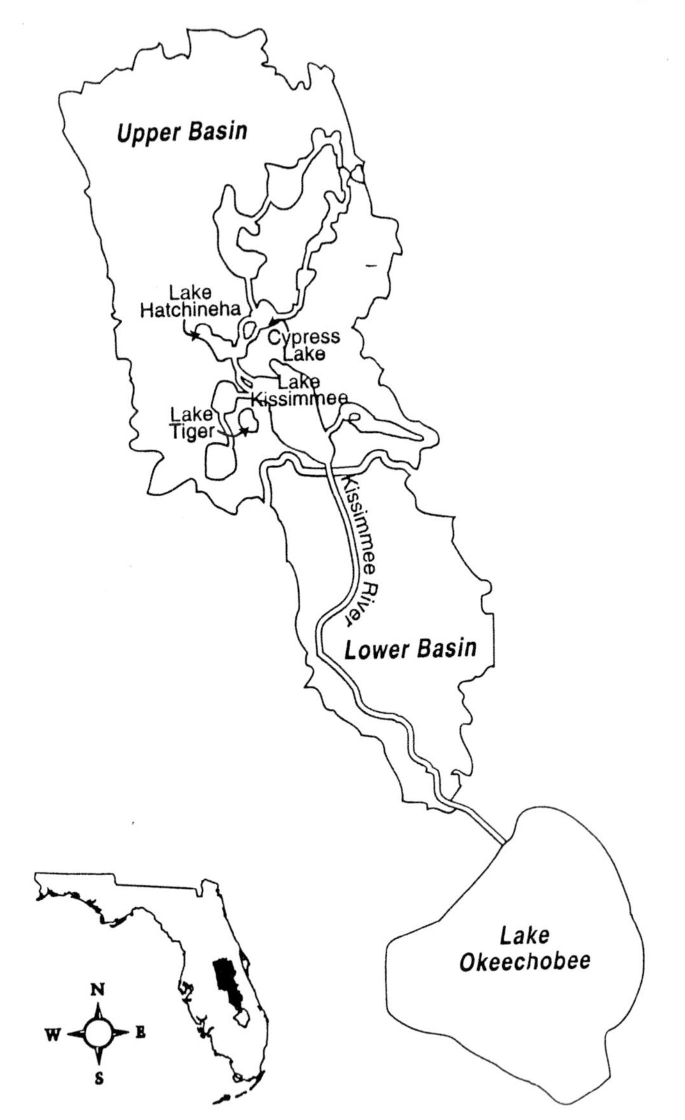

FIGURE 8.6 Map showing the Kissimmee River and its location in relation to the upper and lower basins of the drainage basin. (From Merritt *et al.*, 1996.)

making work for yourself! The construction phase of the project is estimated to take 15 years. Current plans call for the backfilling of over 35 km of the new canal and recarving of 14 km of river channel, together with removal of two water-control structures and associated levees. Figure 8.7 shows a portion of the backfilled dredge channel and reestablishment of the sinuous, natural channel of the river. Backfilling began in the summer

FIGURE 8.7 Photograph of restoration efforts in the Kissimmee River showing portions of the backfilled dredged channel and reestablishment of the river channel to its original, sinuous course. (Photo by South Florida Water Management District.)

of 1999. When restoration is complete, 104 km^2 of the river—floodplain ecosystem will be reestablished, including 70 km of river channel and 11,000 ha of wetland habitat. It is estimated that over 320 species of fish and wildlife will benefit from this effort, including the endangered bald eagle, the wood stork, and the snail kite.

Recovery

An extensive ecological evaluation of the Kissimmee River river–floodplain ecosystem has been started to assess the overall success of the restoration efforts in terms of reestablishment of prechannelization ecological characteristics and plant and animal populations. Vital to the success of this effort is the reestablishment of predredging hydrological regimes within the river channel and surrounding wetlands. Conceptual models for various populations have been developed, and scientists will use these to compare population structure and function following restoration efforts with those of prechannelization populations. Inherent in these assumptions is that reestablishment of the floodplain ecosystem will eventually lead to the development of physical and chemical conditions that allow development of preexisting animal and plant communities. High priority is being given to recreational, economic, and natural heritage valuable organisms such as game fish, wading birds, waterfowl, and threatened and endangered species.

Following channelization, the aquatic macroinvertebrate community shifted from one characteristic of a riverine system to one of a more lake-like community. Thus, success in evaluating the response of this community to restoration efforts will depend upon reestablishment of several ecologically important factors such as streamflow, substrate composition, food quality and quantity, and water quality—all of which influence invertebrate communities. Macroinvertebrates, as we have discussed elsewhere, form an important link between primary production and higher trophic levels, and the reestablishment of macroinvertebrate populations will play a key part in the overall success of the recovery program.

Fish diversity in this region is historically low. Thus, evaluating the success of the restoration efforts will not be based so much on recovery of preexisting fish species and numbers, but on overall ecosystem functioning. Restoration will hopefully restore the higher concentrations of dissolved oxygen extant prior to channelization. The model for this uses nutrient cycling, the movement of larvae, juvenile, and adult fishes, and macroinvertebrates.

Two groups of birds have been selected to evaluate restoration: waterfowl and waders. Population data, regularity of occurrence, and nesting and feeding activities will be considered in the conceptual model developed to evaluate the overall success of the restoration efforts for these animals.

Evaluation of the success of the restoration efforts on the structure and function of the Kissimmee River ecosystem will depend on evaluation of several ecosystem components, their connectivity and interactions, and how well these reflect historic conditions. Determining these will require a major, long-term, intensive research effort by a large team of scientists.

Recommended Reading

Lydeard, C. and Mayden, R. L. (1995). A diverse and endangered aquatic ecosystem of the southeast United States. *Conservation Biology* 9:800–805.

Meyer, J. L. (1990). A blackwater perspective on riverine ecosystems. *BioScience* 40:643–651.

Merritt, R. W. and Lawson, D. L. (1992). The role of leaf litter macroinvertebrates in stream-floodplain dynamics. *Hydrobiologia* 248:65–77.

Toth, L. A. 1993. The ecological basis of the Kissimmee River restoration plan. *Florida Scientist* 56:25–51.

Webster, J. R., Wallace, J. B. and Benfield, E. F. (1995). Organic processes in streams of the eastern United States. pp. 117–197. *In* C. E. Cushing, K. W. Cummins, and G. W. Minshall (eds.), *River and Stream Ecosystems, Ecosystems of the World* Vol. 22. Elsevier, New York.

CHAPTER 9

Warm-Water Rivers of the Midwest

INTRODUCTION

The small streams and rivers of America's heartland, which are the focus of this chapter, have fewer advocates and have received less study than cold-water streams with their highly valued salmonids. Most originate at low altitude and have low gradients. Long runs and pools predominate, but shallow, gravelly riffles can be found where the geology and gradient combine appropriately. Cascades, waterfalls, and boulders are rare. Substrate varies from place to place, of course, but often includes sand, silt, and mud. Water temperatures are cold in the north, then grade into cool and warm waters as one proceeds south.

In their pristine state, and where they have not been greatly altered, the streams and smaller rivers of the broadleaf forests of the Midwest are the prototype of the RCC model. Small streams are heavily shaded, and so periphyton production generally is less important and allochthonous inputs more so, with consequences for functional groups described in Chapter 5. Downstream, as rivers widen, periphyton and macrophytes play a larger role, and direct leaf inputs play a smaller role.

This chapter begins by describing the setting, which is always more complex and heterogeneous than our mental image. We will revisit the River Continuum Concept, because our understanding of the functioning of forested headwater streams has been greatly advanced by studies in this region. In addition, for those whose vision of a stream is a cold, stony-bottom trout stream, it may be interesting to contrast warmer, slower streams that often have sandy and muddy bottoms. Finally, rivers of this region have been greatly modified by human activities. We will end this chapter with a lesson on river history and an examination of the threats and opportunities that confront us today.

DEFINING THE MIDWEST AS A REGION

The boundaries of the "Midwest" are approximate. For our purposes we will include the region from Minnesota to Missouri on the west, from Ohio to Tennessee on the east, and south from Michigan and Wisconsin to the northern border of the gulf states. These are jurisdictional boundaries, but they correspond roughly to natural boundaries. The western edge is the Mississippi River, and also approximately the limit of nonirrigated agriculture. The eastern edge is marked by the rise of the Appalachian Mountains. To the north, along a line that roughly bisects Michigan and Wisconsin, broadleaf forest gives way to mixed hardwood and pine. To the south, roughly along the northern border of the gulf states, ecosystems begin to reflect their subtropical climate.

The vegetation of this region originally was mostly broadleaf forest, although this graded into coniferous forest at the northern boundary and into open woodland and prairie on the western boundary. Along the eastern margin, the Appalachian Mountains add further geological, elevational, and vegetational diversity. Today the land looks very different, of course. Much of the region we will discuss is known as the corn belt. This region is justly known for its agricultural richness due to fertile soils, adequate precipitation, and a climate of cold winters and warm-to-hot summers.

The zoogeography of fishes adds another important element to our thinking about Midwestern rivers. Fishes disperse though river networks; as a consequence interconnected rivers share a common pool of fish species, and a different river basin likely contains some different species. In the Midwest, the dominant distinction is between the drainages of the Laurentian Great Lakes and the Mississippi Basin. The boundary of the Great Lakes includes parts of eastern Wisconsin and Minnesota, and hugs the Lakes' lower outline, separating Michigan and a thin slice of northern Ohio from the rest of the Midwest. To the south lies the diverse Mississippian fish fauna; to the north, the less diverse Laurentian fish fauna. As glaciers retreated throughout the upper Midwest, many species of fishes spread northward, colonizing areas that had limited freshwater habitat only 10,000 years ago. Today, Michigan has 128 species of native, riverine fishes; Missouri, 185; and Tennessee, 286 (Fig. 18.1). Time and history explain part of this northerly decline in fish diversity. Heterogeneity of terrain and isolation of stream drainages, which promote speciation, are additional factors.

Multiple glacial invasions over the past few million years shaped the northern half of this region and substantially influenced the southern half. The most recent glacial retreat profoundly affected the landscape and its drainage patterns. As the last glaciers retreated, large volumes of meltwater accumulated in a land that was depressed by the enormous weight of the ice, forming lakes much larger than the present extent of the Great Lakes. Blocked by the ice of the retreating glaciers, these waters drained south into the Mississippi. Then, only 5000 or so years ago, as the land rebounded and the ice retreated farther northward, the Great Lakes broke though into their present drainage via the St. Lawrence River, and the landscape began to resemble is current form.

The consequences of this glacial action, along with the underlying geology, are difficult to summarize briefly and impossible to overemphasize. Lower Michigan contains some of the deepest deposits of sand and gravel left by retreating glaciers in North America. The rivers themselves are young, their channels having existed for only some thousands of years. River valleys in the upper part of the state are hilly, and river beds are rich in sand and gravel, due to the presence of moraines. But around Saginaw Bay, Michigan, and the western end of Lake Erie, once submerged under the ancestral lakes, the landscape is flat and the soils consist of lake deposits of clay and sand. Below the southern extent of glaciation, in southern Ohio and beyond, streams are much older, and as a consequence in some areas have cut down almost to base level.

Landscape history also strongly influences the hydrology of streams because geological deposits determine the permeability of soils. In regions of deep glacial deposits, rainwater infiltrates easily, resulting in extensive groundwater aquifers. Hydrographs are very constant in these superstable groundwater streams. In regions with extensive lake deposits of clay, soils are very impermeable, and stream runoff is very responsive to rain events, or "flashy."

Many additional variables add to the layers of complexity that seem to make every stream unique and vex the biologists who attempt to choose similar streams as replicates in their field studies. Stream temperatures increase in the southerly direction, from cold to cool to warm. Precipitation is greatest to the east of our region and less to the west. The southernmost extent of glacial deposits occurs roughly in central Ohio, but geology exhibits great local variation as well. Broad trends exist in vegetation, climate, and geology, and this provides some commonality to the streams of a region. But, as anyone who knows two small streams in the same general area can attest, the differences between two nearby streams can seem as great as between two streams hundreds of kilometers apart.

MIDWESTERN STREAMS AND THE RIVER CONTINUUM CONCEPT

Rivers of the Midwest, at least in their ideal state (which, of course, they aren't—more on this shortly) are excellent models of the RCC. Augusta Creek, a tributary of the Kalamazoo River in western Michigan, has been extensively studied as part of a major scientific endeavor to test and refine this conceptual model. According to the RCC (see Fig. 5.2), headwater streams (1st- through 3rd-order) are forested and shaded. Leaf litter and other inorganic inputs provide much of the energy to the food web; aquatic plants, including algae, are comparatively unimportant. As the stream widens into a 4th- through 6th-order river, more sunlight reaches the stream surface and bed, supporting algae and macrophytes. Rivers of 7th-order and above are deep and turbid; their energy supply is a mix of inputs transported from upstream and from the floodplain. As Chapter 5 discusses in detail, the RCC predicts longitudinal differences in the relative abundance of organisms that shred leaves, scrape algae, collect fine particles, and so on.

Studies from a variety of Midwestern streams suggest that the RCC works pretty well in this region. Low-order streams do receive a significant amount of terrestrial organic matter—roughly 600 $g/m^2/year$. That's about 1.3 pounds, some 60% of which is leaves; the rest is branches, fruits, and so on. This won't be news to anyone who rakes leaves each autumn from their wooded lot. The magnitude of these inputs decreases as stream size increases, fitting expectations of the RCC. Measurements of algal production are found to be much lower in 1st-order streams, when compared to 4th-order. Satisfyingly, the functional composition of the invertebrates also supports the RCC. The biological production (a measure of the total growth and reproduction of a population) of shredders was found to be highest in the headwaters and to decline with increasing stream size. The evidence for other groups is supportive, although not as strong. A review of many published studies found a general tendency toward more scrapers associated with streams of intermediate size. Filterers and gatherers of small particles increased their proportion of total invertebrate production as stream size increased. Overall, the RCC performed well.

COLD-, COOL-, AND WARM-WATER STREAMS

With the exception of mammals and birds, the biota of streams lack the capability to regulate their internal temperatures. The activity and metabolism of fishes and insects are governed by the temperature of the water. Through evolution, organisms adapt to the environments in which they find themselves. Enzymes and metabolic pathways function best within a certain temperature range, and so cold-, cool-, and warm-water fishes differ markedly in their ability to be active, to grow, and to survive in different water temperatures. Freshwater can't cool below 0°C, of course, and this lower limit is experienced by many fishes of temperate regions. But the warmest temperatures that stream fishes will encounter for a sustained

Table 9.1 A Classification of Cold-, Cool-, and Warm-Water Streams

Stream classification	Maximum tolerance[a]	Examples
Cold	<22°C	Sculpin, trout
Cool	22–26°C	Suckers, minnows, dace
Warm	26–38°C	Suckers, sunfish, catfish

[a] Maximum tolerance refers to the maximum extended (over weeks or months) temperature at which the fish species can feed, grow, and holds its place in the community.

period differs markedly on a north–south gradient. Summer streamwater temperatures might easily reach the mid-30°C range in southern Illinois, the high-20°C range in southern Michigan, and only the low-20°C range in northern Wisconsin.

Different thermal environments result in different fish faunas (Table 9.1). Some warm-water fishes are limited in their northern extent by inability to tolerate prolonged exposure to low temperatures, including temperatures well-above freezing. A number of warm-water species of southern states, as well as introduced subtropical species, are unable to extend their distribution very far north. Cold-water fishes are outcompeted, at the southern limit of their ranges, by warm-adapted fishes. So important is water temperature to fishes that they can detect a difference of less than a degree, and in the laboratory will spend most of their time within a 4°C range, which defines their thermal preference.

A number of factors in addition to north–south location affect stream temperatures. Geology is important: stable, groundwater-fed streams tend to be cool in summer, warm in winter, and rather constant overall. Rivers that rise and fall in response to rains also tend to exhibit more seasonal variation in temperature (Fig. 9.1). Shaded streams tend to be cooler than those flowing through open meadows. Lakes and wetlands absorb a lot of heat during the summer, and so a river flowing through a chain of shallow lakes may also become warmer. These environmental factors allow for considerable local variation in summer stream temperatures.

Human activities can alter stream temperatures, sometimes dramatically. Simply clearing away the streamside vegetation can result in summer maxima that are 5°C or more above the natural condition. If summer flows also are reduced, perhaps as a consequence of dams of other river engineering, producing a sluggish, shallow stream, warming may be much greater. Dams and reservoirs can have complex effects on the thermal regime. On a small- to medium-sized river, a dam will produce a shallow lake that usually will warm more than the river, and its surface outflow will warm the river below it. Brown trout are excluded from some sections of Michigan's Au Sable River (Chapter 6) in this manner. High dams on large rivers sometimes release water from the bottom of the reservoir, and this can result in abnormally low summer temperatures. (However, few trout fishers will complain about this practice, as it results in superb tailwater fisheries in a number of western rivers.)

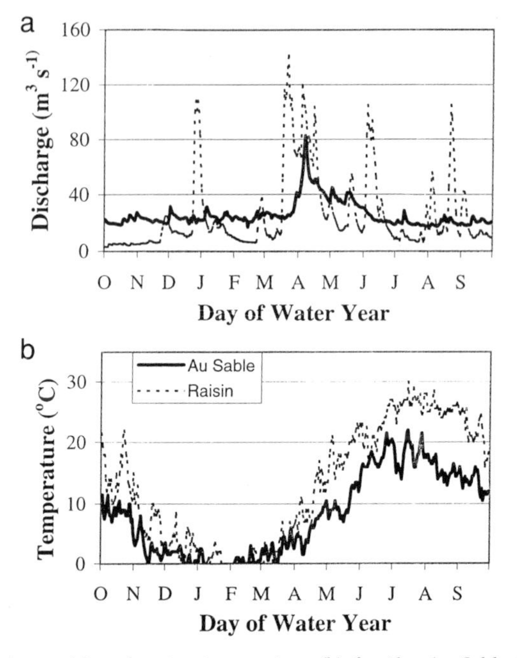

FIGURE 9.1 Discharge (a) and water temperature (b) for the Au Sable River, a stable, groundwater-fed stream, and the Raisin River, a more flashy, surface-fed stream, both in Michigan. Note that water temperatures are less variable, and cooler in summer, in the groundwater-fed stream.

Climate change is another potentially very serious driver of altered river temperatures. It may take 100 years, and the precise extent of warming is still uncertain, but an increase of 2–4°C now seems very plausible. Shifts in species distributions are highly likely, although basin boundaries will limit dispersal. We'll take a closer look at this topic in Chapter 22.

HISTORY

The footprint of human activities looms especially large in the Midwestern United States, and so this chapter provides an opportunity to examine the diverse ways that humans have changed the landscape and the watercourses. A story about Illinois, "Cornbelt, U.S.A.," is illustrative. Reconstruction of presettlement (1820) vegetation reveals Illinois to have been 59% prairie, 38% forested, and 3% water. Land-use change over the ensuing 160 years converted most of this to agriculture. As of 1980, only 19% of Illinois' forests and 0.01% of the original prairies remained. About two-thirds of Illinois' wetlands also were lost during this time period. Plowing the prairies, clearing forested land for agriculture, and draining of wetlands are primary drivers of this transition. "Swamp-busting" gained momentum with the Swamp Land Acts of 1849, 1850, and 1860, and river regulation grew substantially after the 1927 Mississippi Flood, which spawned the 1928 Flood Control Act. Not only were wetlands drained, but many streams were straightened and turned into ditches, the better to carry away runoff. In low-lying areas of wet soils, many farmers installed drain tiles. These pipes, typically 10 to 15 cm in diameter, have a series of holes on their upper side, allowing water to enter the pipes and be carried away. Walk along a "farm stream" and you may see these pipes sticking out of the stream bank, perhaps 1 meter beneath the land surface. These are the exit points for a vast, underground network that carries away rainwater quickly, but also produces unnatural, scouring high flows within the channel.

An extensive hydrologic history of the Illinois River provides another view of human-induced changes in the river (Fig. 9.2). From inspection of annual hydrographs spanning a century we can clearly see that the pattern of annual runoff has changed dramatically. In the more recent period, river levels are noticeably more erratic, major floods are higher and more frequent, and there are fewer years with low, stable water levels during the summer growing season.

Intensive agriculture and altered hydrology are a recipe for erosion, and the Midwestern states are the textbook example. Exposed soil is more vulnerable to erosion by wind and water than is vegetated soil. The combination of intensive agriculture, farming to the stream bank, and fall

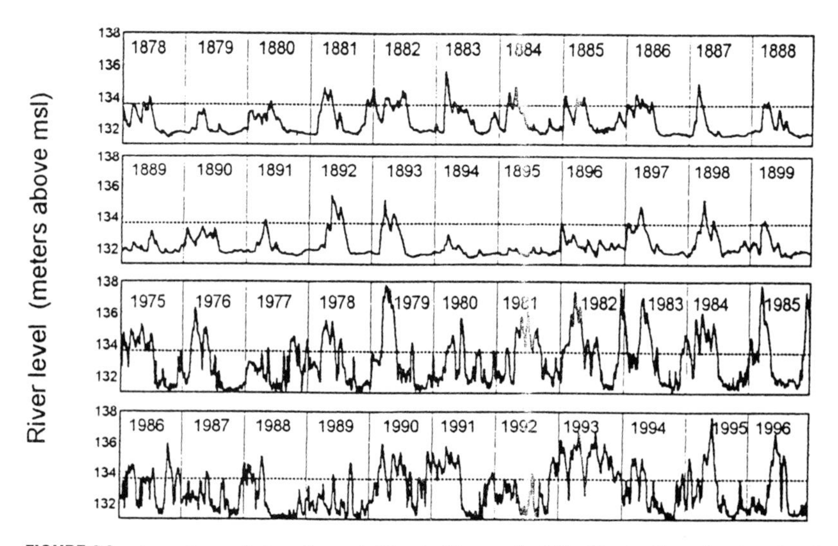

FIGURE 9.2 A century of river flow at Illinois River mile 137. Fluctuations in river level (stage) in meters above mean sea level (msl) before water diversions and modern navigation dams (1878–1899) and after many alterations in the watershed and river (1975–1996). Each block shows daily level from 1 January to 31 December. The horizontal line indicates a flood elevation at which economic damage occurs. (From Sparks *et al.*, 1998,) *BioScience* 48:706–720. © 1998 American Institute of Biological Sciences.)

ploughing (done because it prepares the fields for spring planting and saves the farmer from having to plough in early spring when heavy machinery may become stuck) all lead to high erosion rates. Figure 9.3 shows the estimated rates of soil erosion for different watersheds in the lower 48 states; the Midwest, and particularly the central Mississippi region, experiences very high rates of soil loss. The U.S. EPA publishes periodic assessments of the condition of the nation's rivers, based on state-by-state reporting. These numbers are suspect, in our view, due to inadequate funding of state assessment activities. With that caveat, it still is relevant to note that siltation is the major reported pollutant nationwide and agriculture the leading source of reported impaired river miles nationwide.

A study of the St. Croix River valley from 1830 to the present illustrates the extensive changes that occurred during European-American settlement. This river in western Wisconsin, a tributary of the Mississippi River, has a long history of use by indigenous people, then by French explorers and fur-trappers as a link between Lake Superior and the Mississippi River, and then by settlers moving into the territory that would become Minnesota and Wisconsin. A surprising variety of information sources bear witness to the early appearance of the landscape and its subsequent transformation. When surveyors laid out sections, quarter-sections, and townships, beginning in 1832, they left a wealth of description, including the tree species that marked each corner of the mile-square sections. From this we have a good idea of the presettlement vegetation, which included mixed stands of hardwood and conifers dominated by sugar maple and white pine to the north and sugar maple and basswood to the south (forest covered about half of the watershed's area). Fire probably maintained some oak-savannah regions (16% of the area), and a patchwork of prairie and wet prairie also occurred (another 18%), depending on soil moisture and terrain. Additional wetland systems were associated with the river.

From federal census files, records of lumber and agricultural production, and historical accounts, two periods of rapid change can be seen in the St. Croix valley. From about 1850 to 1880, loggers and then farmers converted the landscape into one that was largely deforested and agricultural. Then, from 1940 to the present, urban areas expanded outward, further fragmenting natural habitat. No direct connections have been made between this landscape change and the river's ecology, but the concepts presented in the first chapters of this book, including the RCC, provide the framework for drawing inferences. In his classic paper entitled "The Stream and Its Valley," Noel Hynes, one of the true fathers of the science of river ecology, wrote in 1975, "in every respect, the valley rules the stream." Surely, as the valley has been transformed, so has its river.

An historical analysis of the Maumee and the Illinois rivers documents substantial decline in the original fish faunas. Originating in Fort Wayne, Indiana, the Maumee flows eastward, entering Lake Erie at Toledo. The largest tributary entering the Great Lakes and the second-largest drainage in Ohio, its 17,000 km^2 basin is 90% agricultural, and much of the remainder is urban. The fishes of the Maumee are well known, and they include 98 species from 28 families (major families include Cyprinidae, 28;

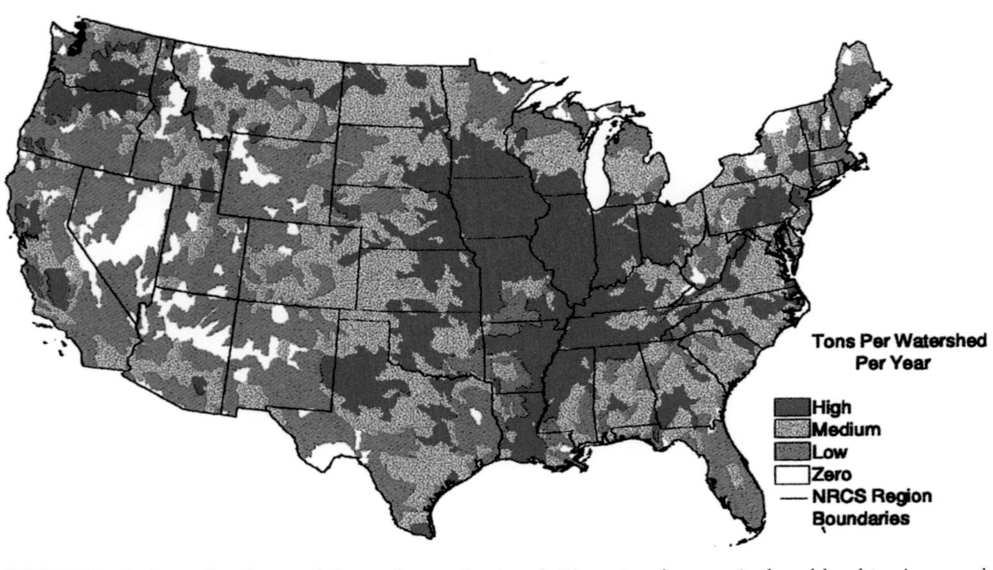

FIGURE 9.3 Estimated sediment delivery due to sheet and rill erosion from agricultural land to rivers and streams for the approximately 2150 watersheds comprising the contiguous United States. Erosion rates are calculated from a model based on the Universal Soil Loss Equation and are not measured values. (From the U.S. Department of Agriculture Natural Resource Conservation Service.)

Percidae, 16; Catostomidae, 13; and Centrarchidae, 12). Overall, it was estimated that 17 species have been extirpated from the river system and an additional 26 species are declining, while 34 are stable and 10 have increased. Declines are greatest in headwater streams, and in some cases they are related to loss of clean spawning gravel.

Originating with the joining of the Des Plaines and Kankakee rivers in northeastern Illinois, the Illinois River flows west and south until joining the Mississippi. This large (72,300 km^2 basin) river drains mostly cropland. Its fish fauna includes 140 species from 27 families (Cyprinidae, 41; Percidae, 18; Catostomidae, 18; and Centrarchidae, 14). The Illinois River has lost 8 species, and an additional 82 species have declined, while 31 are stable. Declines are greatest in midriver reaches, but at least 60% of fishes of headwater and large river reaches also have declined. In 1908 commercial fish catches from the Illinois River comprised 10% of the entire U.S. freshwater catch, second only to the Columbia. By the 1980s, commercial catches were virtually nil. Even the hardy, introduced carp has declined from a commercial catch of about 7 million kg in 1908 to less than 100,000 kg in 1973.

THREATS OF TODAY AND A VISION FOR TOMORROW

Among our most altered and neglected rivers, warm-water rivers of the Midwest offer great challenges for restoration. Human pressures on these rivers are intense. Their watersheds contain some of the richest, most fertile, and most valuable farmland on earth. Cities, towns, and villages dot the landscape, and suburban and exurban sprawl inexorably consumes more and more land. Here and there a river park, a forest preserve, or other relatively undamaged area reminds us of what could be. But farm streams and suckers have few advocates. People have long become accustomed to going elsewhere for their river recreation. Sadly, too little is being done on behalf of warm-water rivers.

There is no reason that a well-focused effort to improve the health of Midwestern rivers could not succeed. River-friendly agriculture, river-friendly development, and protection of the remaining best streams are the key requirements. Agricultural best management practices, or BMPs, include allowing a vegetated buffer strip to grow along the stream's edge to reduce the runoff of sediments and nutrients from fields into the stream. Many more techniques, such as reduced tillage and winter cover crops, also can reduce erosion. Non profit organizations and government agencies are working hard to promote BMPs, and they're finding a receptive audience in many landowners who want to be good land stewards. Much more can always be done, of course, and more progress is needed in helping us understand how suburban land-use practices might be modified as well.

Riparian buffer strips are among the most practical and effective means of protecting rivers. They reduce the delivery of sediments to streams in a number of ways. Buffer strips displace sediment-producing activities away from the stream, trap sediments moving with runoff, and stabilize stream banks against erosion. A vegetated buffer strip serves as a sink for nutrients because plant roots take up the phosphorus and nitrogen

FIGURE 9.4 A photo of Big Darby Creek, Ohio. (Photo by Steven Flint, The Nature Conservancy, Dublin, Ohio.)

before it reaches the stream. Temperature is moderated by shade provided by trees along the stream bank; leaves provide an energy source, and woody debris creates habitat. There is some debate concerning how wide of a riparian buffer strip is needed. Prescribed buffer widths generally fall in the range of 50–100 m and are greater where slopes are steep and soils more prone to erosion. There are many benefits, for a relatively small amount of land set aside.

Finally, if you question whether any river in the Midwest is worthy of protection and restoration, you might visit the Big Darby, designated by The Nature Conservancy as a "Last, Great Place." This 125-km long creek (Fig. 9.4) runs through a landscape of mostly corn and soybeans, along with scattered horse farms, about 30-km west of Columbus, Ohio—hardly the place one expects to find a jewel of a stream. However, trees line enough of the river banks, and also surround many farms as windbreaks, to offer some protection from erosion and supply woody debris to the channel. The Darby's water is reasonably clean, and the creek has enough gradient and coarse substrate to provide frequent riffles. This exceptional warm-water stream supports 40 types of mussels, including a dozen threatened species, and many fish species including the Tippecanoe darter, needlenose gar, flathead catfish, and more. The threats are likely to accelerate, but the Darby has on its side a consortium led by The Nature Conservancy that includes more than 30 public agencies and private groups. They are leading the way for others to follow.

CHAPTER 10

Desert Rivers of the Southwest

INTRODUCTION

Desert rivers and streams—even the names seem like oxymorons. They are large expanses of dry streambeds one moment, floodwaters that move large boulders moments later following intense rainstorms upstream, and back again to dryness soon after. These unique ecosystems provide startling contrasts to rivers and streams found in more mesic (wetter) regions, described in previous chapters, and provide us with many insights into the ecology of flowing-water ecosystems.

GENERAL DESCRIPTION

Major streams in arid regions receive their water from areas of high elevation, often many kilometers upslope, where precipitation is high and usually persistent. This precipitation is often seasonal and thus flows are usually timed to annual runoff events occurring far away. This is also characteristic of major tributaries and results in the distinctive flow regimes of desert rivers and streams. Let's look at some of the features that characterize desert stream ecosystems and contrast them with conditions found in mesic regimes; again, we will use the RCC to provide a model for comparison.

The upper reaches of desert streams (Fig. 10.1), stream orders 1 to about 6, usually do not contain perennially flowing water; instead, the streambeds in the upper reaches exist as dry beds, with water flowing only intermittently during and after significant rainfall in the headwaters. Some of the low-order streams may contain perennial flow, but the water gradually disappears into the sediments downstream. The hydrograph, or water flow history, is flashy, that is, water appears intermittently as a result of periodic, significant rain storms in the headwaters. During these brief periods, flow can be of flood proportions, moving large amounts of sediment downstream and scouring the stream bottom. And, because water is present for only brief periods, macroinvertebrate communities are limited or absent.

Physical conditions can be very different from stream to stream and even within a single stream basin. Air temperatures are high and streams receive ample sunlight; thus water temperatures tend to be higher than in

FIGURE 10.1 Photograph of upper, low-order reaches, illustrating lack of permanent water. (Photo by S. G. Fisher.)

mesic streams. Even the orientation of the streambed can cause local variations; if the bed is well shaded by cliffs, temperatures are ameliorated from those up- or downstream where the full impact of the sun is felt.

Riparian vegetation is scant and usually is found only along the edges of the flood channel; the streambed itself is devoid of vegetation. Vegetation is also sparse inland of the streambed and this has major implications in terms of the contribution of terrestrially derived allochthonous detritus to the stream ecosystem. When the streams contain water for any significant period of time, the water is usually clear and shallow; thus, with the absence of shading riparian vegetation, the potential for high primary production is present, and this is usually found in these situations. Filamentous algal growth is rapid and prolific, albeit for only a short time.

As we continue down the stream continuum to about 5th- to 6th-order and higher reaches, perennial water is usually found in the streambeds of large watersheds (Fig. 10.2). Even this flow, however, is not always continuous for the whole year. Continuous flow is usually found during spring and early summer, but as the season progresses, flow dwin-

FIGURE 10.2 Photograph of Sycamore Creek. (Photo by S. G. Fisher.)

dles and often disappears underground into the porous, sandy stream bottom, leaving a series of pools separated by dry reaches. Evaporation is about ten times as great as precipitation in this region, and the flow in the perennial streams is fed from runoff from higher mesic regions.

Although visible surface water flow may disappear as summer progresses, water flow through the stream channel has not ceased; it only means that flow connecting the exposed pools is moving underground, an important concept in the functioning of desert stream ecosystems. Thus, for practical purposes in terms of the RCC, 1st-order desert streams, in terms of biology, are really found in 6th- to 8th-order channels in the geomorphological network.

In the mid-reaches, order 5–7, where permanent flows occur, some similarities may be found with the RCC model. Streambeds are open to direct sunlight, and luxurious growths of filamentous green algae occur; these reaches are usually autotrophic. Macroinvertebrate communities are dominated by FPOM-feeding gathering-collectors, but grazers are absent because of the absence of rocks and other surfaces which support the algal growth (diatoms) upon which they feed. This region probably reflects the tenets of the RCC more closely than reaches up or down stream.

In the lower reaches of desert streams, water flow again is largely intermittent. In fact, as discussed below, water is at such a premium that the lower reaches of any large rivers containing significant flows largely have been diverted into artificial distribution systems, thus disrupting the development of natural ecosystems. These reaches certainly bear little resemblance to those predicted by the RCC.

ENERGY SOURCES

As might be expected in an environment devoid of riparian terrestrial vegetation, especially plants with deciduous leaves, the input of allochthonous POM to desert streams is not significant. That which does occur is intermittent and limited to interspersed patches where enough moisture is held to support plant growth. Another problem in desert streams that relates to its retention and processing of allochthonous POM is the lack of retention devices within the streambed-boulders, tree snags, etc.—that retard the downstream movement of any allochthonous POM that finds its way into the wetted streambed. These devices in mesic streams inhibit the downstream movement of POM, retaining it throughout the streambed where it can undergo processing and become suitable food for shredders. And, as might be expected, desert stream insect communities are notably lacking in shredder species, although collectors are present but have found other sources of FPOM other than degradation of CPOM by shredders.

Without significant inputs of allochthonous energy, desert streams are almost entirely dependent upon instream primary production for energy resources for secondary producers. Owing to clear water, high sunlight, and the abundant nutrients (more about this later), most desert streams produce rich growths of algae. Usually this includes dense mats of the filamentous green alga *Cladophora* (Fig. 10.3), as well as diatoms and cyanobacteria that

FIGURE 10.3 Photograph of *Cladophora* mats in Sycamore Creek. (Photo by S. G. Fisher.)

coat both the filamentous algae and the inorganic sand grains and stones on the stream bottom. Desert streams have high rates of primary and secondary production, and, in general, are *autotrophic*, that is, they produce more oxygen by photosynthesis than they respire (see Chapter 3). Their daily rates of primary production are 2–3 times as high as similar mesic streams. Secondary production rates are more than double the highest rates found in streams from several other places around the world.

Flash floods play a significant role in the seasonal pattern of primary production of desert streams, as well as in the dynamics of the invertebrates, as we shall see. Floods inhibit primary production by scouring algae from the streambed; the more severe the flood, the greater the impact upon primary production. And, obviously, if more floods occur during the seasons of high primary production, the greater also will be in the impact on overall season primary production.

One of the most fascinating stories of the ecology of desert streams has to do with the cycling of organic matter through the various organisms. FPOM feeders overwhelmingly dominate the energy flow of desert streams. They are small, numerous, and grow rapidly; thus, their turnover rate is extremely high and annual secondary production (weight of insects produced) exceeds the weight of insects present at any one time by 60 or 70 times, as compared to a figure of two to five times for more mesic streams. But there is more to the story. Because the insects rate of assimilation of their food is very low, it has been calculated that they must ingest four times their body weight daily to maintain their population sizes. Comparing this to the rate of primary production by the algae, for one flood-recovery sequence, the insects were consuming six times more material than was being produced! How can this be? Because of the low assimilation, it means that much of the algae consumed by the insects was passed through their bodies relatively unchanged—to be utilized by other insects. Much of the information for this chapter has been gleaned from the research of Stuart Fisher and his colleagues at Arizona State University.

NUTRIENT DYNAMICS

The input, transport, and cycling of nutrients within desert stream ecosystems is quite different from what occurs in mesic systems, where precipitation and groundwater inputs result in fairly continuous annual flows. In desert streams, input and transport dynamics are periodic and associated with the infrequent intense rainstorms and resulting flooding events. In fact, the uptake and cycling of elements by the stream biota during flash-flood events bear little resemblance to what occurs during base flow the rest of the year.

In terms of mass transport of dissolved nutrients by the stream, by far the majority occurs during the flood events. Concentrations of nitrogen and phosphorus, especially the former, are elevated in both precipitation and the resulting runoff. These high concentrations of nitrogen, most of which is retained within the watershed, play a vital role in controlling the growth of primary producers in the permanent flowing reaches of these streams. Studies of primary production by algae in relation to the concentrations of available nitrogen and phosphorus in desert streams indicate that nitrogen is more likely to be limiting. Studies find the decrease in nitrogen concentrations in the water to be closely related to uptake by algae during photosynthesis, whereas uptake of phosphorus appears to be controlled by other factors. However, in some headwater regions of desert streams, phosphorus may be the limiting nutrient.

All in all, the dynamics of dissolved nutrients in desert streams is an exciting field of study. A whole new aspect of the cycling of dissolved nutrients has been opened by scientists who have extended their scope of observations to include the hyporheic community, the intriguing and little known assemblage of microorganisms that exist below the visible stream bottom. They have found that the interchange of water between the open stream and this subsurface region substantially influence photosynthesis, respiration, and the exchange of dissolved nutrients.

SUCCESSION AFTER FLOODS

One of the most fascinating stories of the ecology of desert streams has to do with the dynamics of nutrients and organic matter in regard to flood frequency. The severe, although infrequent, flooding of desert streams significantly affects the biota, essentially eliminating all organisms except the fishes. Algae and macroinvertebrates are swept away as the stream bottom is dislodged and washed downstream by the raging floodwaters; the system is "reset" to an early stage of succession. Fishes, being larger and more mobile, appear to be able to seek out refuge behind large rocks or other places of haven from the torrential flows.

Surprisingly, recovery to preflood conditions is relatively rapid. This is due to many factors. Favorable growing conditions exist for most of the year, and most of the organisms have short life histories, enabling them to develop rapidly. Further, most organisms are small and mobile, again enabling them to reach and recolonize denuded localities soon after the floodwaters recede.

Favorable nutrient regimes in the water remain, and the algae are usually first to reappear; first the diatoms and then the green algae and cyanobacteria. In the low flow conditions present following the floods, dense mats of algae develop. Macroinvertebrates appear soon after; their successful recolonization is related to the fact that the adults can fly, enabling them to rapidly reach denuded areas from surrounding, unimpacted populations. Short life cycles and favorably warm water temperatures promote rapid growth and maturation of these populations. However, all of this is just a prelude to the next flood and repeat of this cycle.

Impact and recovery of desert streams from floods is quite different from their mesic counterparts. Flooding of desert streams occurs at different frequencies, usually only a few times per year. Flooding is fairly frequent in mesic streams, although all floods are not of the catastrophic nature as are floods in desert streams. Thus, biological communities in mesic streams are less impacted and more adapted to frequent increases in flow. The physical characteristics of mesic streams also make them less susceptible to devastating impacts. Further, because the biological communities in mesic streams are more continuous and not usually totally eradicated by flood flows means that there is a ready population of colonizers to replace downstream losses.

DESERT FISHES—ENDEMISM AND ENDANGERMENT

The story of fishes in streams of the desert southwest is extremely interesting, especially in terms of their history of distribution, present occurrence, and displacement by nonnative species. A brief synopsis of this story is pertinent to this chapter.

Since before the turn of the twentieth century, biologists had noted and documented the uniqueness of the fish fauna of this area. Five families, 18 genera, and 32 indigenous species were listed as early as 1895, of which over 78% were endemic to the Colorado Basin, with a larger percentage of species peculiar to a single river basin than anyplace else in North America. Recent studies have not materially changed these numbers; endemism still is around 75%. Long isolation and the special conditions of these streams have resulted in the presence of a distinctive collection of fish species in the Colorado River Basin. The decline and disappearance of various fish species had been reported throughout the twentieth century, usually related to evidence of the presence of introduced fish species. Fish collections in the Salt River near Tempe, Arizona, in 1900 documented 14 native fish species. Twenty years later, only 7 native species remained, and 2 nonnative species were found. By 1940, the numbers were 3 natives and 6 introduced; in 1960, 3 natives and 8 introduced; and by 1967, only 2 native species remained and 20 introduced species were found. Truly, a remarkable example of the replacement of a native fish population by one of nonnatives!

By the 1960s, exotic fish species had essentially replaced the indigenous fishes of the lower Colorado River. Although this change was not apparent to the casual observer, what could be observed were the physical changes taking place as Glen Canyon, Flaming Gorge, Navajo, and other major dams authorized by the Colorado River Storage Project Act of 1956 were built. A poisoning of the Green River system above Flaming Gorge Dam in order to prepare the system for a trout fishery got out of hand, and fishes were killed throughout Dinosaur National Monument.

As a result of these observable perturbations, the public finally began to protest and people began to use the Endangered Species Act as a focal point to halt the rate at which native species were disappearing. Eight species of fishes are native to Grand Canyon National Park, and six of these are restricted to the Park. The humpback chub, bonytail chub, Colorado pikeminnow, and razorback sucker are either listed as endangered or proposed for listing (see Chapter 7). Of these, the humpback chub presently has a reproducing population, but the other three are either exceedingly rare or absent.

Although dam-induced modifications of the Colorado River system are obvious contributing factors to the changes in the fish fauna of this region, biologists suggest that these are not the primary cause. Rather, they suggest that the extinction of the majority of the endemic fish species in the system is the result of the introduction and proliferation of nonnative fishes, which is enhanced by river alteration. Furthermore, the prognosis for native fishes likely is poor because the nonnative species continue to flourish in the modified river.

LARGE DESERT RIVERS

Little has been said about the ecology of large (stream order > ca. 8) desert rivers—and for good reason. They essentially no longer appear in their natural settings because of the extensive man-made alterations to their channels driven by the cultural demands of growing populations (e.g., green lawns in the desert, decorative water fountains, etc.). We will briefly describe two large desert rivers, the Middle Rio Grande River in New Mexico and the Salt River in Arizona.

The Rio Grande

As mentioned above, most perennially flowing desert rivers receive their water from sources far distant, and the Rio Grande is no exception. The Rio Grande is not entirely typical of the hot desert region we are discussing in this chapter, but it is the only river of any size which retains some of its natural ecological characteristic and which exists in a desert system, albeit, perhaps, a cool-desert. Its headwaters are in the San Juan and Sangre de Cristo mountains in southern Colorado, and it flows over 3220 km to its outlet in the Gulf of Mexico. Other major tributaries originate in the Jemez, Sandia, Manzano, and Magdalena mountain ranges of New Mexico. The section we are concerned with flows about 750 km through Great Basin Grassland, Semidesert Grassland, and Chihuahuan Desertscrub plant communities in the temperate, semiarid climate of its valley in the mid-section of New Mexico where nearly 40% of the state's population lives. Total elevational change for the entire river is about 3600 m; in the mid-section it is about 240 m.

Moderate temperatures characterize the valley's climate, being semi-arid in the northern part and arid as the river flows south. Maximum air temperatures, usually occurring in July, range from 21–24°C; minimum air temperatures are about 4°C. Average annual precipitation is about 25 cm throughout the valley, most of it coming from summer storms. Mean monthly discharge ranges from a little less than 28 cms during base flow in September through February to over 113 cms during peak runoff in May. The midsection of the Rio Grande was historically composed of primarily warm-water aquatic habitat, with shifting sand bottoms. Cooler habitat with gravel and cobble stream bottoms were found further upstream in the vicinity of the entrance of the Jemez River.

Historically, this river valley has experienced much change. Cottonwood—willow forests have been eliminated by land clearing, tree harvesting, water diversion, and agriculture. Currently, agriculture is responsible for about 90% of all river water use. Many diversion dams and ditches funnel water away from, and back to, the main riverbed. Livestock have greatly reduced riparian communities by grazing and through stream bank erosion. The remaining riparian vegetation along the Middle Rio Grand Valley today consists of large cottonwood stands confined to banks where they can survive the highly controlled and physically altered aspects of the river. These riparian forests have been given the name "bosque," and include not only the cottonwood trees but other associated vegetation.

Early fish surveys of the area in the 1850s found the historic, native fish fauna of the Rio Grande to contain somewhere between 17 and 27 species. Big river species were present, including the longnose gar, shovel-nose sturgeon, gray redhorse, blue sucker, and freshwater drum. Most of these large species had disappeared from this section of the Rio Grande by the end of nineteenth century. American eel elvers were also found migrating from their hatching in the Atlantic Ocean. Today, 36–63% of the native fish species have been extirpated from the Middle Rio Grande, and 13–19 nonnative species have been introduced. Exotics now comprise 38–76% of the total number of fish species present. Common species today include rainbow trout, carp, white sucker, black bullhead, yellow bullhead, channel catfish, and largemouth bass.

Aquatic macroinvertebrates characteristic of the Rio Grande consist of the major insect orders Diptera (midges and flies), Ephemeroptera (mayflies), Plecoptera (stoneflies), and Trichoptera (caddisflies). The chironomid midge larvae are a significant food source for the fishes in the river, and the macroinvertebrates, in general, have a wide range of habitat preferences.

A key ecological process in this ecosystem is the transport of allochthonous organic detritus originating from the riparian vegetation and upstream reservoirs. Little autochthonous organic matter in the form of periphyton or macrophytes is produced by rivers such as the Rio Grande, because of the shifting nature of the sandy bottom. However, periphyton is found in the tailwater areas below Cochiti Dam, and primary production peaks in early spring and late summer, thus providing a source of detritus for invertebrate collectors when these communities die and a source of nourishment for grazers while they are growing. Planktonic algae originating in the quieter waters of ponds and marshes also contributes to the FPOM pool in the river. Part of the allochthonous energy source is used by shredder invertebrates, which, in turn, are fed upon by other predators. Microorganisms decompose the remaining detritus, resulting in high concentrations of nutrients for plant growth. Sediment-borne nutrients also contribute to a rich riparian zone when they are deposited on the floodplain. Flooding, in turn, releases and transports nutrients to downstream communities.

Benthic macroinvertebrates and fish are the main aquatic consumers found in the Middle Rio Grande. Longitudinal variation in the macroinvertebrate and fish communities through this section of the Rio Grande indicate that changes in hydrology, channel morphology, and water quality are important influences on the structure and function of the aquatic ecosystem; trends in organic matter input do not appear to be limiting.

The Salt River

The Salt River presents an excellent example of the cultural impacts to available natural water ecosystems. In *How to Create a Water Crisis*, Frank Welsh describes the history of the Central Arizona Project, of which the Salt River and the Salt River Project (SRP) play an important role. Most of what follows comes from his account of how the Salt River was destroyed because of the alleged water shortage of this area. We say "alleged," because while desert regions certainly are short of surface water, groundwater supplies may provide an alternative source. Moreover, it is a cultural decision to build cities in the desert and insist on lawns and other amenities more typical of areas of ample moisture. But the surface water of the Salt River was the solution that many seized upon.

The Salt River in its upper reaches (Fig. 10.4) appears to be typical of permanent streams in this region—a verdant riparian corridor lining a

FIGURE 10.4 Photograph of Salt River prior to diversion into Salt River Project canal system. (Photo by C. E. Cushing.)

sparkling water course in an otherwise dry, arid landscape. It rises in the Usery Mountains near the eastern border of Arizona. On its western journey to join the Gila River southwest of Phoenix, it is interrupted by four major dams: first Theodore Roosevelt Dam, then Horse Mesa Dam, Mormon Flat Dam, and Stewart Mountain Dam. Further downstream, the Salt River is joined from the north by the Verde River. Surprisingly, the Salt River ends here; it just vanishes, replaced by a dry bed winding through downtown Phoenix. What happened to the water? Just below the confluence of the Salt and Verde, Granite Reef Dam, maintains water levels for diversion into two concrete lined canals to supply drinking and irrigation water to people in southwest Phoenix, Tempe, Mesa, and Chandler. Today, water flows in the original Salt River bed only when heavy rains force dumping of water from the upstream dams, sending flood waters racing down the original river bed. This has happened twice since 1970, both times causing major damage and loss of human life.

What has the SRP done to the ecology of the Salt River itself? Construction of the four upstream dams has converted several miles of flowing water and their typical biological communities into standing reservoirs. This, as we have seen in other instances, results in the replacement of an aquatic community of organisms characteristic of rivers and streams by a community typical of lakes. The riparian vegetation is completely absent around the reservoirs, another significant loss of an important ecological community. New species of fishes have been introduced into the reservoirs; buffalofish were stocked in Roosevelt Lake after it was constructed, and flathead catfish now dominate the fish community in the

Salt River upstream from the lake. Surely, this was not typical of the fish fauna prior to construction of the dam and intervention by man.

But the greatest ecological impact is downstream, below the dams. Imagine a typical flowing stream with its mix of substrate, riffles and pools, and riparian vegetation being replaced by a concrete-lined ditch! It certainly doesn't take a professional stream ecologist to realize that the ecological dynamics of the Salt River have been changed drastically, and that it no longer even functions as a natural ecosystem.

All of this to satisfy an alleged "water crisis" where, as Dr. Welsh points out, none really existed. With all of the underground water available in aquifers, it merely points out that it is the laws and the politics that are not working, not the supply of water.

Recommended Reading

Fisher, S. G. (1986). Structure and dynamics of desert streams. *In* W. G. Whitford (ed.), *Pattern and Process in Desert Ecosystems*, pp. 119–139. University of New Mexico Press, Albuquerque.

Welsh, F. (1985). *How to Create a Water Crisis*. Johnson Books, Boulder, Colorado.

CHAPTER 11

Special Riverine Systems

INTRODUCTION

In the preceding five chapters, we've described some of the major types of rivers and streams in the country. Obviously, we have not covered all of the kinds of flowing water ecosystems that exist; many other examples come to mind and in this chapter we will briefly characterize some of these unique rivers and streams.

THERMAL SPRINGS

Thermal springs and streams (Fig. 11.1) occur in regions where geological conditions have resulted in heated magma being found close to the earth's surface—such places as Yellowstone National Park, areas of New Zealand, Arkansas, California, and other places throughout the world.

FIGURE 11.1 Thermal spring, Norris Geyser Basin, Wyoming. (Photo by C. E. Cushing.)

Hot springs are generally the result of water flowing through hot rock formations and eventually reaching the surface as fumaroles, geysers, or other outlets. Water temperatures can be extremely hot; the thermal springs in Arkansas flow at a fairly constant temperature of about 64°C throughout the year. In Yellowstone National Park, temperatures of some springs are as high as 91°C; at the altitude of Yellowstone, this is nearly boiling. Thermal waters are often highly charged with dissolved minerals due to the dissolving power of hot water and the fact that they are under high pressures underground. Thus, at spring sources, and for varying distances downstream, one may observe deposits of minerals that precipitate out as soon as both pressure and temperature are reduced when exposed to air.

It is surprising to find life existing at these high temperatures, but nevertheless, it is there, and in abundance. Bacteria appear to be the only living organisms existing in water up to 75.5°C. Algae are the life-forms most adaptable to existing under high temperatures, and their presence provides the brilliant colors found in many of the large, hot pools that form the source of the thermal springs. Cyanobacteria, such as *Synechocyetis* and *Phormidium*, have been found flourishing in waters as hot as 60°C! As waters cool to below 40°C, diatoms begin to appear. Protozoa, water mites, and midges are also found near the sources of these springs where the water has cooled to around 50°C. As the water further cools downstream, the normal occupants of streams begin to appear—mayflies, stoneflies, and fishes.

These thermal springs have an interesting impact on the streams into which they enter, depending, of course, upon the temperature of both the receiving water and that of the thermal spring itself. If the entering spring is relatively hot compared to that of the receiving stream, the fauna downstream of the juncture can be eliminated or reduced in numbers for varying distances downstream.

SPRING STREAMS

Picture a full-sized river or stream emerging from the ground—not a small, 1st-order springbrook, but a large, mid-order-sized flowing body of water. Unusual, yes, but they occur in several places in the United States. Perhaps the best known are the famous spring rivers in northern Florida where the underground geology is conducive to the transport of underground aquifers to the surface. These systems occur elsewhere, and the example we will describe is the Metolius River, located in central Oregon.

The Metolius River (Fig. 11.2) arises from the ground at the base of Black Butte, a volcanic mountain just west of Bend, Oregon, on the eastern side of the volcanic Cascade Mountain Range. It flows for about 43 km before enter-

FIGURE 11.2 The Metolius River, Oregon, at its source, an example of a spring stream that originates as a large stream. (Photo by C. E. Cushing.)

ing Lake Billy Chinook and then the Deschutes River, a tributary of the Columbia River. The watershed is 94% forested with conifers typical of the Cascade Range. Eight subwatersheds contribute water to the Metolius River.

The Metolius at its source has a flow averaging about 42 cms (range 37 to 48 cms) and is equivalent in size to a 4th-order stream. Water temperature is quite constant, averaging 8.5°C, and seldom exceeding 9°C. Most of the river can be characterized as riffle-run. As would be expected from a spring source stream, the water chemistry also is fairly constant. Nitrate nitrogen concentrations at the source average 0.07 mg/L, total phosphorus 0.1 mg/L, and orthophosphate 0.1 mg/L. These values increase downstream to 0.035, 0.08, and 0.08 mg/L, respectively, as the Metolius receives the input of the streams draining the eight subwatersheds.

The Metolius has little overhead canopy, and what was originally there has been reduced by the cutting of large riparian trees. In addition, most of the large woody debris in the stream itself was removed in the early 1900s when attempts were made to float timber to downstream mills.

The open nature of the stream and its size is reflected in the rich macroinvertebrate communities present. Macroinvertebrate populations average approximately 9200 organisms/m^2, comprised of 49 different species, of which 28 are mayflies, stoneflies, and caddisflies. The four most numerous taxa are *Baetis* (21%), *Drunella* (16%), *Rhithrogena* (7%), and chironomid pupae (5%).

Sixty percent of the macroinvertebrates are collectors, 12% are scrapers, 8% are predators, and 4% are shredders. This is fairly representative of the predictions of the RCC for functional group representation for the mid-reaches of a river.

The Metolius River is a popular destination for anglers. Sport fishes present include rainbow, brown, brook, and bull trout; mountain whitefish; and kokanee salmon. Several species of sculpin and suckers are also present, along with several minor species.

Cold-Desert Spring Streams

A group of unique spring streams (Fig. 11.3) occurs in the cold-desert, shrub-steppe region of the western United States—largely in western Utah, Nevada, southern Idaho, eastern Oregon, and southeastern Washington. These regions are characterized by low winter and high summer temperatures, low annual precipitation (ca. 16 cm, most of which falls in winter), and a regional vegetation dominated by big sagebrush (*Artemisia tridentata*) and an understory of perennial bunchgrasses; cheatgrass now occupies much of the area that has been disturbed. Research on these ecosystems has been largely limited to three sites in southeastern Washington and one in southeastern Idaho; yet several interesting features about their ecology have been discovered.

Physical and Chemical Characteristics

Generally, these spring streams are closed; they arise from seepage areas, flow for various distances, and disappear into the arid soils. Perennial flow

FIGURE 11.3 Rattlesnake Springs, Washington, a typical cold-desert spring stream. (Photo by C. E. Cushing.)

is ordinarily present, but the stream terminus usually recedes as the summer season progresses. Base flow is usually fairly constant, seasonal temperatures vary from near freezing in winter to around 20°C in summer. As expected, solar input is quite high, due to the relative absence of riparian vegetation.

In contrast to hot desert streams (see Chapter 10), spring streams experience flash floods during the winter, as Chinook winds over the frozen ground cause accelerated snow melt. As a consequence, recolonization of these two systems depends on contrasting processes. Because floods occur in winter, recolonization of cold desert spring streams is less dependent upon aerial colonizers than hot desert streams. Because the cold desert systems are usually short and enclosed, their floods scour the entire length of the streams, thus precluding the presence of upstream sources of drifting organisms for recolonization. Disturbance survivors appear to be the most likely source for recolonizers in cold-desert spring streams.

Chemical characteristics are not unusual, although total dissolved solids can be fairly high, up to 220 ppm. Dissolved oxygen is usually adequate because of the autotrophic nature of these streams.

Biological Characteristics

Primary production is relatively high in these small spring streams. This would be expected given their openness to the sun, a rich population of algae, watercress, and other instream macrophytes, and adequate concentrations of nitrogen and phosphorus. Input of allochthonous POM can be relatively significant in reaches where waterflow is perennial and riparian vegetation is well developed, but negligible along reaches where flow is intermittent and little riparian vegetation is present. In either case, studies have shown that grazing insects in these spring streams are more dependent upon instream sources of primary production (periphyton) than allochthonous sources.

One of the anomalous aspects of these cold-desert streams is the relationship between their productivity and the winter flash floods. Because they are autotrophic systems, these streams produce more organic matter than they consume and thus, on an annual basis should export more OM than they import. Yet, if one measures import and export over an annual cycle that does not include a flash flood (they occur about every 4 years), the system appears to be importing and storing more OM than it exports. In a flood year, however, export exceeds import. Thus, it appears that these closed systems have the ability to store the excess production in the form of detritus until a flood comes along and flushes the system.

Secondary production of aquatic insects is intermediate between those found in warmer regimes and those characteristic of cold—mesic regions. Collectors were the dominant functional group of insects present and attest to the large amount of detrital storage present.

Cold-desert spring streams are usually 1st-order streams and rarely flow far enough to receive additional tributaries that would raise them to a higher order. They do not fit the original RCC model because they usually occur in open situations and are autotrophic, rather than being shaded and heterotrophic. Insect functional groups are dominated by collectors; shredders are virtually absent—again unlike the RCC model.

Most of the information on cold-desert spring streams has been taken from the results of research conducted by one of us; more details can be found in Cushing (1996).

ALPINE STREAMS

There are two types of streams that can be found flowing in alpine situations, essentially those regions above timberline. One type originates in seeps and bogs, lake outlets, small springs, or melt from snowfields. The second type originates as glacier melt. We will have a look at both types.

Alpine streams originating from melt from snowfields, lake outlets, springs, or seepage areas are gems of nature (Fig. 11.4). They are usually crystal clear, cold, and flow through a treeless area of scenic alpine tundra. Riparian vegetation can be fairly abundant, especially as the streams approach timberline. The growing season is usually quite short, limited to the brief period between the disappearance of the ice-cover in late spring–early summer, when water temperatures begin to warm up, to the onset of winter conditions in early fall. Despite this brief period of time, flourishing communities of algae, macroinvertebrates, and fishes can be found in many of these systems; *Eurothocladius*, a midge, is commonly found in cold, snow-fed high streams.

As with cold-desert spring streams, alpine streams are generally low-order streams and occur in open situations; thus, they depart from the RCC

FIGURE 11.4 An alpine stream, Edith Creek; Washington. (Photo by C.E. Cushing.)

model in that they are open to sunlight. However, this does not assure that they will have high primary productivity; the low water temperatures can inhibit algal growth and the short growing season results in their being autotrophic for only a short time during the annual cycle.

The second type of stream that can be found in alpine situations results from the melting of glaciers and found flowing from the glacier's terminus (Fig. 11.5). These unique streams have special characteristics which shape their biological communities. As expected, the waters are extremely cold (at or near freezing) year round, and they are milk-white in color because of the heavy load of suspended silt resulting from the grinding action of the glacier on the rocks over which it flows. Given these two factors, cold temperatures and a heavy silt load, it is not surprising to find that the biological communities found in these streams are quite depleted and highly restricted in diversity. Extreme low temperatures limit the opportunity for organisms to grow and reproduce; only those organisms able to complete their life cycles in such extreme conditions can survive. Then there is the problem of finding enough food. The heavy silt load

FIGURE 11.5 The White River, Washington, a river emanating from Emmons Glacier. (Photo by C. E. Cushing.)

inhibits the growth of algae by preventing light from reaching the streambed so that photosynthesis can take place, and it also acts as an abrasive scouring the rocks of any algae that attains a foothold. Thus, macroinvertebrate development is restricted not only by cold temperatures but also by the lack of food.

The simple communities that are found in these streams are mostly composed of a few species of chironomids, quite frequently the genus *Diamesa*. They are often found within a few meters of the stream's emergence from the glacier and are confined in distribution to cold water. They are collectors, feeding on the sparse amount of detritus found on the stream bottom. As the stream progresses downstream, *Diamesa* is joined or replaced by other chironomids and eventually by populations of flatworms, mites, mayflies, and stoneflies. Whereas glacial streams do not become rich in fauna for some distance downstream, alpine streams may be rich in fauna right up to their source. Most organisms complete their life cycle in one year.

TROPICAL STREAMS

Although most visitors to tropical areas plan their trips during the "dry season," there is considerable variation in rainfall during most months in the "wet season." Tropical streams (Fig. 11.6) provide some interesting contrasts with temperate-zone streams because of their different types of seasonal changes and considerable variation from one year to another. Considerable research has been done on the streams in the Luquillo

FIGURE 11.6 The Quebrada Prieta, Puerto Rico, a typical tropical stream. (Photo by A. P. Covich.)

Experimental Forest in Puerto Rico, and we will use the results of these long-term studies to provide a brief description of the ecological characteristics of tropical streams. These ecosystems have provided some unique insights as to how these streams compare with those on the mainland.

These streams are generally small, draining, narrow steep-gradient watersheds; they traverse through several types of forest in their passage from headwaters to the sea, and the streambeds are boulder-lined and well-shaded. Open, sunlit pools occur where storms have toppled large trees, opening the streambed to sunlight. The streambeds are subject to high discharge rates following the frequent storm events, and these events shape much of the ecology of these streams. The large runoff events flush the high accumulations of terrestrial CPOM (more about this later), dilute nutrient concentrations, erode banks, fill the crevices (refugia) in the rocks with silt, and dislodge many bottom-dwelling organisms. There is little variation in water temperature, ranging from 18 to 23°C annually.

Large streams, where they occur, are wide and open to the sun; thus, they develop luxuriant populations of periphyton. Nitrogen concentrations are low and often limiting, and phosphorus concentrations are often below detection limits. Despite this, as mentioned, periphyton usually is present where sunlight reaches the streambed and sufficient nutrients are present. Often, preferential grazing fishes, snails, and insects keep some types of algae to a low level of abundance, but this thin film of diatoms and green algae grow back quickly and are rapidly consumed.

Throughout the entire lengths of headwater streams, terrestrially produced OM is the main source of energy. Primary productivity is low because of the heavy shading, but, as mentioned above, periphyton is an important source of food in selected locations. The relatively warm water temperatures enhance the rapid colonization of bacteria and fungi and breakdown of the terrestrial OM. In places where hardwood dams have been created by tree blowdown, these sites create sources of high-quality leaf litter because the detritus is retained in the streambed and has a chance to undergo processing by bacteria and fungi. In most other sections of the streams, the detritus is flushed out frequently by the high discharge rates.

In the headwaters, food webs are composed of relatively few links. All of the functional feeding groups are present, but the organisms are quite different from those found in temperate mainland streams. Rather than insects being the major organisms present in terms of food web dynamics, tropical streams are dominated by decapod crustaceans, mainly atyid shrimp. These perform the role of both herbivores and detritivores. Fishes (eels, mullets, and gobiids), two species of grazing gastropods, and more than 60 species of aquatic insects comprise the rest of the food web. This seems like a wealth of insects, but they play a lesser role to the 11 species of atyid shrimp. Insects are important detritivores in the 1st- and 2nd-order headwater streams. Although production rates of insects are low as compared to temperate streams, the turnover rates of the population are rapid. Incidentally, although mayflies, caddisflies, beetles, dipterans, and other common insects are found, stoneflies are absent in these streams.

CAVE STREAMS

One other unique habitat where flowing water is encountered is in caves, which may contain volumes of water varying from small trickles to fairly large streams. Little is known ecologically of these systems, but some characteristics are known.

Algae are absent due to the absence of light. Most food webs are thus based on either chemosynthesis or detritus; with the latter being the most prevalent. Organisms feeding on this detritus share some physical characteristics—they are usually blind, or at best have vestigial "eyes," and are usually white or almost colorless.

RELATIONSHIP TO THE RCC

Most of the above examples represent small flowing systems, those typical of the 1st- through 3rd-order reaches of the RCC. Because they are unique, they do not always follow the predictions of the RCC in terms of physical–chemical characteristics or populations of biota present. This is to be expected for a multitude of reasons—water temperature, environmental setting, etc. The Metolius River probably most closely matches the predictions of the RCC, even though the lower order reaches are absent.

This chapter completes the section of the book devoted to the description of the various kinds of rivers commonly encountered. We've visited rivers from different geographical regions, talked about different kinds of streams, and briefly discussed some unusual types of streams. We'll now turn our attention to a description of the biota found in rivers and streams.

PART III

The Biota of Rivers

In this part, our goal is to acquaint the reader with the major groups of plants and animals that inhabit streams and rivers and their margins. We emphasize the biota's ecological roles rather than present a detailed description of all of the species, their distributions, and their characteristics. For the vertebrates, excellent field guides exist. For most of the groups that make up riverine food webs, however, it would not be practical to present photographs and range maps of every species, nor would that help the reader understand their ecological roles. In describing this flora and fauna, it will be necessary to use some scientific, or Latin, names and classifications. Hence, we offer here a brief description of how plants and animals are named and how they are classified.

Biologists, by habit, tend toward neatness; chaos is their sworn enemy. Thus, an entire field of biology, known as *taxonomy*, has developed to satisfy this urge to have every organism in its proper pigeonhole; its practitioners are known as *taxonomists*. Taxonomy is the organized system of classification by which groups of organisms of similar characteristics are placed into smaller and smaller groups based on recognized similarities of such things as body shape, ecological affinities, and, of more emphasis today, biochemical (DNA) affinities. The highest or broadest level of classification is kingdom—plants, animals, fungi, protozoans, and bacteria. A kingdom is further divided into larger and then finer categories. At the highest levels, division into categories is based on widely shared traits, like a backbone, or a group of traits, like hair, warm-bloodedness, and nursing of young. As further subdivisions are developed, the "splitting points" become finer and finer, reaching such defining separations as, for example, the number of hairs on an insect's leg! The end result is to provide every organism with a binomial (two-part) name consisting of a genus and species, which identifies it from all other organisms. *Homo sapiens*, of course, is the binomial name for humans; *Oncorhynchus mykiss* identifies the rainbow trout, and no other fish.

Unfortunately, specialists don't always agree on what end point defines a species or even on the highest taxonomic groupings. There were only two kingdoms (plants and animals) when the authors were students; today most college students learn of six (bacteria are split into early and modern). In all likelihood this number will increase as modern methods reveal more of the murkiest aspect of the tree of life, its bottom roots. Biologists are continually revising, renaming, and regrouping organisms as new evidence is found for making these changes. The point of this discourse is to make the reader aware that our taxonomic classifications may not agree with names you find in another source. For the purposes of this book, however, we will use those names currently acceptable to most people in the field today.

The major taxonomic categories, illustrated for a mayfly, are as follows:

Kingdom	Animalia
Phylum	Arthropoda
Class	Insecta
Order	Ephemeroptera
Family	Baetidae
Genus	*Baetis*
Species	*tricaudatus*

Be aware that at each level of classification, there are further breakdowns, such as superorder, subfamily, etc., that help taxonomists with their quest for order. We will not concern ourselves with these levels here, but you may see them in various publications.

This classification is particularly useful for animals, but be aware that botanists use division instead of phylum. Another problem will appear when we discuss algae. Under the old, two-kingdom model, all were plants, united by their ability to photosynthesize. Today, some algae are in the plant kingdom, some are protozoa (or, more properly, protists), and some are bacteria. For simplicity, we will treat algae as plants, but be aware that this is only a matter of convenience.

CHAPTER 12

Algae

INTRODUCTION

Algae, sparkling green or shining brown, are truly the "grasses of the waters." In one form or another, they are present in varying numbers and kinds in almost any river or stream you encounter—sometimes more than you might want, but certainly always present.

We will begin Part III of this book, devoted to descriptions of the various animals and plants found in streams, with the algae—after all, as we have learned previously, they are the main aquatic organisms which produce the basic food sources for other organisms through photosynthesis.

Algae are, for the most part, microscopic plants found free floating (*phytoplankton*) or attached to solid objects (*periphyton*) (Fig. 12.1); but they can also aggregate into life-forms visible to the naked eye. We will emphasize in this chapter the periphytic forms found attached to rocks, sticks, logs, and other solid objects found in rivers and streams because they contribute to riverine food webs in small- to medium-sized river systems.

We do, however, need to acknowledge the role of phytoplankton because they are important in some flowing waters. Algae can be found suspended in streams of any size, from the smallest to large, nearly lakelike rivers. The question is, in which of these ecosystems are they truly "functioning" as algae? If you collect suspended algae in smaller streams ranging from 1st- to about 6th-order, the species you collect will be the same as those found as periphyton, i.e., the suspended algae are actually cells detached from the bottom. As such, although they may be alive, they are

FIGURE 12.1 Periphyton coating on rock. (Photo by R. Lowe.)

actually only in transport until they sink to the bottom or die, and there is debate as to whether they should be considered "living and functioning" in this form or whether they are merely FPOM. Another argument against considering them as part of the actual "living" system in this form is that they do not reproduce and grow as do true phytoplankters. Now, as rivers become larger, lakes, reservoirs, and other still-water habitats drain into them and contribute a true phytoplankton to the flowing system. Now the question becomes, "Do these lake forms thrive and reproduce as they are carried downstream?" In many cases the answer is "yes, they do," and thus can be considered truly phytoplanktonic and "living" members of the ecosystem. In fact, as we learned in Chapter 5, the depth, turbidity, and soft stream bottoms in these larger rivers are not conducive to the growth of periphytic algae, and primary production (photosynthesis) is accomplished by the phytoplankton. So, suffice it to say that true, actively functioning phytoplankton algae are restricted to the larger, slow-flowing, deeper rivers, and we'll devote the rest of this chapter to the periphytic algae most commonly found in streams and rivers.

Because algae in flowing waters are subjected to a current, in order to remain in place they must either find places of refuge or develop organs of attachment. The former are usually found where currents are negligible or absent—within the periphyton mat, in mucilaginous colonies that resist the eroding nature of the current, in rock crevices, or in backwaters where the current is slack. Filamentous forms have the ability to attach to solid objects and withstand being washed away by fast currents (Fig. 12.2). Living in flowing water can be beneficial to plants; the current is usually well aerated and provides a continuous new supply of nutrients. At the

FIGURE 12.2 Filamentous green algae. (Photo by C. E. Cushing.)

same time, the current continually breaks off pieces of filament or washes away parts of the periphyton community. Thus, most algae growing in flowing water have rapid growth rates enabling them to replace those parts lost to the current.

THE BASE OF THE FOOD WEB

In our basic food, or energy, pyramid that was mentioned in Chapter 3, we said that the basis of all energy found in rivers and streams was the result of photosynthesis by aquatic plants or by terrestrial plants and input as by-products. Of the three aquatic plant groups (algae, macrophytes, and bryophytes), the algae are by far the most important. They appear in uncountable numbers, in many life-forms, in all flowing environments—truly they are ubiquitous in rivers and streams.

LIMITING FACTORS

Despite their prevalence, aquatic plants do need certain conditions in which to flourish, and without these, their occurrence may be quite limited. These conditions, or factors, include such things as temperature, light, nutrients, etc., and define the limits under which a particular species can grow—the upper and lower limits within which they can exist. These are called *limiting factors*. Algae need light in the wavelengths suitable for photosynthesis (called PAR or photosynthetic active radiation) to carry on the synthesis of organic matter. They need a temperature regime that is also conducive to photosynthesis; colder temperatures slow down the process while warmer temperatures enhance it. And, just as do your lawns and plants, they need an adequate supply of nutrients, mainly phosphorus and nitrogen, to grow. As we shall see, the relative amounts of light, temperature, and nutrients, together with flow and stream bottom characteristics, govern what algae will be found and where.

MAJOR CLASSES

The three major classes of algae important in streams and rivers are the *diatoms* (class Bacillariophyceae), the *green algae* (class Chlorophyceae), and the *cyanobacteria* (class Cyanophyceae). We need to say a few words about the latter group. As mentioned earlier, advances in our understanding of evolutionary relationships result in changes in the taxonomy of organisms, and nowhere is this more true than for the cyanobacteria. For years, they were called blue-green algae and classified as algae, albeit very primitive ones. Today, all taxonomists classify them with the bacteria, rather than plants, using the name *cyanobacteria*—really a combination of a color and a type of organism. We will use their current name when discussing this group, even though the public may know them better under their former name.

Let's take a look at some of the characteristics of these three major groups.

Diatoms

Walk up to almost any stream or river and pick up a good-sized rock. What do you notice first? It is slippery, at least on its upper surface. Next you may observe that it and its neighbors all have a brownish color to them. Why? Well, both the slickness and the color are because the rock is covered by millions and millions of single-celled algae called diatoms. Diatoms are truly some of the most beautiful and fascinating of all of the algae, and indeed of many other plants and animals. This is because of their unique construction. They are single-celled plants composed of two halves, called *frustules* or *valves*, which fit together much like a pill box (Fig. 12.3). Now this is not what makes them unique. What sets them apart is the fact that each of the frustules is composed of glass, pure silica, and the frustules of each species are characterized by unique striations and patterns in the glass walls; Figs. 3.3, 12.4, 12.5, and 12.6 show some of these. Within the frustules are the organic cell components. These are usually golden-brown in color and impart the same color to the rocks and other solid objects picked from the stream bottom. Another characteristics of diatoms is that they secrete a mucilaginous coating, and it is this coating that makes the rocks and sticks slippery—and makes wading treacherous!

The diatoms are divided into two groups based on the symmetry of their markings. Those that exhibit radial symmetry are termed *centric*, and those that exhibit bilateral symmetry are called *pennate*. Diatoms can

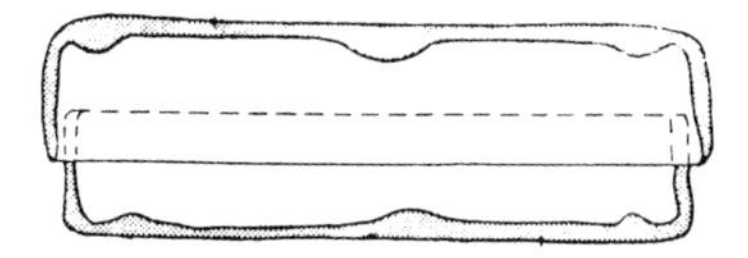

FIGURE 12.3 Diatom frustule construction, showing the upper and lower valves. (From South and Whittick, 1987, with permission.)

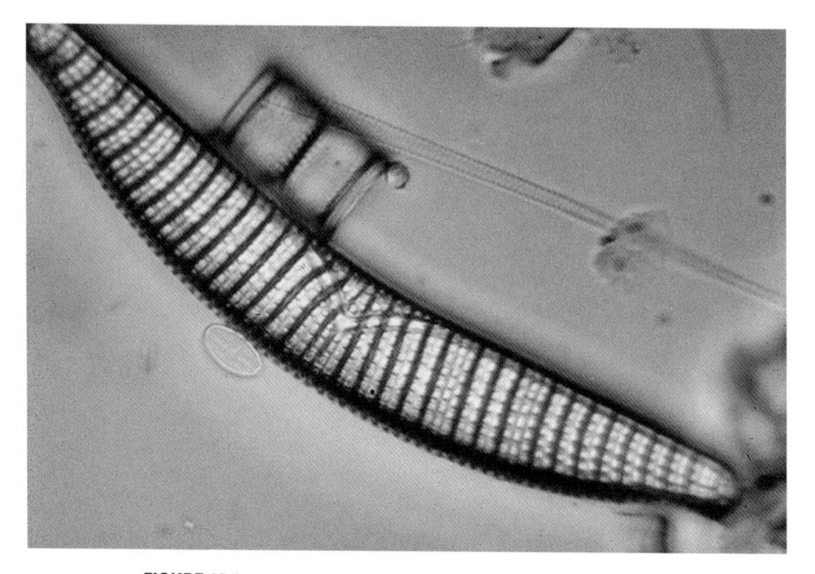

FIGURE 12.4 The diatom *Epithemia* sp. (Photo by R. Lowe.)

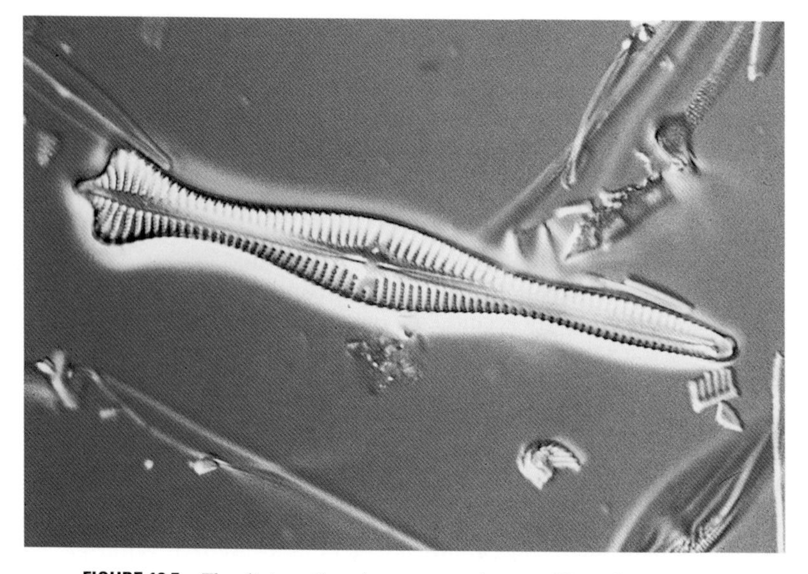

FIGURE 12.5 The diatom *Gomphonema acuminatum*. (Photo by R. Lowe.)

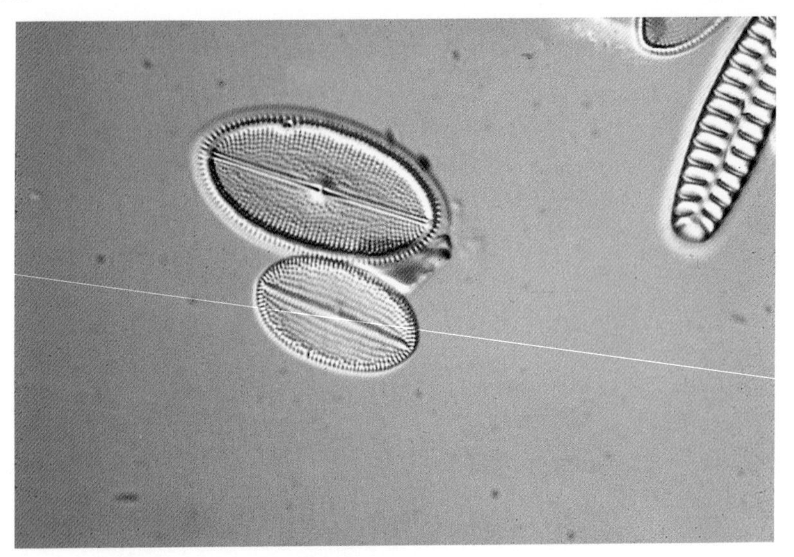

FIGURE 12.6 The diatom *Cocconeis placentula*. (Photo by R. Lowe.)

reproduce either asexually or sexually. The latter usually occurs after a series of asexual reproductions, which may result in a progressive decrease in cell size; sexual reproduction restores maximum size to the cell.

Diatoms can occur individually, in colonies, or in chains, depending on the species. They are usually typical of clean-water environments, and ordinarily exhibit two growth blooms. The first occurs in spring as water temperatures warm and when nutrients are plentiful. Populations then decline during midsummer, but peak again in fall, as nutrients again become plentiful from the death and decay of green algae populations, which are more abundant in midsummer when water temperatures are higher.

The rich coating of diatoms found in most streams is one of the major sources of energy in river and stream ecosystems. They also are important contributors of oxygen because of their high numbers and photosynthetic activities.

Diatoms occur by the literal billions in lakes and oceans. When they die, the cells settle to the bottom, the cell contents decay, and the glass frustules are left to slowly accumulate over eons of time. Eventually, these beds of pure silica can become meters thick, and it is the mining of these deposits that provides industry with one of its most important abrasives, diatomaceous earth; they are also used for other commercial applications.

Green Algae

For purposes of discussion, we can divide the green algae into two general groups, although the second group will be of most interest. Why? Because we can see them without a microscope.

The first group, as you might guess by now, are the microalgae, which cannot be seen without the aid of a microscope. These are the unicellular forms such as the genus *Chlamydomonas*, which can be found in the periphyton. These algae are small and cannot be seen by the naked eye; however, in some standing waters they can occur in rich blooms that result in visible discoloration of the water.

The second group of green algae are the macroalgae, those forms which can be readily seen as one strolls along a river or stream. Macroalgae are somewhat of a step up from the unicellular forms; they are single cells, little different from the unicellular forms, but occur in colonies held together within some kind of matrix. In this form, they may become large enough to be seen without a microscope. Two different life-forms occur; those growing as a *thallus* (Fig. 12.7) and those growing as *filaments* (Figs. 3.2 and 12.2). Have you ever picked up a rock or stone from a river or stream that had what appeared to be a bright green, lettucelike growth attached to it? In most cases, these were green algae belonging to the genus *Prasiola*, often called stream lettuce. If you were to put a small piece of this under a microscope, you would find that it contained groups of four cells, lined up in regular rows and columns. This is a good example of the thallus growth form.

More prominent and easily seen along rivers and streams are the filamentous forms occurring as bright green tufts on solid objects in the river or stream or as filaments up to a meter long waving in the current. Most people call filamentous green algae "moss," but they are wrong. As you learned in Chapter 3, true mosses are small plants with leaflike bracts

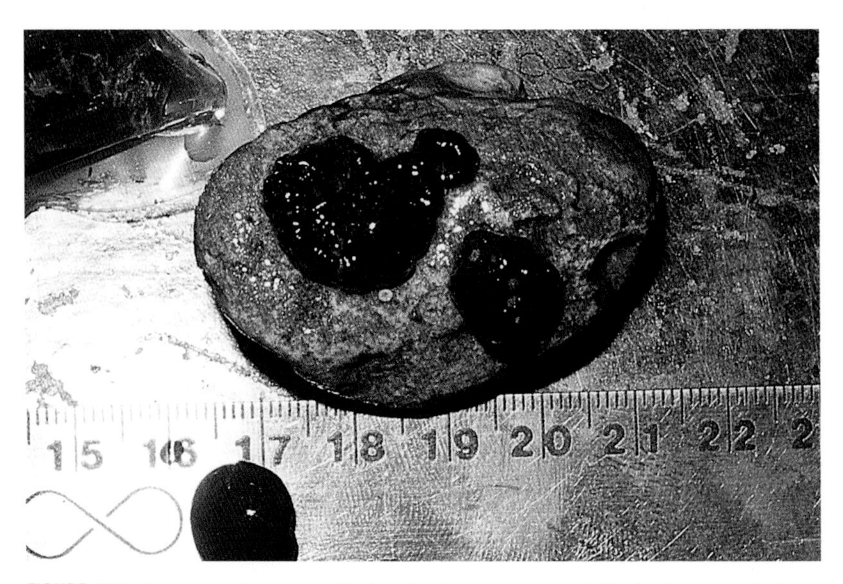

FIGURE 12.7 The cyanobacterium *Nostoc*; this is an example of the thallus growth form. (Photo by W. Dodds.)

found in the cooler reaches of streams. Filaments of green algae may be either branched or unbranched, depending on the genus; unfortunately, the branching requires a microscope to be seen, so differentiation in the field can be difficult. *Ulothrix* is one of the most common unbranched forms found in rivers and streams, and can occur in large bright-green

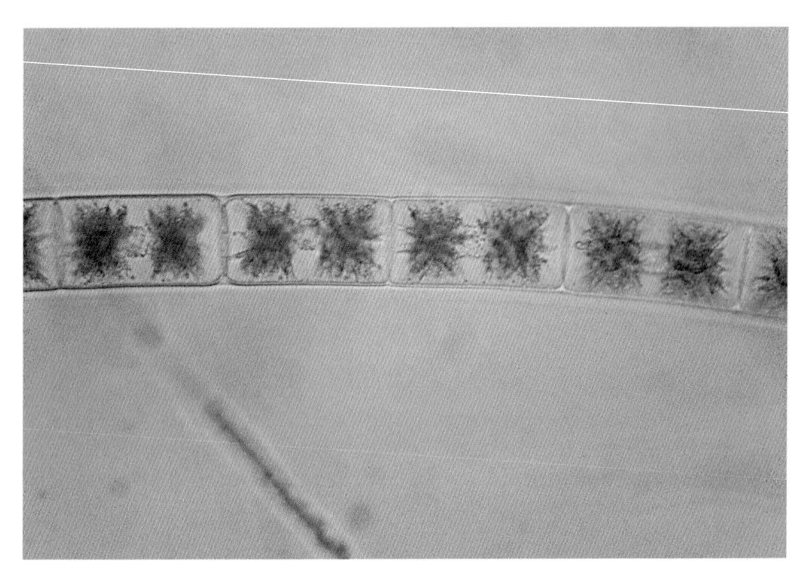

FIGURE 12.8 The unbranched green alga *Zygnema* sp. (Photo by R. Lowe.)

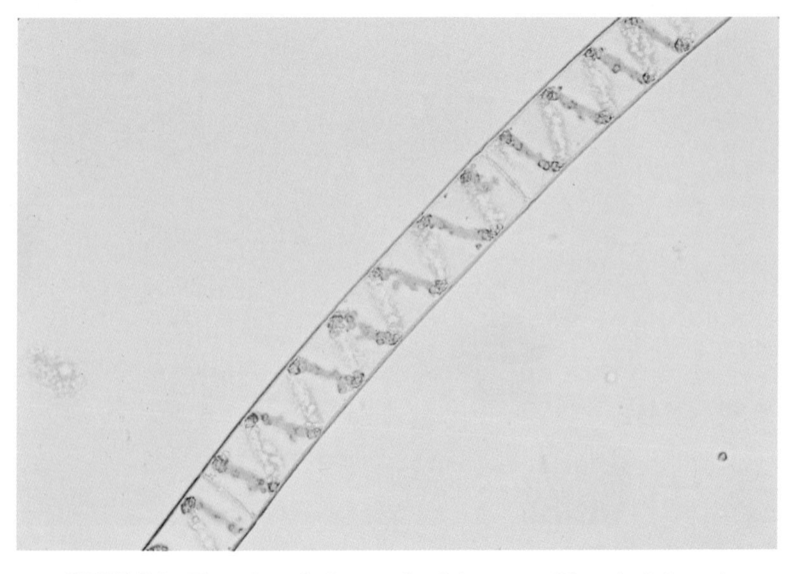

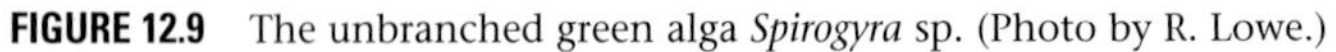

FIGURE 12.9 The unbranched green alga *Spirogyra* sp. (Photo by R. Lowe.)

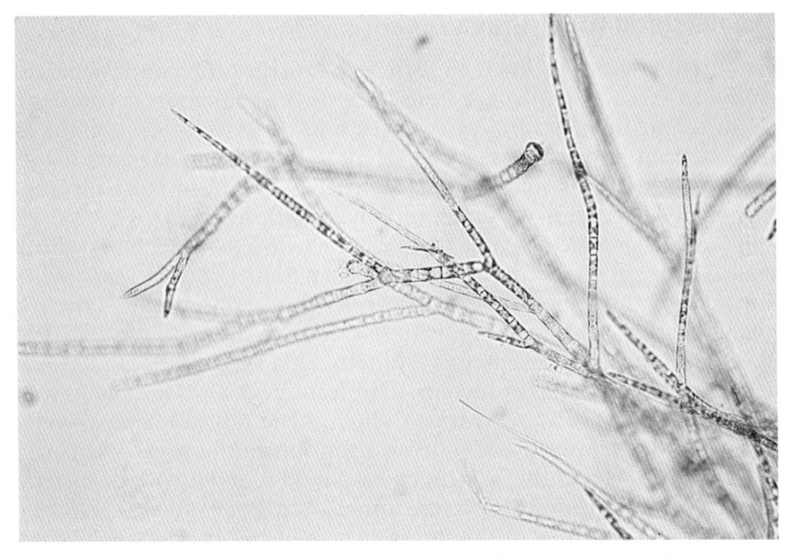

FIGURE 12.10 The branched green alga *Stigeoclonium* sp. (Photo by R. Lowe.)

fronds in the shallows of cold, swift-flowing rivers in spring; *Zygnema* (Fig. 12.8) is another common unbranched alga. *Spirogyra* (Fig. 12. 9) is another unbranched filamentous genus found in quieter and warmer conditions than that of *Ulothrix*. Common branched genera include *Stigeoclonium* (Fig. 12. 10) and *Oedogonium*. In fact, it is difficult to differentiate individual cells of *Ulothrix* and *Stigeoclonium*; the only difference is that the former is unbranched and the latter is branched. *Cladophora* is another common branched filamentous green algae found in cold rivers and streams; it can often be identified by its coarse texture.

Ecologically, the green algae are rich sources of photosynthetically derived oxygen in flowing waters. Their role as food organisms, when they are living, is not as important as are the diatoms whose growth form is more available to grazers. The green algae, however, are important to stream communities in two ways. While living, as either a thallus or filament, diatoms attach to green macroalgae; if you observe a mature filament of green algae under a microscope, you will usually find it to be almost entirely covered by diatoms. Their other important function occurs after the macroalgae die and decay; at this time, they decompose and break up into particles of FPOM which, as we've learned, is one of the two main sources of food for other organisms in river and stream ecosystems.

The filamentous forms of green algae can vary in their seasonal patterns. *Ulothrix* can often be found during winter when other forms find water temperatures too low, and *Cladophora* and *Stigeoclonium* are often prevalent in spring as soon as water temperatures and light regimes are suitable. High summer water temperatures are more suitable for the growth of *Oedogonium*.

Cyanobacteria (Blue-Green Algae)

Cyanobacteria occur in riyers and streams as part of the periphyton community; as you will remember from Chapter 3, periphyton is part of the biofilm found on the surfaces of rocks and other solid objects on the stream bottom. Although not as prevalent as the green algae and diatoms in most periphyton communities, cyanobacteria can be important components of stream ecosystems under certain conditions.

Cyanobacteria are some of the most ancient organisms found on the planet. The fossil record shows them to be at least 3 billion years old. As compared to other types of algae, their cell organization is quite simple, as befits such an ancient organism, with few internal cell features—they even lack a nuclear region. Despite this, cyanobacteria are important photosynthetic organisms and have one important feature which is lacking in other algae—the ability to fix, or "extract," atmospheric nitrogen and convert it into a form that can be used by other organisms in the stream. This is an important feature, and the cyanobacteria can do it under both oxygenated and unoxygenated conditions. In the former, nitrogen fixation is performed by specialized cells called *heterocysts*; under the latter conditions, nitrogen fixation is done by the entire vegetative cell. There are two general types of cyanobacteria in terms of form—filamentous and nonfilamentous; only the filamentous forms have the ability to fix nitrogen.

The nonfilamentous cyanobacteria are generally spherical or elongate in shape and the cells may occur as solitary cells or may aggregate into colonies. *Gloeocapsa* is a common, simple single-cell species that has considerable mucilaginous material surrounding the cell. *Merismopedia*, on the other hand, forms uniform rows and columns of individual cells held together by mucilage.

Filamentous cyanobacteria occur either with or without the specialized cells. Common filamentous species that do not have the specialized cells are *Oscillatoria* and *Lyngbya*. Both form mats and have the special ability to glide.

The most important cyanobacteria in rivers and streams are the filamentous forms, which contain heterocysts and have the ability to fix nitrogen. They are also the forms that cause the most problems in nutrient rich (eutrophic) standing waters, where they form large unsightly mats and dense blooms. The genera most often responsible for these problems are *Anabaena, Aphanizomenon,* and *Microcystis*—often referred to as Annie, Fanny, and Mike. One of the genera most often encountered in flowing waters is *Nostoc*. This genus is usually found as small, firm, dark-green nodules growing on the surface of rocks or other solid objects (Fig. 12.7). They are slippery because the filaments grow in a dense coating of mucilage. They may be anywhere from pinhead size to lumps the size of walnuts.

Cyanobacteria are important in ecosystems not only for their ability to produce oxygen by photosynthesis, but especially where they convert atmospheric nitrogen into a chemical form available to other organisms in the community. These algae occur in a wide variety of environmental conditions; from Antarctic lakes with permanent ice cover to thermal springs where water temperatures reach 74°C; *Spirulina* (Fig. 12.11) is often found

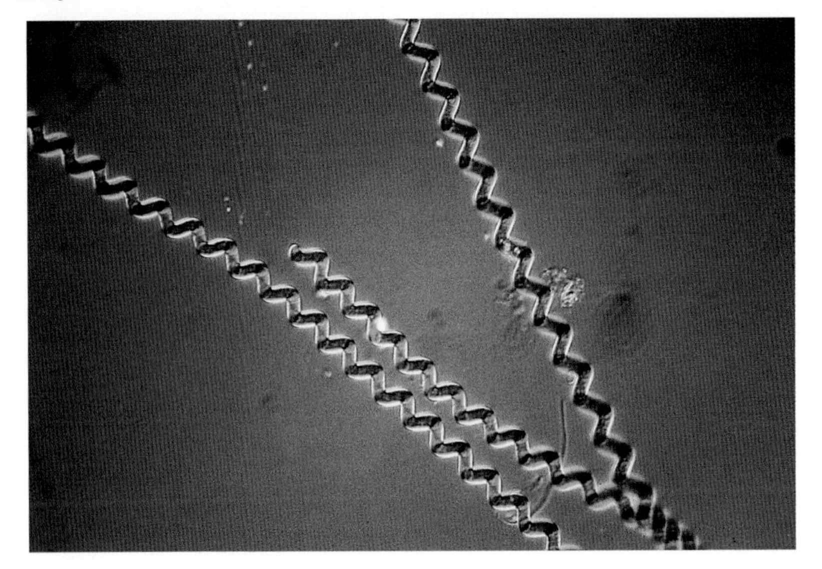

FIGURE 12.11 The cyanobacterium *Spirulina* sp.; often found in thermal hot springs. (Photo by R. Lowe.)

in thermal springs. In fact, it is these algae that give some of the brilliant colors to the thermal springs in such places as Yellowstone National Park.

Although the diatoms, green algae, and cyanobacteria are the groups most commonly encountered in rivers and streams, and also the ones most important to ecological functioning, other taxonomic groups of algae occur, and, under some circumstances, can be quite noticeable.

The red algae (division Rhodophyta), because of their color, can be quite obvious where they occur in high numbers. They occur as filamentous or as a thallus; usually in marine environments, but some forms are found in freshwaters. *Batrachospermum* (Fig. 12.12) and *Lemanea* are common genera found in rivers and streams. They vary greatly in color from violet to yellow-green and blue-green to brown and blackish.

The yellow-brown algae (division Chrysophyta) are also represented in river and stream communities. The most important member of this group are the diatoms, which we discussed in detail above. Other freshwater members of this group are mostly planktonic and thus will not be encountered in rivers and streams. One genus, *Vaucheria* (Fig. 12.13), forms mats in slow-moving water such as ditches. In terms of visible chrysophytes which readers might see, the most prominent are kelp, although don't look for these in rivers and streams!

In this chapter we've presented an overview of the important algae in river and stream ecosystems, with emphasis on those forms that you can observe without the aid of a microscope. It is probably more important for you to just know that they are there, unseen mostly, but providing the important functions of oxygen production by photosynthesis and a food base for

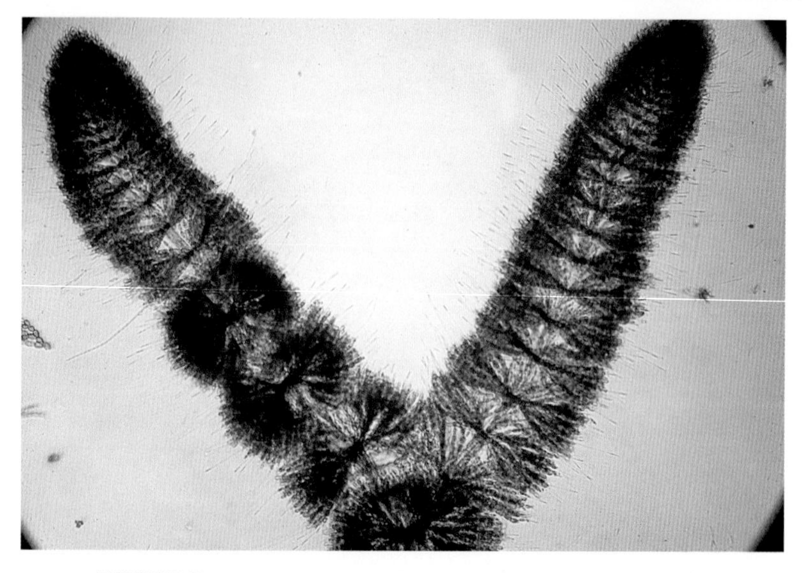

FIGURE 12.12 The red alga *Batrachospermum* sp. (Photo by R. Lowe.)

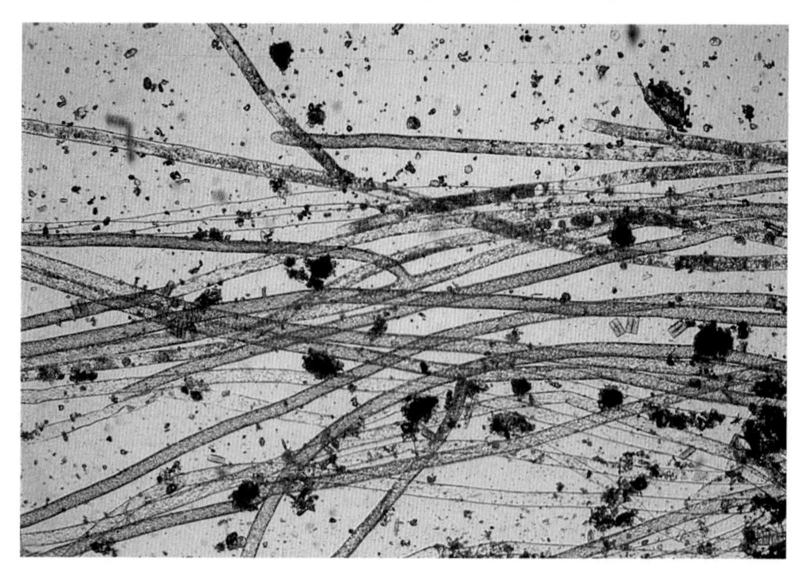

FIGURE 12.13 The yellow-brown alga *Vaucheria* sp. (Photo by R. Lowe.)

grazing organisms. In the next chapter, we will still be looking at plants, but these will be the macrophytes, those plants easily seen by the naked eye.

Recommended Reading

Sze, P. (1986). *A Biology of the Algae*. Brown, Dubuque, Iowa.

CHAPTER 13

Higher Plants: The Macrophytes

INTRODUCTION

In the last chapter, we learned about the microscopic plants found in streams. In this chapter we will discuss the larger aquatic plants, those easily seen with the naked eye. They are either attached to rocks and stones, free-floating, or rooted on the stream bottom; the latter may be entirely submerged or have emergent plant parts. Collectively, we call these *macrophytes*, meaning large plants. Included in this category are flowering plants, mosses and liverworts, some species of encrusting lichens, and a special group, the Charales. We won't be saying much about the lichens or Charales because they are seldom encountered. One other group that is sometimes included as macrophytes are some algal species with large growth forms; these include the filamentous green algae and others that were described in Chapter 12.

Macrophytes, although fairly numerous in terms of species, exhibit few adaptations to living in fast currents. They are most successful in places where the current is reduced and the resultant stress on the plant is lessened.

Scientists group macrophytes in different ways, emphasizing such things as method of attachment and location of plant parts in relation to the water column. We will use the grouping that provides the easiest and quickest way to place any macrophyte you find in the field into a readily identified group. The three groups are (1) attached to solid objects, (2) entirely free-floating, and (3) rooted on the stream bottom. The latter group has two types—those entirely submerged in the water column and those with emergent leaves and flowers.

Macrophytes attached to solid objects are usually found in the colder, headwater regions of streams where the largely rocky stream bottom offers sites for attachment. Representatives of this group include bryophytes (mosses and liverworts), some flattened lichens, and two peculiar families of flowering plants found in tropical waterfalls. The mosses and liverworts are the forms most likely to be encountered in North American streams. Bryophytes, particularly mosses, require high levels of dissolved carbon dioxide; this is one reason for their restricted distribution to areas high in dissolved carbon dioxide, such as headwater springs and turbulent regions. They attach to objects by rhizoid holdfasts (Fig. 13.1). Submerged lichens are uncommon, but mosses occur worldwide. *Fontinalis* is a common moss genus in small headwater streams, although it also occurs in large rivers. Water temperature and availability of dissolved nutrients appear to limit the distribution of mosses, although much remains to be learned about the ecology of these plants. It is interesting that although most macrophytes require high levels of sunlight to thrive, many mosses flourish in shaded conditions; *Fissidens* is a genus that is shade-adapted.

The depth of water also influences the presence or absence of certain species of mosses. Where rocks are rarely submerged, a variety of terrestrial mosses and liverworts can be found in shady places. True aquatic mosses, such as *Fontinalis*, occur below the water level. Thus, as one looks at a large reach of stream, the presence of a mosaic of species is apparent depending on the amount of light and the depth of the water.

The only seed-producing plants attached to rocks that you are likely to encounter are members of the family Podostemacea. Representatives are found in swiftly flowing water, in the vicinity of waterfalls, and in other torrential situations. The plants are attached to the top of rocks by their

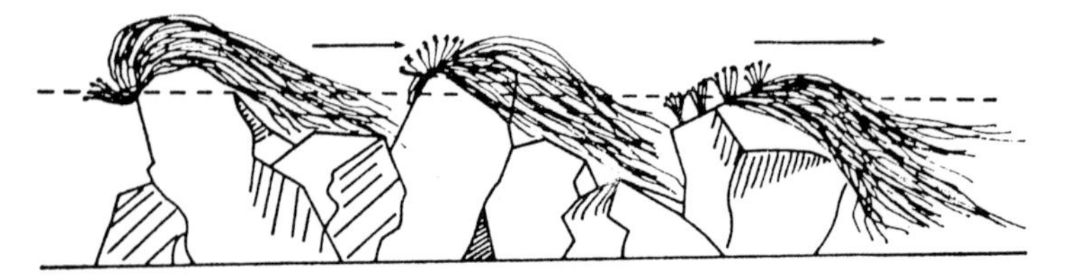

FIGURE 13.1 Diagram showing the position of plants attached to rocks. (From Hynes, 1970, with permission.)

rhizoidlike roots, which enter into the irregularities in the rock surface. The vegetative parts trail in the water or may emerge above the water surface in a wet mass; only the flowering stems rise into the air.

The second group of macrophytes, those free-floating in the water column, are the least important of the three groups in terms of their ecological role in temperate stream ecosystems. The small duckweeds, e.g., *Lemna* sp., can be found in the relative quite backwaters of small springs and streams where they often form floating mats of individual plants. Duckweed can persist through the winter as long as freezing does not occur. Free-floating plants assume greater importance in warmer latitudes. One in particular, the water hyacinth, *Eichornia crassipes,* can proliferate to form dense mats that choke waterways, inhibit navigation, and deoxygenate the underlying water. It is a pest species in many parts of the south and is spreading to other regions where it finds suitable growing conditions.

The rooted macrophytes are the final group we will discuss. These plants are readily observed in rivers and streams, and as mentioned above, we will describe both those plants which occur entirely submerged and those which have parts that emerge from the water.

First, some generalities that apply to both groups. It has been noted that plants occurring in running water differ from those of the same species found in still water; they may have smaller leaves, shorter petioles, and shorter growth sections of the stems. They may also produce fewer and/or smaller flowers and thus, fewer seeds. The reduced seed-production means that those plants in flowing waters tend to rely more on vegetative reproduction than on new production via seeds. The reasons for these reductions are unclear but thought to be related to the current's effect on the leaves.

Rooted plants, either submerged or emergent, occur in regions of the river or stream where the current has slackened enough to allow the development of fine-grained "soils" or places where they can root. These plants have developed a means of firm attachment to the stream bottom by adventitious roots, rhizomes, or stolons, and have tough, flexible stems and leaves. Because rooted macrophytes thrive only where there is soil for rooting and the current is reduced, it follows that they cover only a small percentage of the stream bottom. Obviously, the amount of macrophyte cover on the stream bottom varies considerably with locale, and in many stream habitats they play a small part in the overall ecology of these streams.

Several physical and chemical factors influence the presence or absence of various species. Among these are water hardness, light, and nutrients. A study conducted in Europe found that different species in the same genus may grow under widely different conditions of water hardness; some preferred soft water, others hard water, and some grew in both conditions. Light and nutrients pose the same anomalies—closely related species may differ in their ability to grow under differing conditions of illumination and water fertility. In nutrient-poor waters, phosphorus is probably limiting, but nitrogen and potassium also may be limiting under these conditions. Unlike algae, rooted macrophytes have the advantage of being able to extract nutrients from the substrate in which they are rooted, in addition to absorbing nutrients from the water column by their aboveground parts.

The growing season for macrophytes can be quite long where the water temperature stays above freezing. This commonly occurs in small spring streams fed by underground water sources whose temperatures are considerably above freezing. Such forms as the floating duckweeds (*Lemna*) and watercress (*Nasturtium*) can thrive under these conditions.

What roles do the macrophytes play in the ecology of stream ecosystems? Obviously, the plants produce oxygen by photosynthesis and, in places, produce considerable biomass. But is this plant biomass an important source of food for other consumers? Studies have shown that direct feeding on freshwater macrophytes is very limited; few aquatic invertebrates have the ability to graze them effectively. Some insects with piercing mouthparts can feed on the juices within macrophyte stems. Macrophytes also secrete dissolved organic chemicals into the water, thus adding to the pool of DOM that is important in other aspects of the stream ecosystem. But perhaps the most important contribution of macrophyte biomass to the overall energy flow within a stream is after the plant dies and decays; the breakdown of the tissue provides an important source of FPOM to the stream where it is utilized by the gathering- and filtering-collector invertebrates.

What can we say about the distribution of the various types of macrophytes along a river continuum? It is difficult to be specific about exactly which plants are found in which parts of a stream because local conditions can greatly modify any generalizations that may be attempted. Some distributions, however, can be predicted. Mosses and bryophytes are common in the cold, headwater reaches of streams, along with duckweeds and watercress in slow-flowing backwaters. Fast-water angiosperms such as *Ranunculus* may also be present. Pondweeds (*Potamogeton* sp.) and milfoil (*Elodea* sp.) may be found in the mid-reaches of a river where the current slackens and soft substrates are present. A greater variety of macrophytes may be present in the lower reaches where the current is much slower and soft substrates are present. Common genera found in this region are *Potamogeton, Myriophyllum, Ranunculus*, and *Elodea*.

We've mentioned several genera that you might expect to see in your visits to streams and rivers; now let's take a closer look at some of these groups so that you will be able to recognize them in the field.

ATTACHED PLANTS

Mosses (Musci) and Leafy Liverworts (Family Hepaticae) (Bryophyta)

The water moss *Fontinalis* (Fig. 13.2a) is a plant with three rows of leaves and a long stem floating out from its point of attachment, and it is one of the largest mosses. Another common genus, *Fissidens* (Fig. 13.2b), has a shorter stem and forms matlike communities whose leaves spread from two sides of the stem.

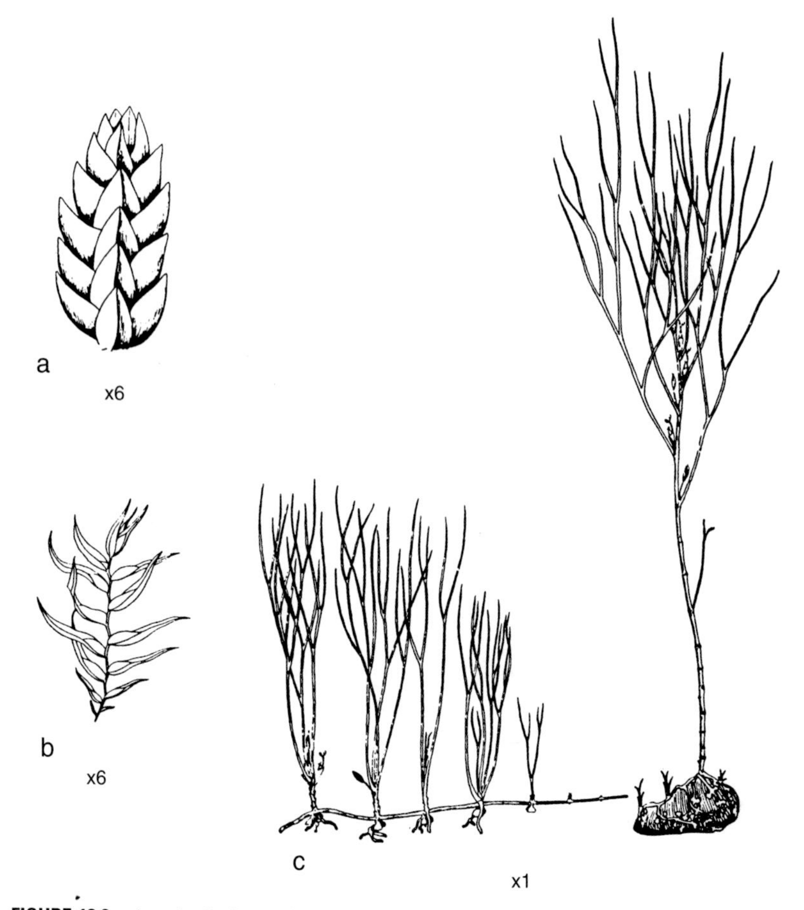

FIGURE 13.2 Attached plants: (a) *Fontinalis*, (b) *Fissidens*, (c) *Podostemon*. (a and b from Fassett, 1940; c from Muenscher, 1944.) (Magnifications are approximate.)

Riverweeds (Family Podostemaceae)

These are much-branched plants fastened to rocks or bedrock in swift-flowing water. They have fine leaves and appear to be cartilaginous. The flowers are very small. Although the plant is widely distributed in the tropics, the only genus found in North America is *Podostemon* (Fig. 13.2c).

FREE-FLOATING PLANTS

Duckweeds (Family Lemnaceae)

The duckweeds are minute, free-floating or submerged plants with flattened or globular plant bodies. There are no definite leaves or stems,

FIGURE 13.3 Free-floating plants: (a) *Lemna*, (b) *Wolffia*, (c) *Eichornia*. (From Muenscher, 1944.) (Magnifications are approximate.)

and they are found either singly or in aggregated colonies. Roots may or may not be present. The flowers are carried in a reproductive pouch on the margin or upper surface of the frond; they are much reduced and usually appear only after a period of hot weather. The most common genus is *Lemna* (Fig. 13.3a), although *Wolffia* (Fig. 13.3b) may be encountered frequently in the eastern United States. The two can be differentiated by the fact that *Lemna* has roots whereas *Wolffia* does not.

Pickeralweeds (Family Pontederiaceae)

These plants are perennials with creeping rootstocks and fibrous roots. The leaves have smooth margins in basal clusters or on branched, leafy stems. The flowers are perfect. They are largely confined to the tropics, but as mentioned above, some species are becoming a nuisance in many waters of the southern United States. The genus *Eichornia* (Fig. 13.3c) is considered to have been introduced, but may have been native to Florida.

SUBMERGED ROOTED PLANTS

Frogbits (Family Hydrocharitaceae)

This family contains one of the most prevalent and widely known aquatic plants, the genus *Elodea* (Fig. 13.4a), commonly known as waterweed and found in countless home aquaria. It is an herb having clustered leaves at the nodes of rhizomes or a few leafy stems. *Elodea* is a perennial species with branched slender stems and with whorled, linear leaves and fibrous roots. They may form extensive floating mats. *Vallisneria* (Fig. 13.4b) is a common related genus found in the eastern United States.

Hornworts (Family Ceratophyllaceae)

The hornworts are submersed, rootless, olive-green to green herbs with a slender main axis from 10- to 200-cm long and with scattered lateral branches. The lower end of the stem is often buried in soft mud. Leaves are whorled and divided into slender, often stiff and brittle segments. The leaves are much crowded toward the apex of the plant, giving the appearance of a "coontail," which is the common name given to the only genus in this family, *Ceratophyllum* (Fig. 13.4c). The plant grows upright early in the growing season, with the stem anchored in the mud; this probably helps with the absorption of nutrients. Later in the season, the plants are found floating in mats enmeshed with filamentous algae. The tips of the branches may break off and sink to the bottom to become "winter buds" and develop into new plants.

Water-milfoils (Family Haloragidaceae)

This family contains another well-known genus, *Myriophyllum* (Fig. 13.4d), the water milfoil. The plants are perennials, with lax and submerged, creeping, or emersed stems. Leaves may be alternate or whorled, and are finely dissected, and may have smooth or serrate margins—even on the same plant. *Hippuris* (Fig. 13.4e), mare's-tail, is found throughout the western, north, and northeastern United States. It has forked, jointed, creeping rootstocks with roots and scales at the nodes. The stems are simple and may be lax and submerged or emersed and erect, with 6 to 12 leaves in whorls at the nodes. *Myriophyllum* has slender, sparingly branched stems mostly rooting freely at the lower nodes. Leaves are whorled or alternate, with the lower leaves usually pinnately dissected into filiform segments; upper leaves are often reduced to bracts.

EMERGENT ROOTED PLANTS

Pondweeds (Family Potamogetonaceae)

This family contains many common species and is found in many aquatic situations. Although there are several genera in the family, by far the largest and most diverse is the genus *Potamogeton*, which contains about 40

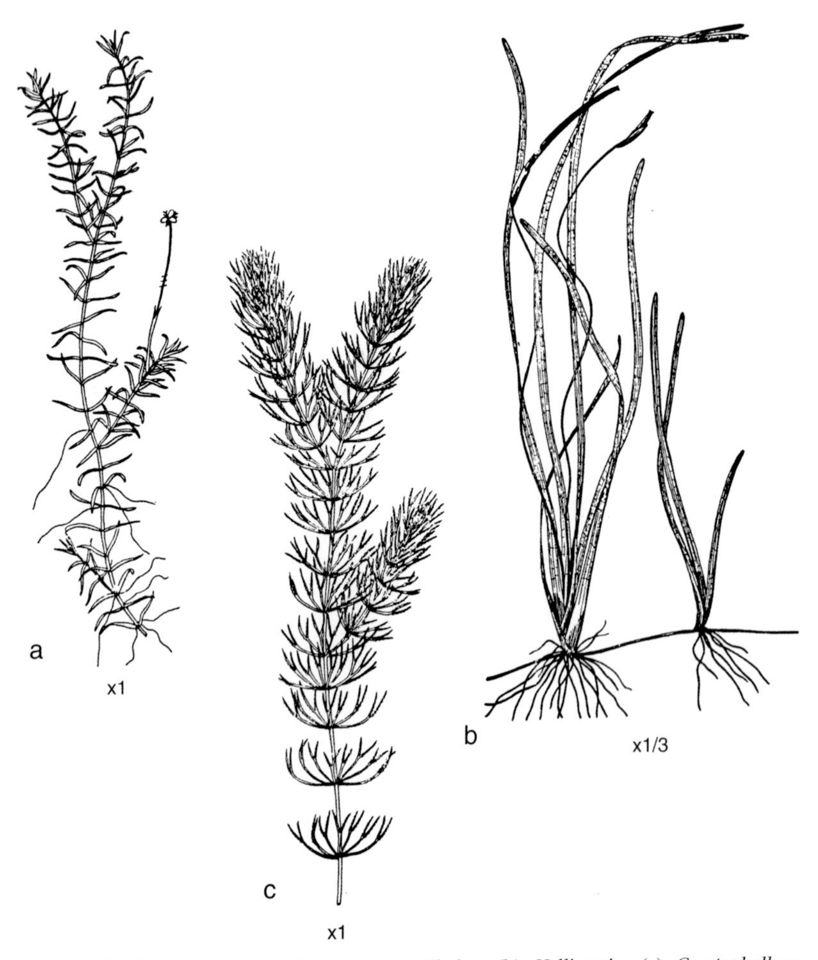

FIGURE 13.4 Submerged rooted plants: (a) *Elodea*, (b) *Vallisneria*, (c) *Ceratophyllum*, (d) *Myriophyllum*, (e) *Hippuris*. (From Muenscher, 1944.) (Magnifications are approximate.)

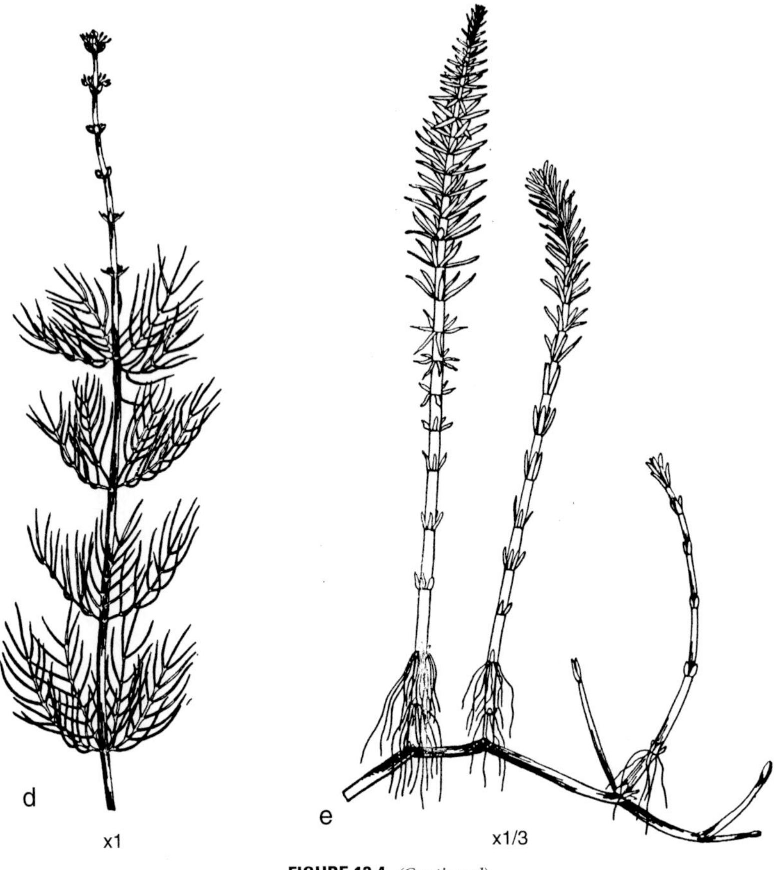

FIGURE 13.4 *(Continued)*

different species. Plants are submersed with creeping rootstocks. Leaves are mainly alternate on erect, jointed stems, or basal and simple.

The large genus *Potamogeton* is probably the rooted plant you are most likely to encounter in rivers and streams. The leaves are quite variable, ranging from thin and grasslike in species like *P. filiformis* (Fig. 13.5a) to broad and "wrinkled" in *P. crispus* (Fig. 13.5b). Leaf form may vary on the same plant, with narrow leaves on the lower, underwater part of the stem and broad, floating leaves on the emergent part.

Ruppia (Fig. 13.5c), widgeon or ditch grass, and *Zannichellia* (Fig. 13.5d), horned pondweed, are two other genera that are widespread in distribution and may be found in flowing waters.

Buttercups (Family Ranunculaceae)

This is mainly a terrestrial plant family with about 250 species, only a few of which are aquatic in the genus *Ranunculus* (Fig. 13.5e). They are annuals or perennials with alternate leaves on erect or creeping stems. The leaf blades are palmately dissected or lobed, rarely lanceolate to linear and entire.

Mustards (Family Cruciferae)

This is the family that contains the plants known as watercress. They are herbs with alternate leaves and a pungent or peppery flavor; people often eat them as salads. The family is largely terrestrial, with only a few aquatic species. The most common aquatic genus is *Nasturtium* (Fig. 13.5f). They are perennials with spreading or creeping angular stems. Leaves are alternate and pinnately compound with 3 to 11 round or elliptical, nearly entire leaflets. They grow densely in the backwaters of springs and springbrooks forming floating mats over the surface of the water.

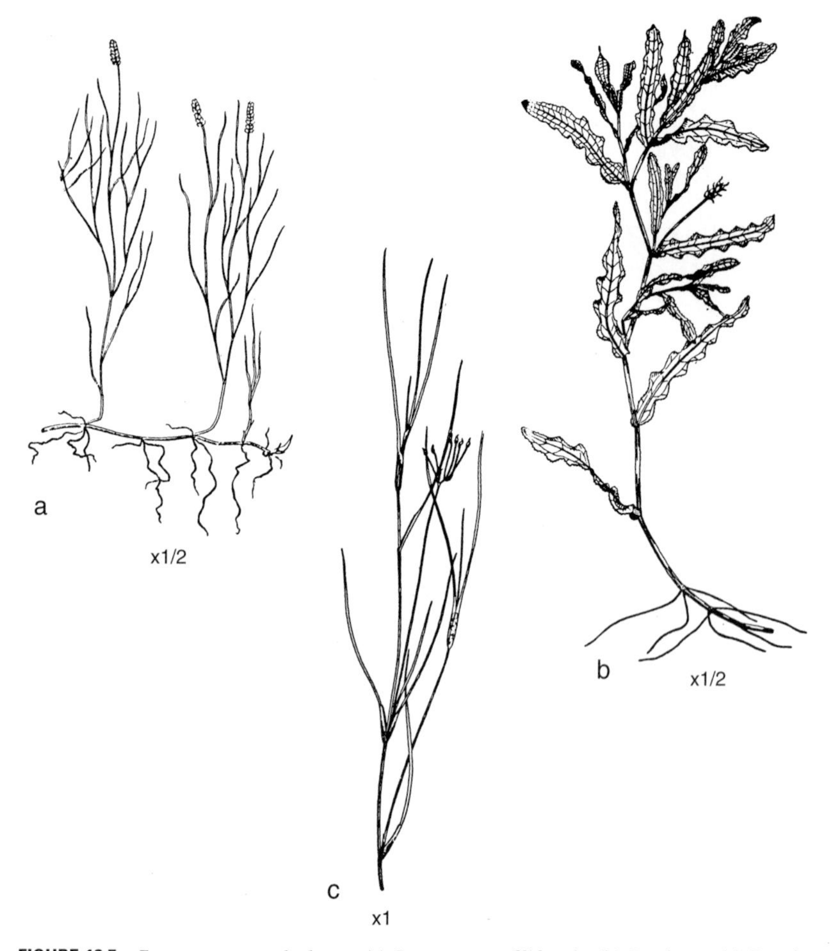

FIGURE 13.5 Emergent rooted plants: (a) *Potamogeton filiformis,* (b) *P. crispus,* (c) *Ruppia,* (d) *Zannichellia,* (e) *Ranunculus,* (f) *Nasturtium.* (From Muenscher, 1944.) (Magnifications are approximate.)

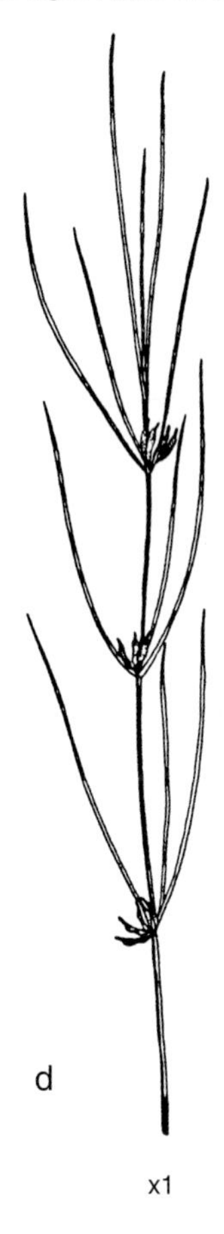

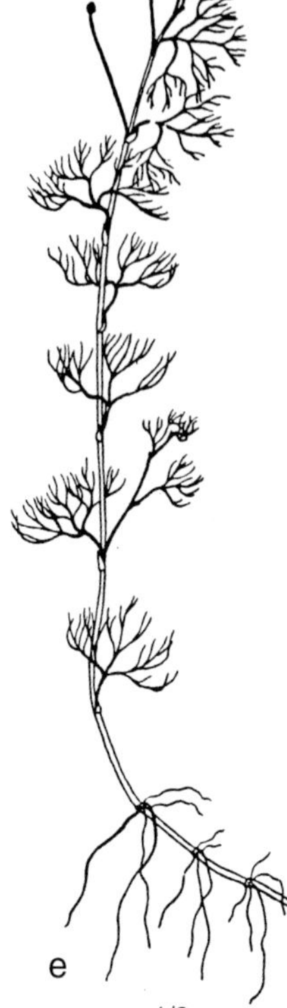

FIGURE 13.5 *(Continued)*

CHAPTER 14

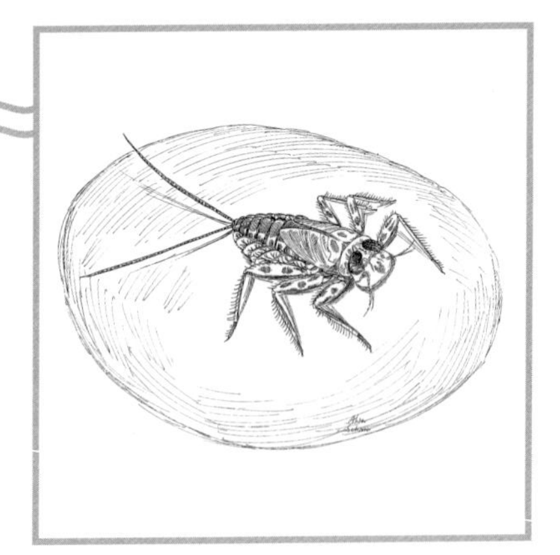

Insects

INTRODUCTION

The insects are probably the most numerous macroinvertebrates of unpolluted river and stream ecosystems. They usually comprise from 70 to 90% of the macroinvertebrates in streams and are of much interest to naturalists, anglers, and others who spend time around flowing waters. Much of this interest is related to the tremendous diversity of insects present in most riverine systems, and the fact that they have an aquatic life stage and an adult stage readily observed on land. Just how diverse are the numbers of insects found in streams? The St. Vrain River, a 1st- to 5th-order stream which drains an alpine watershed on the eastern side of the Rocky Mountains in Colorado, has 210 different taxa! Insects comprise 89–99% of the total numbers and 84–99% of the biomass. Other invertebrates, discussed in subsequent chapters, made up the remainder. The St. Vrain has 31 species of stoneflies, 28 of mayflies, 33 of caddisflies, and 74 of dipterans. The Gunnison River in southwestern Colorado, a major tributary of the Colorado River, contains over 220 species of invertebrates; most of which are rare. Of the 52 taxa that comprise 2% or more of the total numbers, 48 are mayflies, stoneflies, caddisflies, dipterans, and beetles. Third-order Mink Creek, Idaho, contains 16 species of mayflies, six species of beetles, 25 species of stoneflies, over 24 species of caddisflies, and six families of dipterans. This diversity is not found only in Rocky Mountain streams. A colleague furnished the following numbers for the Ogeechee River, a 6th-order, blackwater river in Georgia: 31 species of mayflies, 18 species of caddisflies, 34 species of beetles, 13 species of stoneflies, 59 species of dipterans, 37 species of dragon- and damselflies, three species of megalopterans, and nine species of hemipterans – a total of 208 insect taxa from the main channel and adjoining wetland swamp. To top it off, over 500 species of insects were found in Upper Three Runs Creek, a small stream in South Carolina!

There are 13 orders of insects which have aquatic life stages, and we will describe each of these in terms of their ecology, unique characteristics, unusual interests to man, and other aspects. First, however, let's look at some generalities that apply to all orders of aquatic insects.

Morphology

Insects have three general body regions: the head, thorax, and abdomen (Fig. 14.1a,b). The head bears such structures as the antennae, eyes, and mouthparts. It evolved from the fusion of six or seven anterior segments from the ancestral Annelida–Arthropoda line. In insects where the head is flattened, such as in many stoneflies and mayflies, the sensory structures

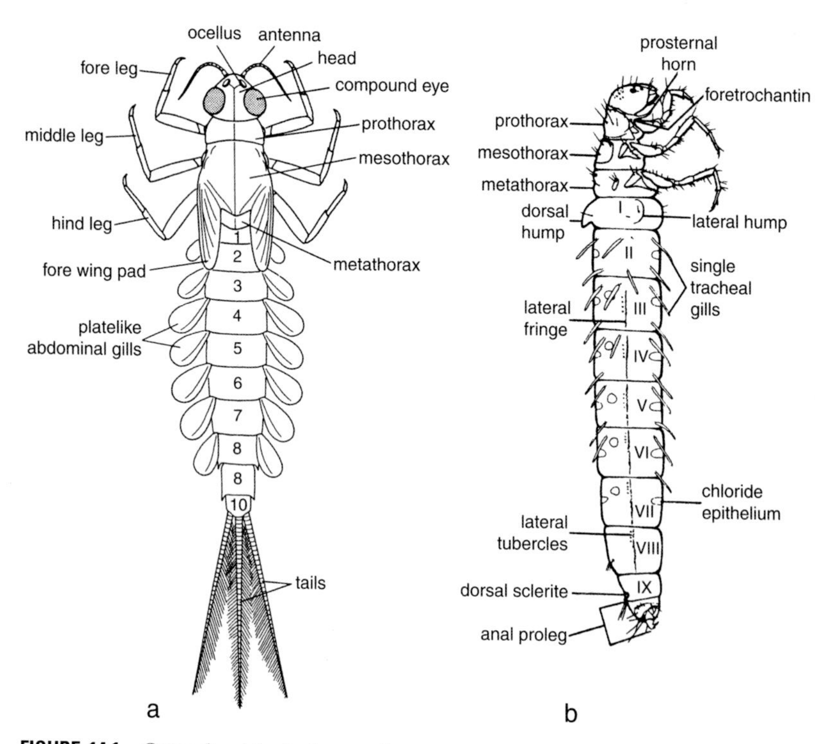

FIGURE 14.1 Generalized body forms of aquatic insects. (a) Nymph, (b) larvae. (From McCafferty, 1981, and Wiggins, 1996.)

(antennae and eyes) are found on top of the head, and the mouthparts are located underneath the head.

The thorax consists of three segments: the prothorax, mesothorax, and metathorax. Each of these three segments bears a pair of legs; the meso- and metathorax also each bear a set of wings (if wings are present). The three pairs of jointed legs have various modifications to enhance swimming, grasping, burrowing, or other activities, and terminate in one or two claws. Aquatic insect adults usually have two pairs of wings, although a single pair is present in some groups (mayflies and dipterans).

The abdomen consists of eleven segments, although the last two are usually fused and difficult to distinguish. Gills arise from the sides of the abdominal segments in some groups (mayflies and caddisflies, for example); in others, like the stoneflies, the branched gills are found on the thorax and first two abdominal segments. Reproductive organs are usually located at the end of the abdomen.

For a more detailed description of morphology and physiology of insects, we recommend Merritt and Cummins' excellent book, *An Introduction to Aquatic Insects of North America*, 3rd ed. Other books listed in the recommended readings at the end of the chapter are also good sources for information.

Life Cycle

Aquatic insects, for the most part, undergo one of two kinds of development, termed *hemimetabolous* and *holometabolous*. Hemimetabolous insects pass through three distinct life stages—egg, immature (often called a *nymph* or *naiad*), and adult, which is terrestrial. Examples of hemimetabolous insects include stoneflies, mayflies, and dragonflies. Holometabolous insects undergo complete metamorphosis and pass through four life stages—egg, immature (usually called a *larva*), pupa, and adult (Fig. 14.2). Note that the difference in the two is the pupal stage which holometabolous insects pass through. The immature forms of hemimetabolous insects closely resemble the adults, whereas the immatures of holometabolous insects do not resemble the adults; it is during the pupal stage that the insects acquire their adult characteristics. Caddisflies, aquatic moths, and dipterans are examples of holometabolous insects. One other type of life cycle, exhibited by the springtails, is called *ametabolous*; it indicates that there is no change in the body form throughout the life of the insect.

Aquatic insects require different periods of time to complete a generation, and *voltinism* is the term used to indicate the number of generations per year. An insect that has a single generation each year is termed *univoltine*, a pattern common in temperate streams. *Bivoltine* and *multivoltine* species have two or more generations per year, respectively. *Semivoltine* and *merovoltine* species take two or more years to complete a life cycle, respectively. The pattern is based largely on temperature and other environmental conditions. The same species can have different patterns in different streams: populations living in lowland, warmer streams can be bivoltine, while another living in colder, high elevation streams can be univoltine.

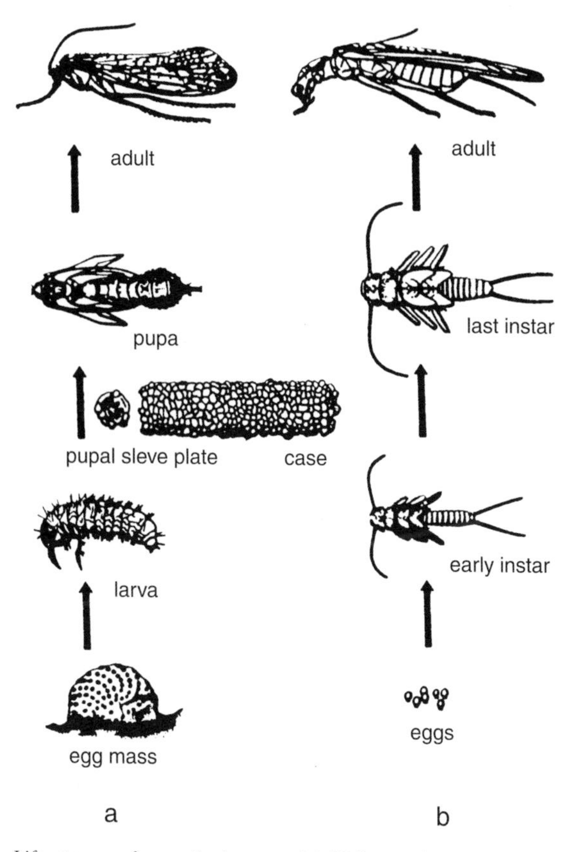

FIGURE 14.2 Life stages of aquatic insects. (a) Holometabolous, (b) hemimetabolous. (From Daly, 1996.)

Basically, each species requires a certain thermal history to develop from an egg to an adult. This is measured in *degree-days*, a degree-day being defined as one day at a certain temperature. For example, if the immature was exposed to 14 days when the water temperature averaged 9°C, this period would equate to 126 degree-days. If the next 20 days warmed up to an average temperature of 11°C, this would be an additional 220 degree-days, or a total of 346 degree-days in our example. When the required number of degree-days are attained, usually 2000 + for many mayflies, the insect emerges as an adult. Thus, those insects which require relatively few degree-days to complete development may have more than a single generation per year (multivoltine); this is true for many dipterans. On the other hand, some stoneflies take two years (bivoltine), or even three (multivoltine), to complete a generation. The interested reader can find a detailed discussion of life cycle characteristics in Hynes' book *The Ecology of Running Waters*.

As an immature develops, it periodically outgrows its exoskeleton, sheds it (called *molting*), and becomes a larger version. This brief act of getting rid of an old skin is called *ecdysis*, and the shed skin is called the exuvium (plural: exuvia). Anglers refer to these cast skins as "shucks." These developmental stages are called *instars*, and insects pass through various numbers of instars as they metamorphose. All of the instars of a population of one species that are of the same generation are called *cohorts*. Thus, a cohort is an age-group; they were hatched at about the same time and mature in step with each other, much like the freshman, sophomores, juniors, and seniors are cohorts in our school systems. For those species with more than one generation per year, there may be more than one cohort present at any one time. Ecologists use cohorts as indicators of production. They can keep track of the gain and loss of biomass (due to growth and mortality, respectively) of a particular cohort as it matures and determine how much biomass is produced as it matures and emerges.

We briefly addressed the subject of *production* or *productivity* in Chapter 6, but perhaps we need to say a few words about it here because of the importance of insects in determining secondary production in streams. Because insects contribute the greatest biomass of all invertebrates found in most river and stream ecosystems, and their densities exceed those of all other animals, it is not surprising to find that the biomass produced by insects during a year contributes a high percentage of the total secondary production. As mentioned in the first paragraph of this chapter, the biomass of insects can constitute over 90% of the invertebrate biomass in a stream, and this can be equally as high when expressed as annual production. Fishes, because of their large size and sometimes rapid growth rate, can also exhibit high biomass and production in rivers and streams, but among the invertebrates, the insects are champions. Much has been published in scientific journals documenting values found for the productivity of insects in many rivers and streams, of all sizes and all locations. Of particular interest is the finding in some early studies that fish production exceeded what theoretically could be supported by macroinvertebrate production. See chapter 6 for an explanation of this apparent paradox.

Habitat

Insects, in relation to their high diversity in rivers and streams, can be found occupying just about every available niche present. They are found on the top, sides, and bottom of stones, deep within the substratum, floating in the current, and even in the surface film. In softer bottoms, insects build tubes within which they maintain a current of water with their gills to obtain fresh, oxygenated water and food. Some work the sediments similar to earthworms. Others build portable cases of inorganic or organic matter and live within these structures as they seek food. Some build fixed retreats on stones, wood, or other solid substrate and let the current bring their food to them. Several of these adaptations were discussed when describing the insect functional groups in Chapter 4.

Aquatic insects can be found in waters ranging from below freezing, where they can survive being frozen within the substrate, to waters reaching temperatures of nearly 50°C. A greater diversity of insects can be found during the season of colder rather than warmer temperatures; indeed, many of the most common forms found in temperate streams grow throughout the winter when water temperatures may be close to freezing.

As discussed in Chapter 2, insect adaptations to living in a current are necessary for successful existence. Aquatic insects have developed several strategies for coping with an environment that is continually trying to wash them downstream. Some avoid the current by dwelling in places of low current velocity—within the substratum, on the lee side of rocks and stones, or other places where the current is less. Others exhibit diverse adaptations to live where the current is stronger: these include construction of weighted cases (caddisflies), attachment to anchored pads of silk (blackflies), recurved claws for clinging (mayflies and stoneflies), development of organs which act as "suction" devices (some mayflies and dipterans), and development of a flattened or streamlined body form to enhance staying out of the current or using the current to force the body onto the substrate (many mayflies). However, some ecologists have questioned the effectiveness of the latter strategy as a means of enabling these insects to live in the slow moving boundary layer of water found just above the surface of rocks and stones; just how extensive a boundary layer exists is a highly controversial subject.

Another aspect of insect life in rivers and streams is the phenomenon known as *drift*. Obviously, life in a current is not conducive to staying in one place easily, and downstream movement of insects is the rule, rather than the exception. Despite many adaptations to cope with life in a current, many insects are dislodged or voluntarily leave the bottom and are carried downstream for varying distances. Studies have shown that this is not entirely accidental and that some organisms appear to enter the drift because of normal activities. Three types of drift have been identified: (1) *catastrophic*, in which the organisms enter the drift because of physical disturbance—floods, pollutants, etc., (2) *behavioral*, where they enter the drift because of normal activities, usually in greater numbers at night, and (3) *constant*, the continual, usually low numbers of organisms found drifting at any time of the day.

Drift is important in the downstream colonization of denuded areas resulting from floods, pollutants, or other disturbances that eliminate established populations. Drifting organisms are vitally important to trout and other organisms that depend on the current to deliver their food while they expend minimal energy in its pursuit.

But if drift is a continual downstream movement of insects, while upstream movement of immature insects is negligible, why, over the millennia, are there still populations of aquatic insects in the upstream reaches of streams—why haven't all of the insects been transported downstream to the oceans or the mouths of rivers? A solution was proposed by Karl Müller, working in Sweden in the 1950s, who suggested that downstream transport of immature forms was compensated for by an upstream flight of egg-bearing adults. Subsequent research has largely verified this theory, termed the "colonization cycle." Another part of the explanation seems to lie with successful reproduction by those individuals that successfully remain in the river's upper reaches.

The nocturnal periodicity of drift is dramatic—numbers in the water column can increase 10 or even 100-fold at nightfall, remain relatively high but variable during the night, and then plummet to low numbers at dawn. One of us (JDA) provided the first strong evidence that predation by fishes causes drift behavior to be nocturnal. Observations from a Rocky Mountain stream revealed that fishes preferentially fed upon larger, as opposed to smaller, mayflies of the genus *Baetis*. Furthermore, small *Baetis* drifted about as frequently by day and by night, but larger size classes became increasingly nocturnal. To some degree this behavior is "hard-wired," meaning that mayfly drift is nocturnal even in the absence of trout, as experimentally determined in the laboratory, for example. However, there is growing evidence that drift is affected by hunger level and can be influenced by the presence of fishes, perhaps sensed chemically. Drift is best interpreted within the context of an insect's behavioral rhythms, feeding opportunities, and the presence of predators.

Food

Many of the aspects of insect feeding were presented in Chapter 4, and it would serve the reader well to review the discussion of functional feeding groups. Actually, there isn't much within the river and stream that doesn't constitute a source of food for one type of insect or another—algae for the scrapers, CPOM for the shredders, FPOM for the gathering- and filtering-collectors, and all, in turn, are fair game for the predators. Thus, aquatic insects are herbivores, detritivores, carnivores, and omnivores, feeding on all categories of food.

Enemies

Insects are preyed upon by many other animals—bigger insects, amphibians, reptiles, mammals, birds, and fishes. Trout have been shown to exhibit a selective predation on some species of insects—literally selecting certain species from among many other available organisms. And, of course, man must be counted among the enemies of insects in streams through activities that include habitat alteration and destruction of habitat, introduction of pollutants and insecticides, and many others.

INSECT ORDERS

This section will discuss various aspects of the ecology, biology, and other interesting characteristics of the orders of insects found in rivers and streams. For many reasons, including their abundance, their significance in the ecological dynamics in riverine systems, and the interest directed toward them by anglers and naturalists, more is known about the five major orders [the mayflies (order Ephemeroptera), stoneflies (order Plecoptera), caddisflies (order Trichoptera), aquatic beetles (order

Coleoptera), and aquatic flies (order Diptera)], and we will describe these orders more thoroughly. Conversely, little is known of the spongillaflies and other lesser known orders; our level of treatment will reflect these attributes. So, with these caveats, let's begin our exploration of the insects living their lives, or at least the immature stages, in rivers and streams. We'll start with the "big five"—water beetles, aquatic flies, mayflies, stoneflies, and caddisflies.

Major Orders

Water Beetles (Order Coleoptera)

Although most beetles, the most species-rich family of insects, are terrestrial, there are over 1000 aquatic and semiaquatic species known in North America. Most are found in ponds and lakes, but several live in rivers and streams.

Beetles are found in a variety of riverine environments, and members of a few families reside mainly in rocky-bottomed rapid streams. The families best adapted to life in rivers and streams are the Elmidae (riffle beetles) (Fig. 14.3a), Psephenidae (water penny beetles) (Fig. 14.3b), Dryopidae (longtoed water beetles), and Ptilodactylidae (ptilodactylid beetles). Members of the families Dytiscidae (predaceous diving beetles) (Fig. 14.3c), Gyrinidae (whirligig beetles), Haliplidae (crawling water beetles), Hydrophilidae (water scavenger beetles) (Fig. 14.3d), and Hydraenidae (minute moss beetles) can also be found in running water. Other habitats frequented by aquatic beetles include emergent vegetation, tree holes and flower cups, and the wetted margins along streams. They are known from hot springs, hyporheic zones, and brackish and intertidal zones.

Metamorphosis is complete, that is, beetles are holometabolous. Both the adults and larvae live underwater, but the final immature instar moves onto land to pupate; the adult returns immediately to the water. Some genera pupate under water, and representatives of some families have terrestrial adults and aquatic immature stages. Most beetles exhibit three, some up to eight, instars, and most species in North America are univoltine. It is not surprising to find so much diversification in such a large group of insects; indeed, to not find such a diversification of life stages would be more surprising. Adults disperse by flight following pupation. Mating takes place near or in the water, and egg laying occurs in water with the eggs being deposited in aquatic plant tissue, on the substratum, on algal mats, or in moist soil adjacent to the water.

Larvae range from 2 to 60 mm in length and are variously shaped. Most aquatic larvae have gills, though a few obtain oxygen through the body surface. Some, however, must visit the water surface to obtain oxygen in the form of a bubble, which they trap and retain on the body surface. Some members of the whirlygig beetles (family Gyrinidae) have a unique adaptation which enables them to swim in the surface film; each eye is divided so that the upper half is used for aerial viewing and the lower half for underwater vision. Adult beetles measure 1–40 mm in length and usually have oval or elongate bodies. They usually use the leathery wing

FIGURE 14.3 Beetles. (a) Elmidae, *Lara avara* larva (photo by L. Serpa); (b) Psephenidae, *Ectopria* sp. and *Psephenus* sp. larvae (photo by K. Krieger); (c) Dytiscidae larva (photo by R. Merritt); (d) Hydrophilidae, *Hydrochara* sp. adult (photo by M. Higgins).

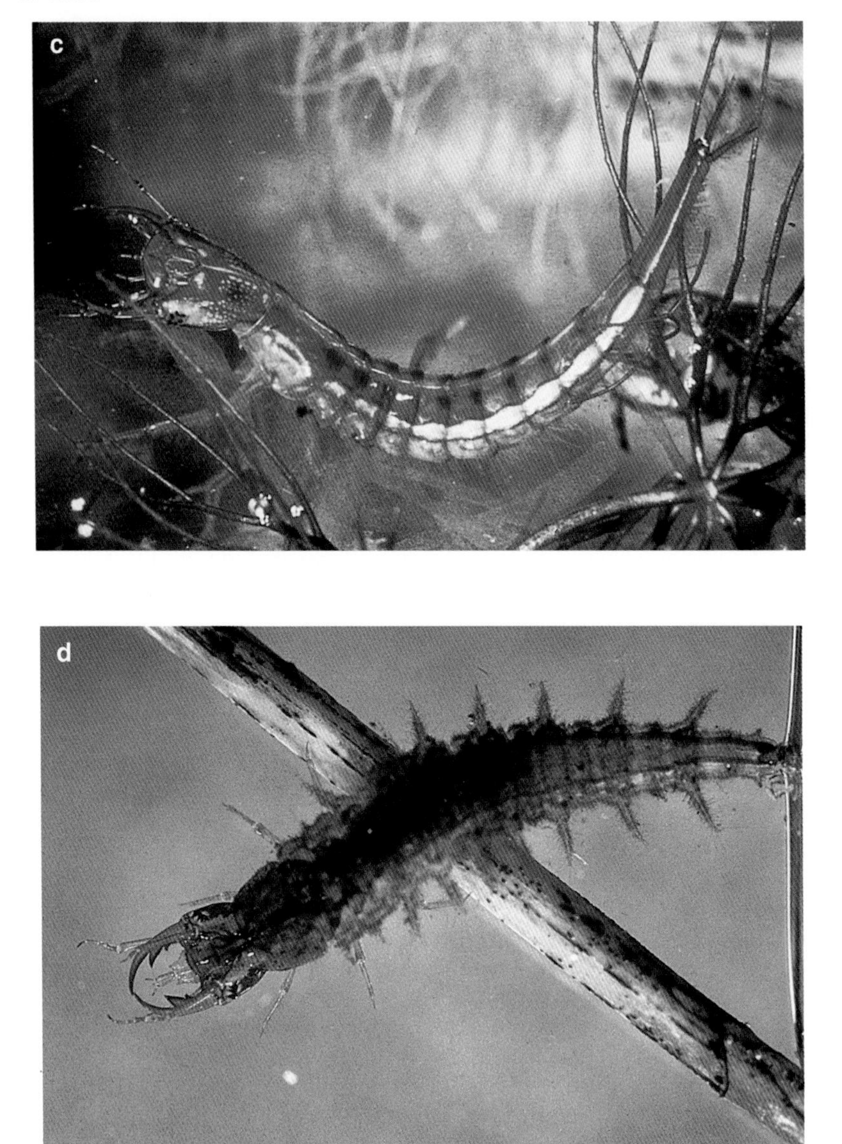

FIGURE 14.3 *(Continued)*

covering, the *elytra*, to trap and hold an air supply in proximity to the spiracles (the name Coleoptera means "sheath-winged").

Aquatic beetles exhibit a wide variety of morphological adaptations. Swimming is usually accomplished by modified hind legs, although the middle legs sometimes are modified. The predaceous diving beetles use both legs in unison, whereas the water scavenger beetles alternate leg strokes. Long grasping claws are characteristic of some beetles that inhabit the bottoms of rivers and streams, and others have well-developed claws or terminal abdominal hooks to enhance clinging to substrates. The water penny beetles have developed disc-shaped bodies for attachment to the bottom.

Also as expected from such a diverse group, feeding habits by beetles are quite varied; many families are predaceous as both larvae and adults. Some are herbivorous, scraping algae from substrates, and one group, the hydroscaphids, is unusual in that it feeds largely on cyanobacteria. Others are detritivores, feeding on CPOM and FPOM. Large dytiscids have been known to attack and eat small fishes and tadpoles, and in one case, kill a garter snake! Beetles use a number of adaptations in their role as predators. Some inject toxic or digestive fluids from their mouthparts. Some emit secretions into the water, which can inactivate or even kill certain fishes. Beetles, in turn, are important constituents of many aquatic food webs; they are fed upon by fishes and waterfowl.

Anglers have begun to develop imitations of beetles as more is learned about how actively trout feed on terrestrial insects which fall or are blown into streams and rivers. Some common patterns include the Black Beetle, Dropper Beetle, CDC Peacock Beetle, Foam Peacock Beetle, and the Foam Beetle.

True Flies (Order Diptera)

Aquatic flies are a large, highly diverse assemblage of insects occurring in a wide variety of habitats as immatures. Many groups have developed physical adaptations to enable them to live in fast-flowing waters, such as the development of suctorial discs on larval body segments and the ability to obtain oxygen from the water directly through the skin. We have already mentioned how blackfly larvae attach to stones (Fig. 2.2). Diptera larvae may be the most abundant insects in many river and stream communities; over half of the known aquatic species of insects belong to this group. There are over 3500 species known from North America, in about 35 families. The family Chironomidae (midges) (Fig. 14.4a) is the largest family of freshwater insects. Other important dipteran families found in riverine systems include the Tipulidae (crane flies) (Fig. 14.4b), Empididae (dance flies), Tabanidae (horse flies and deer flies) (Fig. 14.4c), Blephariceridae (netwinged midges) (Fig. 14.4d), Simuliidae (black flies) (Fig. 2.2), Culicidae (mosquitoes) (Fig. 14.4e), and Dixidae (dixid midges) (Fig. 14.4f). Given the great diversity of this order in terms of numbers of species and variation in physical characteristics, the information presented later is fairly general; space does not allow detailed descriptions of each family. Also note that when using common names of dipterans, such as black flies, the two words are separated. This is because these are "true flies," whereas the mayflies, stoneflies, caddisflies, and others are not, and so their names are not separated.

Aquatic flies are holometabolous. Larvae pass through three or four instars, although black flies may have six or seven. Development takes as little as a week in some mosquitoes, black flies, or other biting midges, and

FIGURE 14.4 Diptera. (a) Chironomide larva (photo by D. Penrose); (b) Tipulidae, *Tipula* sp. larva (photo by R. Merritt); (c) Tabanidae, *Tabanus* sp. larva (photo by M. Higgins): (d) Blephariceridae, *Blepharicers* sp. larva, ventral view showing suctorial disks (photo by R. Merritt); (e) Culicidae, *Aedes eudes* larvae (photo by M. Higgins); (f) Dixidae, *Dixa* sp. larva (photo by S. Solada).

FIGURE 14.4 *(Continued)*

FIGURE 14.4 *(Continued)*

as long as a year for some crane flies. Voltinism varies from several generations per year, to some arctic midges that require several years for a single generation. Development time is usually determined by environmental conditions, especially temperature. Pupation varies greatly, both in terms of how it occurs and where it occurs—on land, in water, in cases of other insects, etc. Just prior to emergence, the pupa is called a *pharate adult*. It then emerges from the pupal skin as a fully mature adult. Some adults need to rest prior to flight while body parts harden and become functional; others fly immediately. Mass emergences may occur—just ask anyone who has been enveloped in a cloud of mosquitoes!

Adults may live up to a month. Mating occurs in flight, on vegetation, on land, or even on water. Egg laying is variable; it may occur over, on, or in the water, or on substrates on land or in the water.

Larvae are elongate and resemble maggots, measuring from 1 to 100 mm in length. The head may be separate and discernible, or poorly developed and difficult to see. Eyes are poorly developed and antennae are variously developed. No jointed legs are present on the thorax; however, some larvae possess *prolegs*: fleshy, leglike appendages on the thoracic segments. Wing pads are absent and the abdomen consists of 8–10 segments (sometimes fused or appearing to be subdivided), sometimes bearing prolegs.

The pupae may be active or quiescent; some remain in the water (mosquitoes and midges) and others (crane flies) pupate on land. Some species pupate within a capsulelike case called a *puparium*; others may be free-living or concealed within a cocoon. Many develop respiratory tubes or thoracic gills for breathing.

Adults, as with the larvae, are highly diverse in body form. All have only one pair of functional wings (Diptera means "two wings") and are terrestrial. Eyes are relatively large, mouthparts are highly variable depending on method of feeding, and tails are not present.

The aquatic flies are important constituents of river and stream ecosystems. They can occur in extremely high numbers, up to several thousand individuals per square meter, although populations this high are usually associated with polluted conditions where they can thrive when other insects cannot. The Chironomidae frequently dominate both numbers and biomass in stream ecosystems, testifying to their importance in these ecosystems. Representatives occur in all of the functional feeding groups; thus, they are herbivores, detritivores, carnivores, and omnivores. Shredders are represented by larvae of the Tipulidae, grazers by the Blephariceridae, gathering-collectors by the Chironomidae, filtering-collectors by the Simuliidae, and predators by the Tabanidae. These are only single examples of families filling these functional niches; some families perform multiple functional feeding practices. Dipterans act as primary consumers in stream food webs, and are fed upon by other invertebrates, fishes, amphibians, reptiles, birds, and mammals.

Members of this order have one other characteristic not shared by the other insect orders. Many of them are vectors of pathogens because their feeding habits involve the need for a mammalian or avian blood meal, during which the disease organisms may be transferred to humans. Dipterans can also be pest organisms, occurring in such large numbers as to even prevent human habitation in some areas; many also bite!

It is obvious that because of their great numbers and diverse sizes and shapes, aquatic flies are widely imitated by flies tied by anglers. All life forms—larva, pupa, and adult—have been copied, but probably the group most represented in fly collections are the midges. Often, when trout are rising but no obvious adult hatches are on the water, an angler who fishes a small midge imitation just under the water surface will be successful. These patterns imitate the small midge larvae coming to the surface to emerge—hence the general term "emergers" for this group of patterns. Popular immature patterns include the CDC Midge Pupa, Mosquito Larva, Peacock Chironomid, Griffith's Gnat, and Rising Midge Pupa. Dry flies such as the Adams, Black Gnat, and Mosquito are used to imitate adults.

Mayflies (Order Ephemeroptera)

Mayflies are truly the "ballerinas" of the insect world; dainty, ephemeral insects performing their dance like mating rituals above the water. Their Latin name refers to the ephemeral nature of their adult life. Although this may not be true of all members of this order, it certainly characterizes the general notion people have of them as a group. There are between 17 and 22 North American families of mayflies, depending upon which taxonomist you believe; the taxonomy of this group is unsettled at the family level. There are probably over 700 species. The main families are the Acanthametropodidae, Ametropodidae, Baetidae (Fig. 14.5a), Baetiscidae (Fig. 14.5b), Behningiidae, Caenidae, Ephemerellidae (Fig. 14.5c), Ephemeridae (Fig. 14.5d), Heptageniidae (Fig. 14.5e), Isonychiidae, Leptophlebiidae, Metretopodidae, Neoephemeridae, Oligoneuriidae (Fig. 14.5f), Polymitarcyidae, Potamanthidae, Pseudironidae, and Tricorythidae. The immature forms are aquatic, and the greatest diversity of mayflies is found in warm rivers and streams. As with the stoneflies, mayflies are highly susceptible to pollution and thus are important indicators of water quality.

Mayflies are hemimetabolous and are often multivoltine or semivoltine, although in temperate regions, univoltinism is predominant. Egg development undergoes a *diapause*, or resting, stage in many mayflies found in the temperate zone. This period of dormancy may last anywhere from 3 to over 11 months. Nymphs pass through several instars, depending upon environmental conditions, ranging from 12 for *Baetisca rogersi* to as many as 40–45 for *Stenacron interpunctatum*. Several cohorts may be produced each year. Mature nymphs crawl, swim, or float to the water surface to emerge. Adults have two winged stages, rather than one. The first winged stage is called a *subimago* or *dun*. It is sexually immature and must undergo a final molt before becoming a sexually mature *imago* or *spinner*. Subimagos often appear to be dull, as compared to the glossy imagos. This is because the subimagos are covered with a layer of microscopic hairs which render them waterproof and buoyant and reduces their chances of being trapped in the water film as they emerge from the nymphal exuviae. Adult male imagos form mating swarms above the water surface, often appearing as dancing clouds of insects moving up and down. Individual females enter the swarm of males, and they are joined by a male. They then leave the swarm, mate in the air, and the female deposits her eggs, commonly on the water surface. Females of some species crawl underwater to deposit their eggs on solid surfaces. The swarms of emerging adults can be truly spectacular. Some of the large burrowing mayflies found in the slower currents of large rivers can emerge in such great numbers that their

FIGURE 14.5 Ephemeroptera. (a) Baetidae, *Callibaetis ferrugineus* nymph (photo by M. Higgins; (b) Baetiscidae, *Baetisca* sp. nymph (photo by D. Penrose); (c) Ephemerellidae nymph (photo by R. Merritt); (d) Ephemeridae subimago (photo by J. Wallace); (e) Heptageniidae nymph (photo by H. Daly); (f) Isonychidae, *Isonychia* sp. nymph (photo by D. Penrose).

FIGURE 14.5 *(Continued)*

FIGURE 14.5 *(Continued)*

dying bodies can form drifts that clog car traffic on bridges and make the roads slippery!

Nymphs are usually cylindrical or flattened in body form and range in length from 3 to 30 mm, not including tails. Mayflies are found in a variety of habitats in flowing waters, and their body forms are highly variable and adapted to where they live. There are swimming nymphs capable of moving freely in the water column where currents are not too strong (families Ametropidae and Baetidae, in part); creeping and crawling nymphs which move freely across the substrate (families Ephemerellidae, Baetidae, in part, Tricorythidae); flattened and streamlined nymphs which frequent faster currents by clinging tightly to the substrate (families Heptageniidae, Baetiscidae); and burrowing nymphs which dwell in burrows in soft sediments (family Ephemeridae). The latter are found only in rivers and streams where accumulations of soft sediments provide suitable habitat. Mayfly nymphs have a superficial resemblance to stonefly nymphs. Mayflies have only a single claw on each leg, compared to two in stoneflies; and generally have three *cerci*, or "tails," although some have two, as do all stoneflies. Gills usually occur on the abdominal segments, although some species have gills on the legs and head. The gills on the burrowing mayflies are relatively large and are used to create a current of water through the burrows, thus providing food and fresh, oxygenated water to the nymph. Gills are also modified for other uses. In some species the abdominal gills overlap to form a "sucker-like" disc that helps the nymphs maintain their location in swift currents (Fig. 14.6a). The first pair of gills in some species is modified to cover and protect the remaining pairs (Fig. 14.6b).

Immature mayflies are mainly herbivorous insects, feeding on algae and FPOM. A few forms are carnivorous. Because of their many body forms and the variety of niches they inhabit in stream ecosystems, it is not surprising to find mayfly representatives in all of the functional feeding groups. There are gathering-collectors (e.g., *Baetis* and *Ephemerella*), filtering-collectors (e.g., *Homoeoneuria* and *Isonychia*), grazers (e.g., *Ameletus* and *Heptagenia*), and predators (e.g., *Analetris*). Rarely, they perform as shredders. This indicates that it is likely that mayflies, along with the other invertebrate constituents of stream food webs, are successful in partitioning the available food resources, that is, algae on the rocks for the scrapers, FPOM on the bottom for the gathering-collectors, and FPOM in transport for the filtering-collectors. However, some studies in Colorado streams have reported a lack of evidence for food partitioning by mayflies. Some closely related species have also been found to partition habitat space according to several physical and water quality variables. Although each of the three species, *Drunella grandis*, *D. doddsi*, and *Tricorythodes minutus*, had its own unique niche, there was some overlap. Nymphs, in turn, are fed upon by other insects, fishes, and other vertebrates. Adults have only vestigial mouthparts and do not feed; they live for only hours or a few days, at most.

Mayflies, because of their great diversity in body shape, color, and importance as food items for trout, are undoubtedly the most important insect to anglers in terms of imitation by artificial lures. McCafferty states that over 150 different mayfly species have been imitated by fly tiers in North America alone! Common names have been given to these patterns,

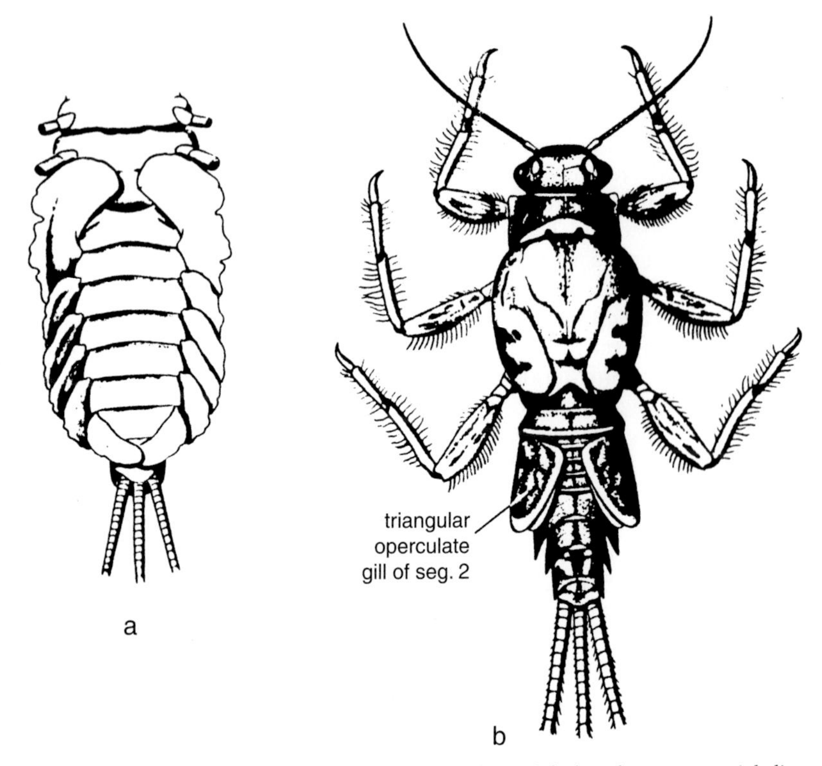

FIGURE 14.6 Gill modifications in mayflies. (a) Gills modified to form a suctorial disc, *Rhithrogenia*; (b) first operculate gill modifed to cover remaining gills, *Tricorythodes*. (From McCafferty, 1981.)

and in 1981 McCafferty published a dictionary of these names in his highly recommended book, *Aquatic Entomology*.

Because mayflies are important fish food items in all of their life stages—nymphs, subimagos, and imagos—imitations of all of these life-forms are found in the fly boxes of most well-equipped anglers. Knowledge of the season when different life stages are present dictates the patterns to be used, patterns which imitate nymphs prior to emergence, and those of subimago (dun) and imago (adults, spinners) after emergence. Anglers also have their own language to denote differing life-forms. Thus, "spinners" denote the dying and dead adults which drift on the surface after mating and death; "duns" are the darker-colored subimagos; and "nymphs," "soft hackles," and "wet flies" refer to the underwater immature forms. Even wing-form is used; mayflies are called "upwings" because their wings are held upright, stoneflies are called "flatwings" because they hold their wings flat over their backs, caddisflies are called "tentwings" because their wings are held folded tentlike over their bodies, and dipteran midges are called "glassywings" because of the transparency of the wings.

Philosophies differ among anglers. One school champions "match-the-hatch," believing that their flies must be precise imitators of the insects present, right down to color, size, and matching body parts. Another school of thought believes that knowledge of fish habits, accuracy and delicacy of presentation of the fly, and overall technique is equally important. The latter tend to use "attractor" fly patterns and general imitations rather than trying for exact duplication. In fact, some of the best known, widely used, and most successful fly patterns do not imitate any specific insect; the Rio Grande King, Stimulator, and the Wulff series are good examples.

Some specific mayfly patterns include the Blue Quill, Beaverkill, Hendrickson, Quill Gordon, Pale Morning Dun, and Trico Parachute among dry fly patterns, and the Hare's Ear series, March Brown, Pheasant Tail, Light Cahill among the nymphs. It is worth repeating that there are more fly patterns tied to imitate mayflies than any other insect. Randle Stetzer, in his 1992 book *Flies, the Best One Thousand*, lists 80 dry mayfly patterns and 80 nymph mayfly patterns, by far the most for any group, and this doesn't count other patterns listed as wet flies, soft hackles, searching nymphs, or miscellaneous dry flies which are good mayfly imitators. In addition, many famous mayfly patterns carry the names of people or places famous in fly-fishing history; see the sections on the Au Sable and Beaverkill rivers in Chapter 6.

Stoneflies (Order Plecoptera)

Stoneflies, as their common name implies, are most frequently found in cold, well-oxygenated, rapid-flowing streams with rocky bottoms. Some occur along the shoreline of wave-swept lakes. Indeed, most stoneflies have fairly stringent environmental requirements; this is reflected in their distribution and value as indicators of water quality conditions (see later discussion). Smaller species have been known to inhabit hyporheic regions for considerable portions of their immature lives—often at considerable distances from the shoreline. Nine families occur in North America [Capniidae, Leuctridae (Fig. 14.7a), Nemouridae, Taeniopterygidae (Fig. 14.7b), Chloroperlidae, Peltoperlidae (Fig. 14.7c), Perlidae, Perlodidae (Fig. 14.7d), and Pteronarcyidae (Fig. 4.2b)], containing over 500 species. Nymphs are aquatic and adults terrestrial, with very few exceptions.

Stoneflies are hemimetabolous; most species are univoltine although some of the larger species, such as some of the Pteronarcyidae, are semivoltine and may take two or three years to mature, depending upon environmental conditions. Ten to over 22 instars may be produced, and some species may spend a period of dormancy for 3–6 months as young instars or eggs if adverse seasonal conditions occur. This occurs in the families Leuctridae, Capniidae, and Taeniopterygidae. Those species that have a dormant period undergo what is termed *heterodynamic* or fast development, whereas those without a dormant period have slow or *homodynamic* development. Nymphs appear to migrate toward the shoreline just before emergence and may increase in drifting numbers. They then crawl out of the water onto rocks or plants when the adults are ready to emerge, leaving the cast skin (exuviae).

FIGURE 14.7 Plecoptera. (a) Leuctridae, *Leuctra* sp. nymph (photo by S. Solada); (b) Taenioterygidae, *Taeniopterys* sp. nymph (photo by D. Penrose); (c) Peltoperlidae, *Tallaperla* sp. nymph (photo by S. Solada); (d) Perlodidae, *Isoperla* sp. nymph (photo by D. Penrose).

FIGURE 14.7 *(Continued)*

Most adults feed and many imbibe water. Stoneflies are poor fliers; many have short wings or none at all. The wings are folded back over the abdomen and held flat. They roost in riparian vegetation and fly in short, fluttering flights when disturbed. Mating does not occur in flight, but upon the ground or on vegetation. Egg laying usually occurs over the water; this together with their weak flying ability, makes them especially vulnerable to trout and other fishes. Some females crawl into the water to deposit their eggs on stones.

One interesting characteristic of stoneflies in the Northern Hemisphere, but not the Southern, is the behavior known as "drumming" during the mating period. Adult males beat their abdomens against the substrate, producing species-specific signals which facilitate mate location. Males drum throughout their adult life, but only virgin females respond.

The length of mature adults range from a few millimeters in the case of some members of the family Capniidae to over 5 cm for some Pteronarcyidae. Adults resemble nymphs to a large degree and are mainly spring and summer emergers. However, some emerge when snow covers the banks of the streams; these are known as "winter stoneflies." Nymphs characteristically have two cerci, and the legs are well-developed and have two apical claws. Mature nymphs have developing wing pads, and small, fingerlike or filamentous gills occur singly or in clusters at various places on the head, thorax, and/or abdomen. Some of the common species perform what appears to be "push ups" to enhance aeration of the gills when oxygen concentrations in the water decrease. Nymphs are not good swimmers, but are well adapted to clinging and crawling on the bottom stones.

Nymphs are herbivorous, omnivorous, detritivorous, or carnivorous, depending on the family. Herbivorous species are mainly shredders that feed on plant detritus accumulations; others graze periphyton. Carnivorous species feed on other insects and invertebrates, ingesting their prey whole. A switch from herbivory–detritivory to carnivory is common as development proceeds, though some species retain the same food habits throughout their immature lives. In two Colorado rivers, *Pteronarcella badia* and *Pteronarcys californica* were herbivore–detritivores throughout development, *Claassenia sabulosa* and *Hesperoperla pacifica* were constant carnivores, and *Isoperla fulva* shifted from herbivory–detritivory to omnivory and then to carnivory. Suffice it to say that stoneflies exhibit a great deal of plasticity in their feeding habits. In terms of functional group classification, some Pteronarcyidae, Capniidae, and Nemouridae are shredders, some Taeniopterygidae, Leuctridae, and Peltoperlidae are collectors or scrapers or both, and Perlodidae, Perlidae, and Chloroperlidae are predators. These groupings are not rigid because of the aforementioned changes in feeding habits with development and the difficulty of placing omnivores into a single category.

One of us (CEC) and a colleague observed a large perlid nymph feeding on black fly larvae on a large stone in the East Fork of the Salmon River. The nymph would traverse the rock horizontally; when it was immediately "downstream" from a black fly larva, it would turn toward it, walk up to it, and seize it in its mandibles. Then it would resume its horizontal movement and repeat the procedure. Apparently, it was receiving some kind of chemical stimulus from the black fly larva to locate it; the perlid sometimes changed course several centimeters below the black fly.

Predaceous stoneflies have been intensively studied in the laboratory by JDA, Bobbi Peckarsky of Cornell University, and others, because they are relatively easy to observe. The rate at which prey of various kinds are encountered, attacked, captured, and consumed helps explain why certain prey are more likely to be eaten. Prey within a certain size range relative to the predator, with soft rather than hard exoskeletons, and with limited escape capabilities are most likely to be consumed.

Stoneflies are preyed upon by larger stoneflies, hellgrammites, salamanders, fishes, and birds. Thus, given their significant numbers in stream ecosystems where they occur, they play important roles in the dynamics of energy transfer through these food webs. One author (JDA) studied the relationships between trout and the structure of insect communities in Colorado streams and found that the presence of trout did not play a major role in the diversity or numbers of stream insects, including stoneflies. In another stream, stoneflies consumed about twice the standing crop of other insects, suggesting that there was an abundance of prey relative to consumption.

The subjects of habitat diversity and space partitioning, or where stoneflies are found and what determines how they utilize the various niches in the physical environment of rivers and streams, are beyond the scope of this book, but the interested reader is referred to the discussion in Stewart and Stark (1988). This book contains a wealth of detailed information on North American stoneflies.

The first artificial fly tied, some 450 years ago, was an imitation of a British stonefly. Today, both the immature and adult stages of stoneflies are imitated in a wide variety of patterns. In his book *Flies, The Best One Thousand*, Randle Stetzer includes 30 stonefly nymph patterns and 20 adult patterns. Included in the nymph patterns are the Montana Stone, Rubber Legs Hare's Ear, Stonefly Creeper, Bitch Creek, Kaufmann's Black Stone, and Pheasant Back Stone. Popular adult patterns include the Stimulator series, CDC Little Yellow Stone, Sofa Pillow series, Henry's Fork Salmon Fly, Fluttering Stone, Egg Laying *Pteronarcys*, and Improved Golden Stone.

Some of the truly fabulous fly-fishing on such streams as the Madison River in Montana, the Crow's Nest River in Alberta, and the Deschutes River in Oregon occurs during annual hatches of large stoneflies. At this time, when the slow-flying females are dropping their eggs on the water's surface, large trout seem to lose their natural wariness and go on literal feeding frenzies.

Caddisflies (Order Trichoptera)

Trichoptera means "hair-wing," and refers to the presence of hairlike setae that cover the wings. There are over 1350 species in 22 families recognized in North America, although, as in the mayflies, the taxonomy is unsettled. Diversity of caddisflies is highest in cool rivers and streams. It is thought that this high diversity is related to their ability to produce silk, which provides an evolutionary advantage for many physiological and physical functions, such as respiration, feeding, and adaptations to their physical environment. Caddisflies are closely related to the Lepidoptera; the adults of both orders closely resemble each other. In terms of the ecology of temperate stream ecosystems, the families Rhyacophilidae (Fig. 14.8a), Glossosomatidae, Hydroptilidae (Fig. 14.8b), Psychomyidae,

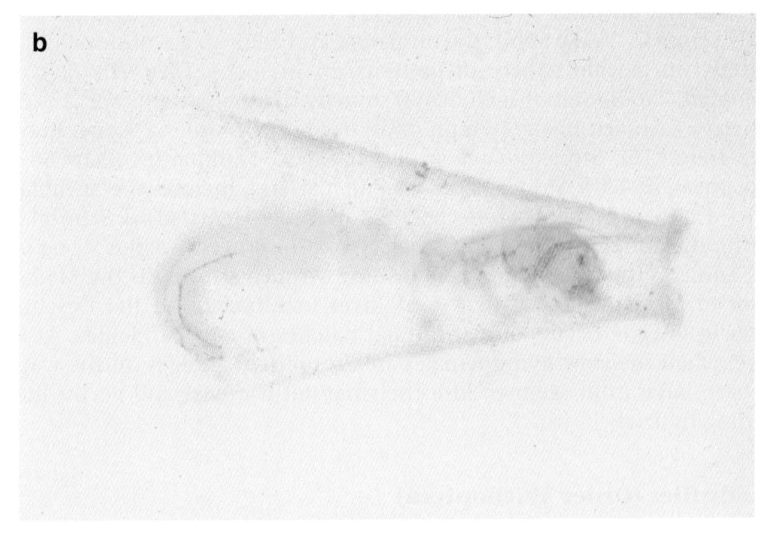

FIGURE 14.8 Trichoptera. (a) Rhyacophilidae, *Rhyacophila* sp. larva (photo by A. Elosegi); (b) Hydroptilidae, *Osyethira* sp. larva (photo by S. Solada); (c) Lepidostomatidae, *Lepidostoma* sp. larva (photo by D. Penrose).

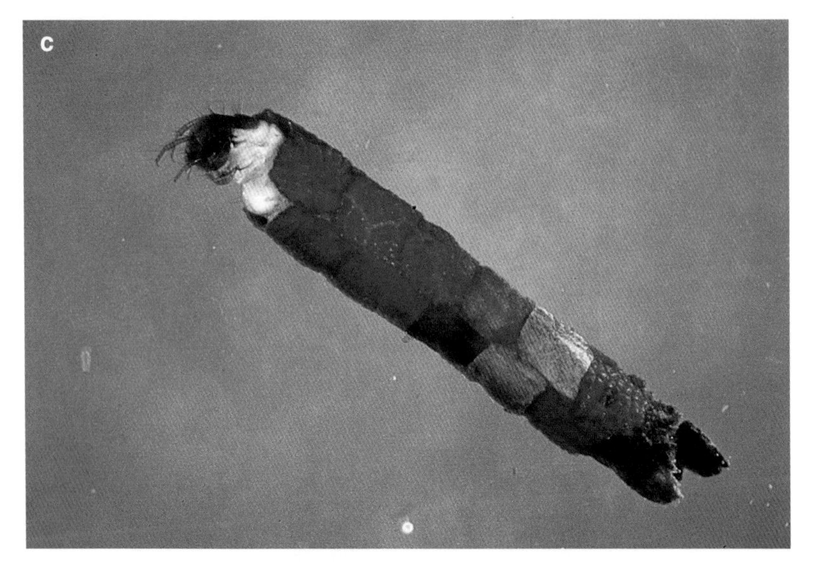

FIGURE 14.8 *(Continued)*

Hydropsychidae (Fig. 4.7), Brachycentridae (Fig. 4.8), Limnephilidae (Fig. 4.5), Lepidostomatidae (Fig. 14.8c), and Leptoceridae play significant roles.

The caddisflies are one of the most interesting groups of insects found in rivers and streams, largely because of their case-building characteristics. All families in this order except two, the Rhyacophilidae and Hydrobiosidae, construct a case of some sort from organic or inorganic material held together by silk. Although differing widely among families, case form is consistent within genera. These cases are transportable for some species, fixed for others. Figure 14.9 shows some representative cases constructed by caddisflies. The cases are of extraordinary variety in terms of shape and construction material. One family (Helicopsychidae) constructs a spiral case resembling a snail shell from sand grains (Fig. 14.9l); indeed, early taxonomists thought it was a snail.

Caddisflies are holometabolous, having both an aquatic larval and pupal stage. Most species are univoltine, although bivoltine and semivoltine life cycles have been reported. Five instars are the norm for caddisflies (7 or 13 in some cases), and the final instar constructs a pupal case of silk within its case or on a solid substrate in those forms not having larval cases. Regardless, the pupal stage lasts from 2–3 weeks for most species, and in some instances, a dormant stage (diapause) may occur for up to 6 months duration. The mature pupa uses large, sclerotized mandibles to break out of the pupal case and then swims or crawls to the surface to emerge as an adult. Mating occurs on the ground or on vegetation, and egg laying usually takes place over water where strands or masses of eggs are deposited by the female. Some species crawl or dive into the water to lay their eggs.

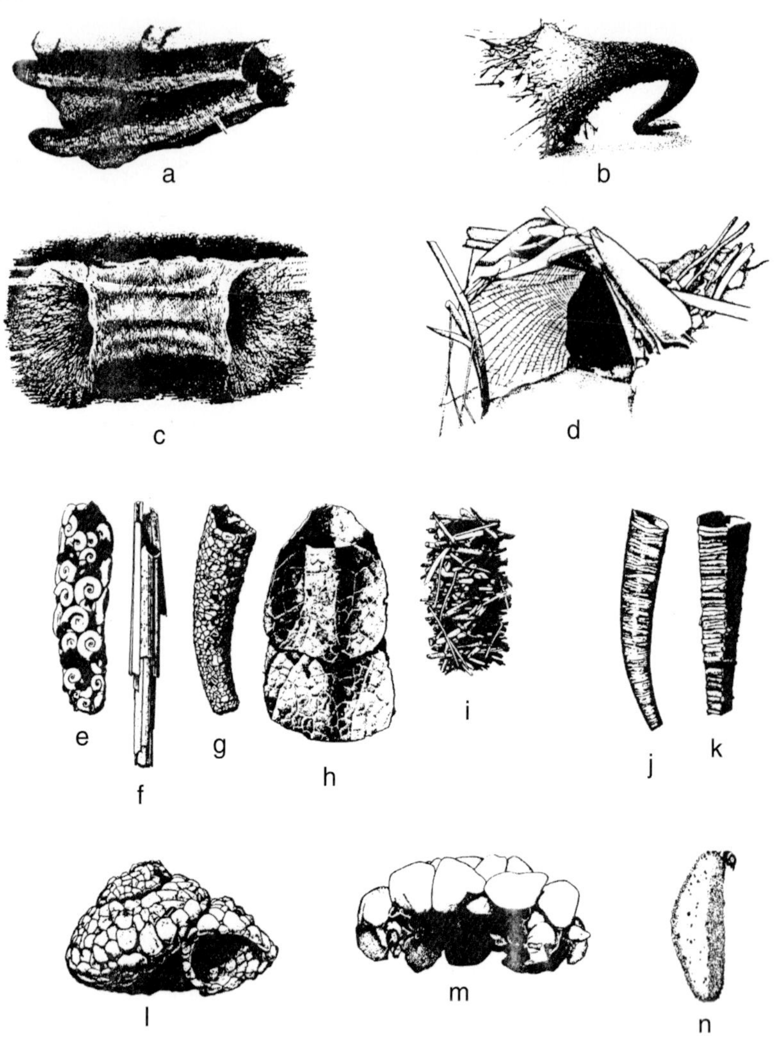

FIGURE 14.9 Representative caddisfly cases. (a) *Dolophilodes* sp. (Philopotamidae); (b) *Neureclipsis* sp. (Polycentropodidae); (c) *Paranyctiophylax* sp. (Polycentropodidae); (d) *Hydropsyche* sp. (Hydropsychidae); (e) *Philarctus* sp. (Limnephilidae); (f) *Arctopora* sp. (Limnephilidae); (g) *Pseudostenophylax* sp. (Limnephilidae); (h) *Clostoeca* sp. (Limnephilidae); (i) *Platycentropus* sp. (Limnephilidae); (j) *Micrasema* sp. (Brachycentridae); (k) *Brachycentrus* sp. (Brachycentridae); (l) *Helicopsyche* sp (Helicopsychidae); (m) *Glossosoma* sp. (Glossosomatidae); (n) *Hydroptila* sp. (Hydroptilidae). (Modified from Wiggins, 1996).

Larval caddisflies are generally cylindrical in shape, resembling caterpillars. The head capsule is heavily sclerotized and antennae are small and inconspicuous. Sclerotized plates are present in various configurations on the thorax. Larvae range from 2 to 40 mm in length when mature and wing pads are absent. The three jointed thoracic legs terminate in a single claw, and anal claws are present at the end of the abdomen. Abdominal gills are present and they may be single or multibranched.

Adults hold their wings folded tentlike over their bodies. Antennae are relatively long and slender, and tails are absent. Adult caddisflies vary in length from less than 2 to over 40 mm, and most live less than a month. They are largely inactive during the day, resting on riparian vegetation, but actively fly at dusk and into the night. Caddisflies are weak fliers, but strong runners.

Given the high diversity of this group, it is not surprising to find that they feed on a wide variety of materials and obtain their food in many ways. Although individual species may feed on a wide variety of material, their method of feeding is usually restricted. For example, a wide variety of plant and animal material is ingested by filter-feeding larvae that construct fixed retreats and spin nets having specific mesh sizes. The caddisflies are broadly grouped into five assemblages based on the type of case they construct, which reflects their ecological role in rivers and streams. Let's look at the categories and how some of the caddisflies feed.

The saddle-case makers, family Glossosomatidae, construct portable tortoiselike shells of sand and pebble grains (see Fig. 14.9m) and move across the surface of the substrate grazing the periphyton. They can be found in large numbers in some streams, literally covering the surface of larger rocks and stones. The purse-case makers of the family Hydroptilidae are free-living as early instars and feed on algae. The tube-case makers are represented by several families and feed in a variety of ways. Members of the family Limnephilidae, one of the largest caddisfly families, function as both grazers and shredders; indeed, they may change feeding habits as they mature, with early instars grazing periphyton and later instars shredding CPOM. A generalization is that if the larval case is made of inorganic material, e.g., sand or pebbles, the larva is a grazer; if it is made of plant material, it is probably a shredder. The net-spinners or retreat-makers are perhaps the most diverse and interesting caddisflies. They function as filtering-collectors and have developed many adaptations to a filter-feeding mode of life. Some, such as the Hydropsychiidae, Psychomyiidae, and Philopotamidae, construct fixed retreats and spin a net which filters food particles from the current (see Fig. 4.7). Others, such as the Brachycentridae, have dense hairs on the forelegs, which they use to filter FPOM from the current after fixing their case to the substrate (Fig. 4.8). Members of the family Rhyacophilidae are predaceous and free-living and ingest other small invertebrates. Thus, as is obvious, caddisflies have representatives in all of the functional feeding groups. The mouthparts of adults are largely nonfunctional and they feed very little.

Because of their diversity and high densities in streams, caddisflies are important members of aquatic food webs. As secondary producers feeding directly on plant material or FPOM, they form a link between the primary producers and the higher level carnivores, such as fishes.

Caddisflies also demonstrate several examples of resource partitioning in streams. For instance, six closely-related species of caddisflies in the family Hydropsychidae were found distributed along the Flathead River in Montana from its headwaters in Glacier National Park to its entrance into Flathead Lake. The species were *Parapsyche elsis, Arctopsyche grandis, Hydropsyche cockerelli, Hydropsyche oslari, Hydropsyche occidentalis*, and *Cheumatopsyche campyla. P. elsis* was dominant in the upper reaches, giving way to *A. grandis, H. cockerelli*, and *H. oslari* in the mid-reaches, and finally to *H. occidentalis* and *C. campyla* in the lower reaches of the river. Two observations stand out concerning the ecology of these six species, all of which build fixed retreats and spin nets to obtain food. First, the ratio of CPOM to FPOM decreased from the headwaters to the mouth, meaning that the coarser detritus prevalent in the headwaters gradually gave way to finer and finer particles downstream. Second, the mesh-size of the nets spun by the caddisflies also decreased downstream, coincident with the decrease in available particle size. Thus, the caddisflies each occupy an ecological niche where the size of the net that they produced was most conducive to capturing the dominant size of food resource available.

The distribution of caddisfly functional feeding groups shows an interesting pattern in relation to available food resources. In eastern deciduous streams, headwater benthic communities are dominated by shredders and collectors, and give way to grazer-collector abundance downstream. Western montane streams generally fit this pattern, except that grazer abundance extends further upstream. This conforms quite closely with the predictions of the RCC.

Given the wide diversity of life-forms and ecological characteristics found in the caddisflies, it has been necessary for us to make broad generalizations. For the reader interested in pursuing more details about this fascinating group of aquatic insects, there is no finer book than *Larvae of the North American Caddisfly Genera (Trichoptera)*, 2nd ed. (1992), by Glenn Wiggins, and we highly recommend it to those who want to learn more.

As with other members of the "big five" (mayflies, stoneflies, beetles, and dipterans), caddisflies play an important role in angling, and this is not surprising given the high densities and variety of body forms found in most trout streams. Immature forms are often called "caddisworms" or "periwinkles," and adults go by such names as "sedges" and "shadflies." Both immature and mature forms are widely imitated. Larval forms are imitated by a number of patterns, often misnamed "nymphs," and are actively fished to imitate larvae drifting in the water column and, especially, pupal forms struggling to get to the water surface to emerge as adults. The Speckled Sedge, American Grannon, Colorado Caddis, and Little Green Rock Worm are just a few of the many patterns tied to imitate larval and pupal caddisflies. Popular adult fly patterns used to imitate caddisflies include the Elk Hair Caddis series, Gray Caddis, Canadian Sedge, Goddard Caddis, and several patterns referred to as Skittering or Fluttering Caddis. The latter are widely used to imitate the flight of egg-depositing females as they flutter across the surface of the water. They are vulnerable to trout during this behavior, and anglers have not missed this opportunity to capitalize on actively feeding trout during these times.

FIGURE 14.10 Collembola, Podinidae (photo by R. Merritt).

Minor Orders

Springtails (Order Collembola)

Although we include Collembola in our chapter on insects, they are generally considered to be distinct taxonomically; however, most authors include them in major publications on insects. The name springtail relates to the presence of a small, forked springlike organ on the underside of the abdomen, which allows them to jump for considerable distances, up to several centimeters.

Although springtails live mainly in the soil, litter, and moist vegetation, some specialized aquatic species are important constituents of the *neuston* (organisms that live on the water film) (Fig. 14.10). The common freshwater families are the Isotomidae, Poduridae, and Smithuridae. Primarily found in standing waters, they have been found in rivers and streams where it is thought that their occurrence here is primarily related to dispersal. Collembolans have been found in the hyporheic zone and in the surface film in caves. They can be common along the side of rocks, just at the waterline.

Little is known of the ecology of these organisms. Springtails are mainly detritivores, consuming a wide variety of microflora and dead plant material. Many feed on diatoms, algae, and other planktonic organisms trapped in the surface film, while others appear to favor a liquid or bacterial diet.

Collembolans have great diversity associated with their development. Instars vary from two to over 50, and the adult in many species is not definitive (ametabolous), thus making the number of instars difficult to measure. They produce several generations per year (are multivoltine).

Water Bugs (Order Hemiptera)

This order includes the true "bugs," and most of them by far are terrestrial. Those truly aquatic species are included in the suborder Heteroptera (the suborder Homoptera includes a few semiaquatic species). There are approximately 400 species in North America. Six families are underwater dwellers, five who live mostly on the water surface, and six families are essentially shore- or riparian-dwelling. The name Hemiptera means "half wing" and refers to the characteristic division of the forewings into a thickened, leathery base and a membranous apical portion. Adults, however, vary in their wing development, ranging from no wings at all to fully developed wings. As with the beetles, which some hemipterans resemble, some species have truly aquatic and terrestrial adults. In the latter stage, they can disperse to other nearby bodies of water. Generally, their body shape ranges from oval to slender and elongate, from 1–65 mm in length.

Most hemipterans are not found in fast-flowing reaches of rivers and streams; rather, they are more common in the slow-flowing, lakelike backwaters. Three families of this group live on the top of the water film; they are the well known water striders or skippers [families Gerridae (Fig. 14.11a), Veliidae (Fig. 14.11b), and Hydrometridae] which skate across the surface of slow-moving water on the water film. In Canada, they are sometimes called Jesus bugs because they "walk on water." Few are found in fast-flowing streams; those that are will be found on the water surface along the stream edge. One genus, *Rhagovelia*, the broad-shouldered water strider, can be found skating on the surface in rapid currents where it has adapted to the

FIGURE 14.11 Hemiptera. (a) Gerridae (photo by R. Merritt); (b) Veliidae (photo by R. Merritt); (c) Notonectidae (photo by R. Merritt); (d) Corixidae, *Graptocorixa californica* (photo by L. Serpa); (e) Naucoridae (photo by A. Elosegi); (f) Belostomatidae. *Lethocerus* sp. (photo by R. Merritt).

FIGURE 14.11 *(Continued)*

FIGURE 14.11 *(Continued)*

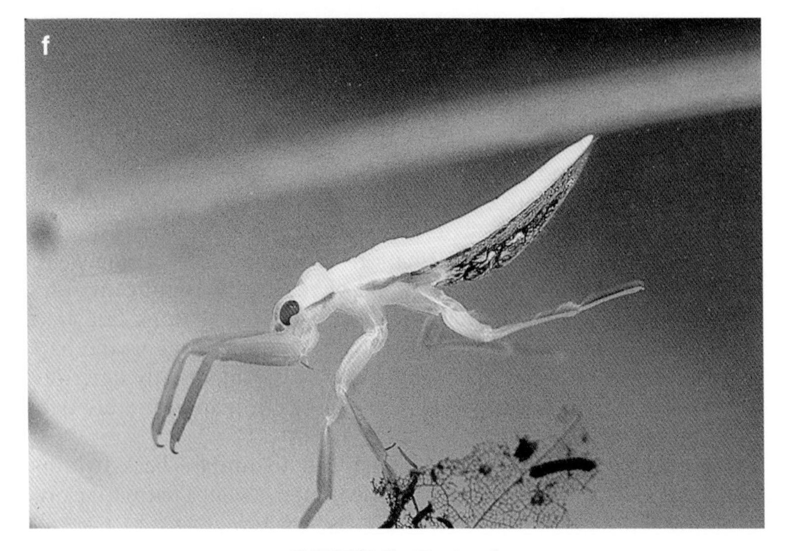

FIGURE 14.11 *(Continued)*

"fast lane." Another common hemipteran is the back-swimmer, *Notonecta* (family Notonectidae) (Fig. 14.11c), which swims upside down with strong beats of its oar-shaped hind legs. It swims in jerky movements and captures slow-moving prey. Notonectids prey heavily upon mosquito larvae, and their use as a biological control in this respect has been advocated. Water boatmen (family Corixidae) (Fig. 14.11d) are the largest group of water bugs. They swim right-side up and both middle and hind legs have swimming hairs to enhance their ability to swim; they are also common in the slower reaches of rivers and streams. Creeping water bugs (family Naucoridae) (Fig. 14.11e) swim, as well as crawl, in quiet water and are rarely found in fast currents; *Ambrysus* lives in small, clear streams. The giant water bugs (family Belostomatidae) (Fig. 14.11f) are spectacular, large bugs fond of slow currents and usually found in debris accumulations on the bottom. They often feign death when disturbed, but can inflict painful bites; they are called "toe biters" by careless waders! Belostomatids are often pests in fish hatcheries where they prey heavily upon the smaller fishes.

Metamorphosis is hemimetabolous. This is characteristic of the univoltine life cycle common in the temperate zone, although reproduction may be continuous in tropical regions and many gerrids are bivoltine. There are generally five nymphal instars, although four occur in some of the smaller water-striders. Nymphs and adults resemble each other and occur in the same general habitats. Adults lay eggs in the spring and immatures develop during the summer and autumn and overwinter as adults. Adults are not known to swarm, but large numbers are often attracted to lights.

Water bugs are predacious; all of the members of this order have mouthparts modified into beaks with which most of the terrestrial forms pierce plant tissue and the aquatic forms extract their food from animal prey. They can also inflict painful bites if you aren't careful when you handle them. The forelegs of many species are modified for capturing and grasping prey. Two claws are present; in the Gerridae, the claws are preapical on the fore legs. Water bugs are active predators on other insects and crustaceans, but have been known to feed on carrion. The giant water bugs have been known to attack fishes, frogs, and water snakes. Corixidae are also able to grind up small bits of food and are classified as herbivore–detritivores, omnivores, predators, or scavengers.

Although considerable information is available for some selected groups of hemipterans, they are not known to play significant roles in most riverine ecosystems. Obviously, they act as predators, but are evidently not as readily preyed upon by other animals as one would expect. This is due to the presence of scent glands, which apparently deter predation. However, the Corixidae do not possess scent glands and are heavily preyed upon.

A few fly patterns have been developed to imitate back-swimmers; these flies include the Graywinged Backswimmer, Water Boatman, Corixid, Moon Backswimmer, and Grousewinged Backswimmer.

Aquatic wasps (Order Hymenoptera)

This order is primarily terrestrial, but the larvae of many aquatic insects are parasitized by the larvae of some wasps. There are nine families and 51 species of aquatic Hymenoptera in North America.

Aquatic wasps frequent the same habitat as their hosts. They parasitize all orders of aquatic insects except Collembola, Ephemeroptera, Orthoptera, and Plecoptera. The adult females of these wasps enter the water and seek out different aquatic insect life stages in which they deposit their eggs; hence, they are called diving or aquatic wasps. They may attack the eggs, larvae, or pupae of aquatic insects, depending on the species of wasp and host. They may also parasitize the nonaquatic life stages of insects, in which case the wasp does not need to enter the water.

These wasps are holometabolous and many emerge as adults from the hosts. Larvae change radically from instar to instar.

Water Moths (Order Lepidoptera)

Several families of moths have aquatic or semiaquatic members, but most are primarily terrestrial, including the family Pyralidae, which has the most truly aquatic species. Eight other families have aquatic or semiaquatic species, four of which are found in North America, for a total of five families on this continent with about 150 species that are aquatic or semiaquatic; 50 species are truly aquatic (Fig. 14.12a).

All immature stages of water moths occur in the water, and adults are terrestrial. Larvae range from 3–75 mm in length when mature. Larvae may or may not have gills, depending on species and habitat. Lepidopterans are holometabolous and most have from 5–7 instars. One study revealed a long overwintering generation and a short summer generation for one species of *Petrophila* in a New York stream. Two or three generations per

FIGURE 14.12 Miscellaneous aquatic insects. (a) Lepidoptera, Noctuidae larva (photo by R. Merritt); (b) Megaloptera, Sialidae, *Silas* sp. larva (photo by R. Merritt); (c) Megaloptera, Corydalidae, *Corydalus cornutus* larva (photo by R. Merritt); (d) Odonata, Gomphidae, *Progomphus obscurus* nymph (photo by D. Penrose); (e) Odonata, Lestidae nymph (photo by R. Merritt).

FIGURE 14.12 *(Continued)*

FIGURE 14.12 *(Continued)*

year have been documented in Northern California for *Petrophila confusalis*, but this species has only a single generation per year in Montana.

Immature moths are herbivorous, feeding on plant tissues. Many are found in slow-moving, depositional areas of streams. Some are leaf miners, others bore into the roots and stems. A few species of the genus *Petrophila* occupy rapid streams. The larvae live in silken tents where they graze algae from the rock surface. Pupae of some aquatic species live underwater in silken cases; they cut an escape slit prior to pupation. Emerging adults float or swim to the surface. Some species in one Hawaiian genus, *Hyposmocoma*, construct cases similar to caddisflies from which they feed on algae, moss, and lichens in rapid streams. Eggs are usually laid on the undersides of floating leaves, though some females deposit their eggs inside leaves or flowers. Rock-dwelling females have swimming hairs on the hind legs and readily deposit eggs at considerable depth on rocks in swift-flowing water. Eggs are laid singly or in clusters.

Immature stages are found on rocky substrates in rivers and streams and on or in aquatic vascular plants. Factors affecting the density and distribution of moths include water velocity, water temperature, dissolved oxygen content, substrate texture, and growth of algae. Some species in the same genus can coexist if conditions are conducive to both, but changes in one or another of the above-mentioned characteristics can limit the distribution of one species, providing a competitive advantage to the other.

Because they can occur in high numbers and feed heavily on plants, they have been suggested as biological control agents for some pest aquatic plants. However, this is unlikely to occur in riverine ecosystems.

Water moths have not been imitated by fly tiers, although several general patterns imitating other insects come fairly close to being good imitations of these insects.

Dobsonflies, Alderflies, and Fishflies (Order Megaloptera)

This is a relatively small group of insects that is closely associated with, and sometimes included with, the Neuroptera. Two families, Sialidae and Corydalidae, have approximately 50 species in North America. The dobsonflies, also called hellgrammites, are well known to anglers who recognize them as excellent trout bait; fly fishermen imitate them with several patterns. The larvae of some megalopterans can be confused with immatures of some beetles and caddisflies, which they closely resemble. Wings are large, as their scientific name implies, but they are weak fliers.

Sialid larvae (alderflies) (Fig. 14.12b) can be very abundant in streams and rivers where they burrow into soft sediments and detritus. They are predacious and feed indiscriminately on other insect larvae, annelids, crustaceans, and mollusks.

Alderflies are holometabolous, and larvae may pass through as many as 10 instars during a life span of one to two years (uni- or bivoltine). Larvae leave the stream prior to pupation and pupate in an unlined chamber dug several centimeters into the soil and litter along the shoreline. Adults emerge in spring and summer and are active during the warmer hours of the day. They are not strong or frequent fliers, and adults apparently do not feed.

Larvae attain a maximum length of about 25 mm, including the median caudal filament, or tail. Legs have two claws, and the first seven abdominal segments have prominent, segmented lateral gill filaments. Adults are small, usually from 10–15 mm long, with black, brown, or yellowish orange bodies. Wings are colored much like the bodies.

Corydalids (hellgrammites) (Fig. 14.12c) occur in a wide variety of aquatic habitats, including streams and rivers. They are predacious and feed on a wide variety of small aquatic invertebrates, similar to the alderflies. They, too, are holometabolous, and have a life cycle lasting 2–5 years with the larvae going through 10–12 instars. Pupation occurs in soil chambers excavated near the shoreline, although some species pupate in dry streambeds and others seek rotting logs or stumps along the shore. Adults emerge from spring to midsummer and are nocturnal fliers.

Hellgrammite larvae are usually larger than alderfly larvae, attaining lengths ranging from 30–65 mm. Other characteristics are similar except that the abdomen terminates in a pair of anal prolegs, each of which bears paired claws and a dorsal filament. Adults are up to 75 mm long and are black, brown, or yellowish in color with pale, smoky, mottled wings.

Megalopterans are often found under large rocks in rivers and streams. They are less well-known ecologically than many other insect orders, probably because they are not as common in collections as are many other groups, notably the mayflies, stoneflies, dipterans, and caddisflies.

Larvae of these insects, especially the dobsonflies, are excellent bait for anglers. However, there is little in the fly tying literature to show that much attention has been given to specifically imitating these insects. Nevertheless, because the immature dobsonflies bear a resemblance to many of the large stonefly nymphs, there is no reason to believe that some of the stonefly patterns couldn't pass as reasonable imitations of dobsonflies. This would include such patterns as Carrier's Stone, Rubber Legs Brown Stone, Giant Stone, and Stonefly Bugger.

Spongillaflies (Order Neuroptera)

These insects are primarily terrestrial and include the lacewing flies and ant lions. One small family (Sisyridae), however, is aquatic and lives as a parasite associated with freshwater sponges. Six species occur in North America, three each in the genera *Sisyra* and *Climacea*. They are widespread in North America, occurring, as would be expected, where sponges occur, but are rare in the southwest and western mountain areas. Spongilids are closely related to the order Megaloptera, and you may find them grouped together by some authors under the order Neuroptera ("nerve-winged").

The larvae of sisyrids are aquatic; the other life stages occur on land. There are three larval instars, and they overwinter as *prepupae,* that is, a larvae within a pupal cocoon. Pupation occurs on land. Metamorphosis is holometabolous, and anywhere from one to several generations per year may occur. No feces are produced by the larvae or pupa because the alimentary canal is closed during this time of development. As soon as the adult emerges from the pupal cocoon, it excretes a fecal pellet that incorporates all of the waste products accumulated up to this time of development. Adults are short-lived and emergence usually occurs during warm weather.

The larvae live on and within sponges, feeding on the sponge tissues through mouthparts modified to form long, hollow tubes. Larvae are small and soft-bodied, usually measuring from 3–8 mm in length. No taillike structures are present and the antennae are long; wing pads are absent. Many *setae*, or spines, are present on the body. Adult bodies are brown with brown wings and are usually from 6 to 8 mm long.

Although they generally remain with a single host sponge during larval development, they can "swim" to another host by flexing movements of their bodies. Little else is known of the ecology of these insects.

Dragonflies and Damselflies (Order Odonata)

Dragonflies and damselflies are more characteristic of standing or slow-flowing waters; thus, their occurrence in rivers and streams is restricted to backwaters and other reaches where the current slackens. They are also found around small springs and seepage areas, but this is usually where some of the water is backed up and slow-flowing because of vegetation. The order is composed of two families; the Anisoptera, or dragonflies, and the Zygoptera, or damselflies. Dragonflies are generally larger and hold their wings to the side when perched; damselflies are relatively smaller and hold their wings along the body when at rest.

There are approximately 450 species of odonates in North America, and they are well known to anybody who frequents wet places. They are known for their striking colors, darting flight, and many myths. As a youngster, one of us (CEC) used to flee from them in our farm pond; the common name "devil's darning needles" indicated that they could sew your lips together! Some fly actively and rarely alight during daylight, others perch as soon as the sun goes behind a cloud, while others seem to fly only during particular times of the day.

Larval dragonflies are robust in body shape (Fig. 14.12d) and their abdomens terminate in short, stout, spinelike appendages, whereas damselflies (Fig. 14.12e) are slender and have three taillike appendages, the gills. One of the most striking morphological characteristics is the presence in the nymphs of a modified, hinged, prehensile labium (Fig. 14.13), with which they can strike out and nab unwary prey. When retracted, this organ forms a masklike covering over the bottom of the head and other mouthparts. The thoracic legs are well developed and two pairs of wing pads are present. Immatures can be found crawling or sprawling on vegetation or other debris, and even burrowing. Some dragonflies have the capability of "jet propulsion," being able to move rapidly by forcibly ejecting water from the rectal chamber. The compound eyes of the adults are very large, and the abdomen is very long and slender.

Metamorphosis is hemimetabolous. Adult damselflies slightly resemble the nymphs, whereas adult dragonflies bear no resemblance to the immatures. Larval development may range from 8–15 instars, although 10–12 is the norm. They may be univoltine, semivoltine, or merovoltine; only rarely is more than one generation per year produced. Adults emerge during warmer weather after overwintering as eggs or nymphs. Some species emerge all at once and soon disappear; others emerge throughout the summer. Adult life may last up to a few months. Diurnal flight activity is associated with feeding, reproduction, or dispersal to other sites. Males may establish mating territories, which they defend vigorously. Mating may take place in flight or while perched, and at various distances from the water. Mating is complicated, and involves the male depositing sperm in secondary genitalia on the second and third segments of his abdomen. He then seizes the female behind the head with grasping organs at the end of his abdomen. She then flexes her abdomen under his and engages her genitalia with the sperm deposited earlier. In position, they form a "head-to-tail" loop (Fig. 14.14). It

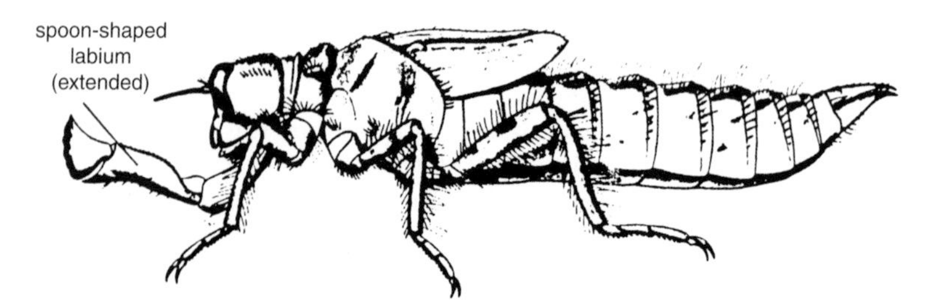

FIGURE 14.13 Dragonfly nymph showing modified labium which can be extended to capture prey. (From McCafferty, 1981.)

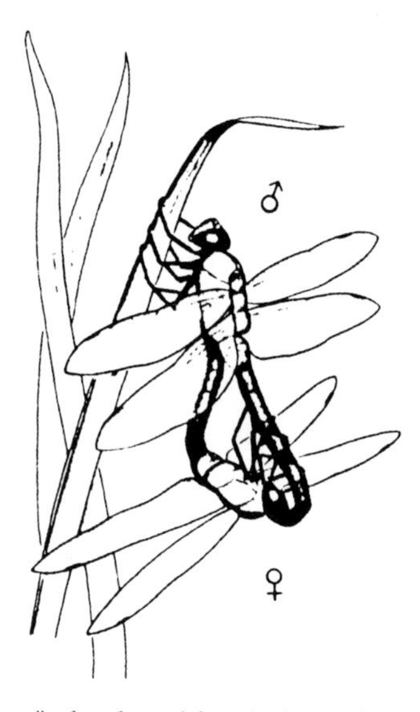

FIGURE 14.14 "Mating loop" of male and female dragonflies. (From McCafferty, 1981.)

is common to see damselflies in tandem, and studies have shown that this is to assure successful mating by the male. Should a second male mate with the female soon after another male, a special adaptation of the copulatory organ allows the second male to remove the prior male's sperm, an unusual form of competition between males for mating success. Eggs are laid by the male and female in tandem and are deposited directly into the water, in emergent or floating plants, or directly onto the substrate of shallow water.

Adults are extremely active predators. They consume other insects, many of which are pest species, which they catch in flight, giving them the common name of "night hawks" or "mosquito hawks." You might observe a dragonfly "working" a mayfly swarm, capturing one after another in what must be a feast for these agile predators. Adult odonates themselves are heavily preyed upon by frogs and birds, such as kingfishers, herons, falcons, blackbirds, and several others. The immatures are especially vulnerable when they are emerging. Immatures are consumed by game fishes.

Anglers have conceived many imitations of these "flying dragons," especially the damselflies. Some popular damselfly patterns include the Giant Damsel, Wingcase Damsel, Swimming Damsel, and Marabou Damsel for the immatures, and the Blue Damsel, Parachute Damsel, and Braided Butt Damsel for adults. Larval dragonfly patterns include the Filoplume Dragon and Lake Dragon. Live nymphs are also used for bait.

Crickets, Grasshoppers, and Locusts (Order Orthoptera)

These are generally terrestrial insects (the name means "straight wing"). Some, however, are hydrophilous (water-loving), and some have adapted to live on emergent aquatic vegetation in shoreline habitats. Six families have semiaquatic species. They are not found where the current is rapid, and those adapted to living on or near the water are usually found at the water's edge or in standing water. Some grasshoppers may dive and cling to submerged vegetation; their legs are flattened to aid in swimming.

Metamorphosis is hemimetabolous and immatures generally resemble the adults. They are mostly univoltine, although some may take up to three years to mature.

Most orthopterans feed on plant tissue; however, some are predators and others feed on detrital particles in the water. They are not important constituents of riverine ecosystems.

Despite their general lack of ecological importance in rivers and streams, knowledgeable anglers know that a grasshopper, live or imitation, plunked down along the edge of a slow-flowing meadow stream can bring a slashing strike from a trout. Thus, several fly patterns have been developed to imitate grasshoppers, including Dave's Hopper, Joe's Hopper, Henry's Fork Hopper, and several other hopper patterns.

INSECTS AS BIOINDICATORS

As mentioned earlier, several orders of insects, especially Ephemeroptera, Plecoptera, and Trichoptera (EPT), require high-quality water for their existence. Thus, biologists have used their presence or absence, in conjunction with the numbers present at a particular location in a stream or river, to develop several indices of water quality.

The simplest of these uses a weighting system of pollution tolerance by the various families of mayflies, stoneflies, caddisflies, and dipterans. By collecting 100 insects at random at a particular site, multiplying the number of each family by their tolerance index, summing these numbers, and dividing by 100, gives an index number which is then compared with a table of value ranges indicative of various levels of pollution.

Other indices have been developed using various combinations of EPT numbers to develop ratios in conjunction with numbers of other, less tolerant organisms. Rather than go into the details of all of these variations, suffice it is to say that aquatic insects, because of their range of tolerances for environmental conditions, provide excellent organisms for use as indicators of environmental degradation.

Further, because of their location in the food web, they are excellent organisms to sample and analyze for the presence of contaminants—DDT, pesticides, radionuclides, PCBs, and other compounds. These chemicals are usually absorbed in high concentrations by unicellular algae and FPOM, and when fed upon by insects, the compounds are concentrated in their tissues and amenable to analyses and comparison with control populations. This is called *bioaccumulation*.

Books have been written on this subject (see Recommended Reading), and a chapter is devoted to biomonitoring in the Merritt and Cummins volume. The interested reader is referred to these publications.

Recommended Reading

Hynes, H. B. N. (1970). *The Ecology of Running Waters*. University of Toronto Press.

McCafferty, W. P. (1981). *Aquatic Entomology*. Jones & Bartlett, Boston.

Merritt, R. W. and Cummins, K. W. (eds.). (1996). *An Introduction to the Aquatic Insects of North America*, 3rd ed. Kendall/Hunt, Dubuque, Iowa.

Rosenberg, D. M. and Resh, V. H. (1993). *Freshwater Biomonitoring and Benthic Macroinvertebrates*. Chapman & Hall.

Stetzer, R. S. (1992). *Flies, the Best One Thousand*. Frank Amato.

Stewart, K. W. and Stark, B. P. (1993). *Nymphs of North American Stonefly Genera (Plecoptera)*. University of North Texas Press, Denton.

Ward. J. V. (1992). *Aquatic Insect Ecology. 1. Biology and Habitat*. John Wiley & Sons, New York.

Ward, J. V. and Kondratieff, B. C. (1992). *An Illustrated Guide to the Mountain Stream Insects of Colorado*. University Press of Colorado.

Wiggins, G. B. (1992). *Larvae of the North American Caddisfly Genera (Trichoptera)*, 2nd ed. University of Toronto Press.

CHAPTER 15

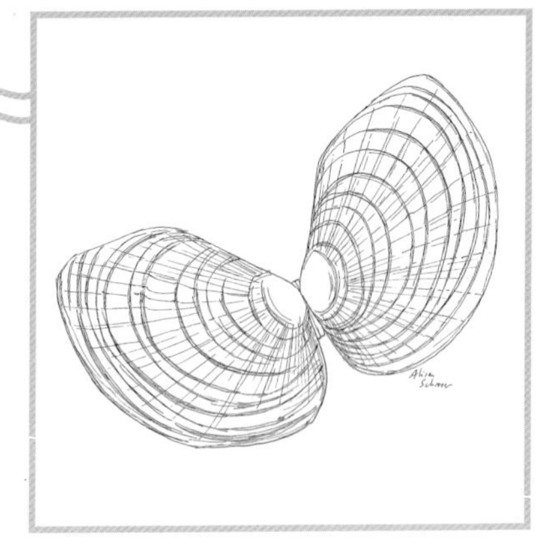

Mollusks

Two groups of freshwater mollusks are common residents of rivers and streams: the univalve Gastropoda (the snails and limpets) and the bivalve Pelecypoda (the clams and mussels). Mollusks are abundant and diverse in marine environments, including some groups such as squid and octopi that are strictly marine. The gastropods comprise almost three-fourths of the roughly 110,000 described species, occurring in the seas, freshwater, and on land. As we shall see, the gastropods and bivalves of river ecosystems are diverse, ecologically important, and perhaps the most seriously imperiled of all animal groups in North America. Their life cycles are interesting and complex, some are colorful (and colorfully named), and some, especially bivalves, are economically important for harvest and as nuisances.

Readers seeking a more detailed account of mollusks should consult *The Ecology and Classification of Freshwater Invertebrates* by J.H. Thorp and A. P. Covich (1991), and *Fresh-Water Invertebrates of the United States* by R. W. Pennak (1989). Much of what follows is derived from these two excellent sources.

SNAILS AND LIMPETS (CLASS GASTROPODA)

Like all freshwater mollusks, gastropods have a soft body inside a calcareous shell. As adults, these univalve animals range from 2 to 70 mm in longest dimension. A muscular *foot* propels the animal, a structure called the *mantle* secretes the shell, and the body consists simply of a head and a visceral mass. Prosobranch snails (349 species in North America) have a calcareous *operculum,* a kind of "trap door," which is pulled in after the foot to protect the individual. Pulmonate snails (150 species) have made a secondary invasion of freshwater from land. They lack an operculum and do not respire with gills, as do prosobranchs. Instead a modified portion of the mantle cavity acts as a lung, an adaptation that reflects the terrestrial ancestry of this group.

The geometry of snail shells helps to distinguish major groups. Limpet (Pulmonata, family Ancylidae) shells are simple cones (Fig. 15.1a). Others have spiral whorls all in one plane (Pulmonata, family Planorbidae, Fig. 15.1b). Yet others, both pulmonates and prosobranchs, have their whorls elevated into a spiral (Fig. 15.1c). Shell details useful to the taxonomist include height of the apex, whether the shell is dextral (aperture to right when viewed with spire pointing away) or sinistral (aperture to the left), and further details of the operculum (Fig. 15.1). Some representative species are shown in Fig. 15.2.

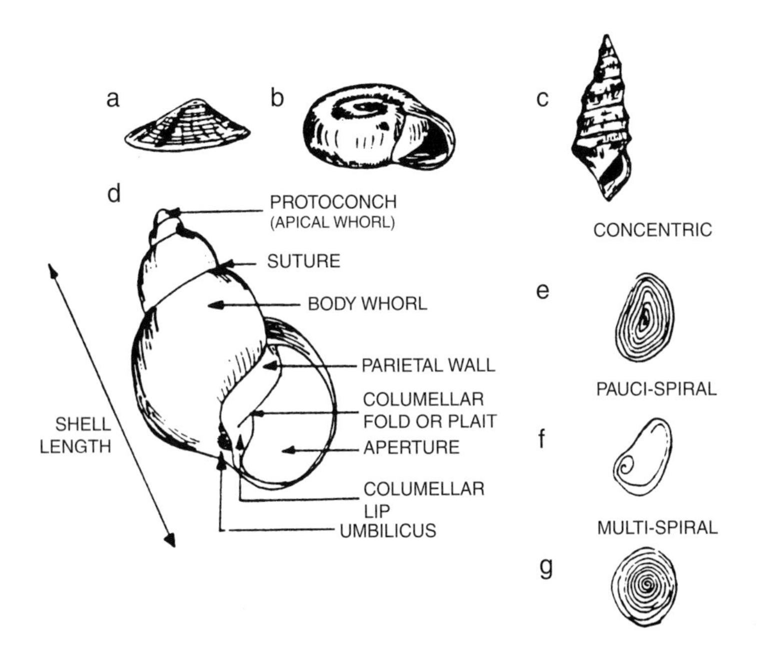

FIGURE 15.1 Basic anatomy of a gastropod shell. Shell type a is conical, type b is planospiral, type c is spiral. Major features of the shell are shown in d; three types of opercula (e, concentric; f, paucispiral; g, multispiral). (From Brown, 1991.)

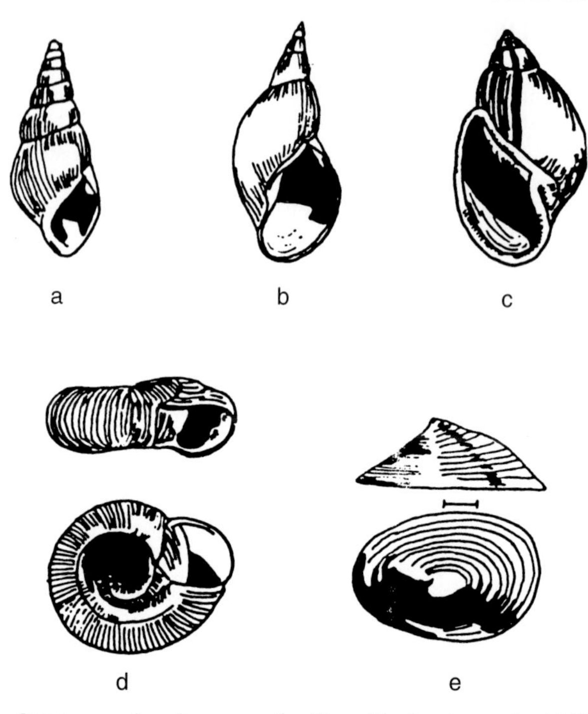

FIGURE 15.2 Some examples of common families of freshwater snails. (a) Pleuroceridae (*Elimnia*), (b) Limnaeidae (*Lymnaea*), (c) Physellidae (*Physella*), (d) Planorbidae (*Planorbula*), (e) Ancylidae (*Ferrissia*). (From Brown, 1991.)

Freshwater snails are herbivorous or omnivorous, feeding on algae and sometimes on organic detritus. Their mouths contain hardened jaws for cutting food items, and further inside the mouth region snails possess a unique rasping structure, called a *radula*. This long, straplike, toothed structure moves rapidly back and forth, grinding food against the roof of the mouth. The radula is continually worn away at its anterior end, and it is replaced by outward growth, rather like a human fingernail. The gastropod digestive system is relatively sophisticated (in comparison, say, to a mayfly nymph); there may be a specialized crop with sand grains to grind food particles, and mollusks produce a variety of enzymes, including cellulases, that can break down the cell walls of algae. Algae and diatoms are frequently observed in the guts of snails, and the rasping radula enables efficient scraping of periphyton from rock surfaces. But detritus and small animals also contribute, more so in snails found in slow currents where organic matter tends to be more available.

Snail grazers can be formidable competitors of grazing aquatic insects. Laboratory studies (conducted in cages termed "snail jails" by the biologists) show that increasing snail densities shift the algal assemblage from filamentous greens to tightly adhering diatoms, and eventually to the relatively unpalatable cyanobacteria. Snails grazing on algae coating the surface of macrophytes may aid the latter by removing these epiphytic algae. Fishes that pick off the snails may facilitate excessive algal build-up on macrophyte leaves, illustrating the complex interactions of food webs and the strong impact snail grazers can have.

River snails occur in a wide range of habitats, from small streams to large rivers, but the majority are found in shallow waters less than 3-m deep. Numerous environmental variables, along with specific adaptations, influence where different snails are found. Certain species or entire families are usually found on sandy bottoms, or rocky substrates, or exclusively in mud. One species, *Fontigens nickliniana*, is known as the "watercress snail," for obvious reasons. A few have colonized caves and, as commonly occurs, have lost their eyes and body pigmentation. Pulmonates and detritivorous prosobranchs predominate in silty habitats of slow current, whereas limpets and prosobranchs are most likely to be encountered in locations of faster current, where they encounter relatively more algae and less detritus.

All freshwater mollusks are limited to waters containing sufficient calcium. Very soft water, defined as having a calcium concentration below 3 mg/L, excludes roughly 95% of all species, and even moderately hard water (< 25 mg/L) excludes roughly half of the freshwater molluskan fauna. The presence or absence of suitable substrate also can be a factor governing species distributions. Last but not least, predators, including those that crush the snail and those that invade it, likely play a contributing role in snail abundance and distribution. Crayfish can chip away at snail shells with their mandibles. The red-eared sunfish has specialized teeth for crushing shells and prefers thin-shelled victims over more armored gastropods. Snails can form a significant part of the diet of suckers, perch, whitefish, sheepshead, and various members of the sunfish family. Leeches are an invasive predator, and at least one British snail exhibits violent shell-shaking, apparently an avoidance reaction, when its mantle fringe comes into contact with certain leeches and flatworms.

Snail sex is complicated. Prosobranch individuals are male or female (there are some exceptions), but each pulmonate individual produces both egg and sperm, usually simultaneously. Individual pulmonates often cross-fertilize, but they also can self-fertilize, and the advantage of their hermaphroditism may be in the ability of a single individual to colonize new habitat. Because newly hatched snails ("spat") can be dispersed in the mud on a bird's foot, the benefits of reproducing without a partner are obvious. And in case this makes the prosobranchs seem boring, we should note that in some species the male's right tentacle is enlarged, serving as a copulatory organ, while others have a specialized penis or no organ. Snails usually lay eggs in a gelatinous mass on some substrate, in small or larger egg masses depending on the species. In the family Viviparidae eggs are deposited in a fold of the mantle inside the snail's body, where they hatch and develop, to be "born" free-living. Snail life spans usually range from less than one year (most pulmonates) to five years, depending on the group.

River snails are diverse in North America, especially in streams of the southeast, numbering about 500 species. The approximately 150 species of Pleuroceridae are found almost entirely east of the Mississippi and are especially rich in the southeastern United States. Although some taxa are widespread, many are quite restricted in their range, and three genera are (or were) found only in the Coosa River, Alabama.

As we shall see also for the bivalves, many species of river snails are threatened as a result of human activities. A 1970 report on the 103 species of freshwater snails in the Ohio River drainage found 41 to be rare and 8 probably extinct. Because more is known about the extent of imperilment of freshwater bivalves, we will focus on that group to better understand the reasons for such high imperilment.

CLAMS AND MUSSELS (CLASS PELECYPODA)

Freshwater bivalve mollusks—clams and mussels—have a shell separated into two symmetrical valves. These enclose a soft body as well as the mantle, which secretes the shell material, and greatly enlarged gills, which filter food from the water column. The two valves are connected by a mid-dorsal isthmus, which secretes the material that forms the elastic connecting hinge. Internal hinge teeth form the fulcrum on which the two valves open and close. Bivalves have a muscular, flexible, and extendable foot that assists in burrowing and movement (Fig. 15.3). Animals usually burrow into the substrate, where their valves and mantle prevent invasion by fine sediments and resist damage from stones. Most are burrowers and hence are referred to as *infauna,* but some, including the exotic zebra mussel *Dreissena polymorpha,* dwell on the surface and so are *epibenthic.* Adult size ranges from 2 to 250 mm in longest dimension for the biggest specimens.

The freshwater mussels and clams of North America are the most diverse in the world (Fig. 15.4). The families Margaritiferidae and Unionidae (collectively the super family Unionacea) occur throughout the world but reach their greatest diversity in North America, particularly in the southeast. Some 281 species and 16 subspecies currently are recog-

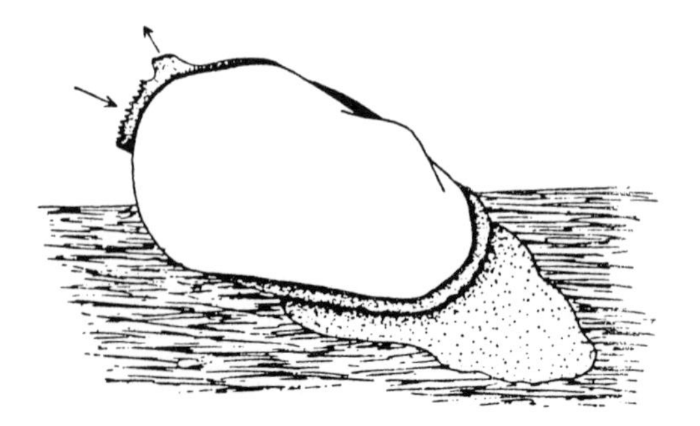

FIGURE 15.3 Female *Lampsilis siliquoidea* in natural position in the river bottom. Arrows indicate inhalent and exhalent siphons. Note muscular foot extended to lower right, and fringe of mantle extending beyond shell. (From Pennak, R. W. (1989). Fresh-Water Invertebrates of the United States, 3rd ed. Copyright © 1989 John Wiley & Sons, Inc.]

FIGURE 15.4 A collage of pearly mussels. North America boasts approximately 300 species of freshwater mussels. (a) *Quadrula quadrula* (mapleleaf), is commercially harvested; (b) *Cycolonaias tuberculata* (purple wartyback) is relatively common; (c) *Epioblasma triquetra* (snuffbox); (d) *Lampsilis fasciola* (wavy-rayed lampmussel) are threatened/endangered in many states; (e) *Epioblasma torulosa* (tubercled blossom) is globally extinct; (f) *Dreissena polymorpha* (zebra mussel) is an exotic species, shown attached to *Potamilus alatus* (pink heelsplitter). (Photos by K. Cummings.)

FIGURE 15.4 *(Continued)*

FIGURE 15.4 *(Continued)*

nized, mainly from the Unionidae. The Sphaeriidae or fingernail clams number 33 native (and 4 introduced) species. Two important exotics, the Asiatic clam *Corbicula fluminea* and the zebra mussel *D. polymorpha* will receive special mention later. Many unionaceans have limited geographical distributions and most are found in rivers, whereas the sphaeriids tend to be widespread, and many are inhabitants of lakes, although some occur in small streams. *Corbicula* is a widespread invasive species, and the zebra mussel is well on its way to colonizing waterways across the country.

Environmental requirements vary with the species, of course, but unionaceans generally are most successful in clean, well-oxygenated water, on substrates of sand or gravel, and where currents are swift enough to remove silt but slow enough for sediment stability. Some species prefer finer sediments; these tend to be species of embayments, deep water, and lakes. Shallow habitats are preferred, generally less than 4–10 m, but bivalves occur in the bottom of deeper rivers if they are well oxygenated. Bivalves, like snails, need calcium carbonate for their shells, and so the remarks made previously about water hardness and snail occurrence apply equally here. Water that is acid or low in calcium excludes most species, whereas alkaline, hard waters are favorable.

Droughts and falling water levels can expose freshwater bivalves to drying, and many species can withstand these conditions for a considerable time, even months. Valve closure minimizes water loss while allowing oxygen uptake, and other metabolic adjustments may take place.

Most freshwater bivalves filter fine particles from the water, including algae, bacteria, and organic detritus. Material is filtered onto gill surfaces, then moved by ciliary action to the mouth, where some sorting occurs prior to ingestion. Enzymes are secreted along the way, and within the stomach a rodlike structure rotates like a mix-master. More size-based particle sorting takes place, and small particles eventually end up in very acid pockets of the digestive system. In lakes, large rivers, and embayments, and in streams below lake outlets, there may be sufficient phytoplankton in the water column for this to be an important component of the bivalve diet. However, many rivers have relatively low quantities of algae and much higher quantities of suspended organic matter (Chapter 5), and so organic matter and bacteria likely are dietary mainstays in most circumstances. The lower reaches of large rivers typically provide an ample supply of fine particulate organic matter for suspension feeding, and so bivalves do well at the lower end of the river continuum. Native unionaceans reportedly cannot filter bacteria from suspension, while the invasive zebra mussel and Asiatic clam can. Another feeding modality, termed pedal feeding, relies on the foot to draw detrital particles into the mantle cavity. Known from some marine species and *Corbicula*, it may turn out to be common in freshwater bivalves that burrow in soft sediments.

Bivalves enhance water clarity by their filtering activity and serve as benthic "sinks" for organic matter and nutrients. Quantitatively, however, this effect in many rivers probably is small. It is unlikely that the filter-feeding activity of riverine bivalves often is great enough to "clean" the water column of algae and organic particles, although this does occur in lakes with dense populations of the Asiatic clam or zebra mussel. In some large, slow-moving rivers, these same invaders likewise have been found capable of effectively filtering the entire water column in the time it takes to pass over a dense bed of clams or mussels.

Bivalves have their enemies, of course, and various fish are foremost among these. Ducks and shorebirds consume small individuals in shallow habitats, as do turtles. Crayfish also feed on small bivalves. Most fishes also prey on smaller species because shell strength increases in proportion to shell size. However, carp and freshwater drum are able to crush bivalves, and channel catfish swallow larger specimens whole. Mammals are also important predators, and more likely to feed on larger individuals. Great piles of open shells next to the river's edge points to a favored feeding location of a raccoon or muskrat. Additional mammalian predators include mink, otter, and humans.

Many freshwater bivalves are hermaphroditic, generally simultaneous and often self-fertilizing. Cross-fertilization requires release of sperm into the water column, usually in response to specific temperature cues. With one exception, North American freshwater bivalves brood their embryos through early development internally, in the gill region. The exception is the zebra mussel, which release sperm and eggs into the water, resulting in external fertilization and a free-swimming larva.

The life history characteristics of freshwater bivalves—the details of reproduction, growth, life span, etc.—are fascinating and diverse. Unionaceans possess a unique, parasitic larval stage, called the *glochidium* (pl. glochidia). In one North American species the glochidium attaches to a particular salamander (the mudpuppy, *Necturus maculosus*); all others parasitize fish. The host provides nutrition to the larva and a means of dispersal. There are several types of glochidia, referred to as "hooked," "hookless," and "axehead." When released by the adult they snap their valves repeatedly, and this action is stimulated by mucus and fish tissues. Glochidia attach to fins, scales, or gill filaments. While not species-specific in their attachment, certain unionaceans typically are associated with certain fish species, due mainly to immune responses that allow some fish species to slough the parasite. Examples are known where restoration of fish host populations was followed by recovery of endangered unionacean populations. For their part, unionaceans attempt to increase the likelihood of attachment by timing the release of glochidia, and by various techniques that anglers can appreciate. Some suspend glochidia on a mucilaginous "fishing" line to enhance encounters with fishes. Others produce a brightly colored, wormlike material, letting it wave from their exhalent siphon, and female lampsilids possess a pigmented extension of the mantle edge that they pulsate above the substratum surface (Fig. 15.5). When a fish attacks, glochidia are instantly released. These fascinating adaptations, combined with high fecundity (reportedly 200,000–17 million glochidia/female/season) help to ensure some success for this improbable life cycle. Even so, successful attachment is just the first hurdle for larval unionaceans. If the host's immune response doesn't eliminate the parasite, then after metamorphosis it must successfully settle into suitable substrate. One estimate is that one in one million shed glochidia become settled juveniles! The previously mentioned high fecundity, coupled with a long life span (easily 10 years, and approaching a century in some instances) helps to counteract these odds. One individual of *Margaritifera margaritifera* from Sweden was reliably aged at 132 years by counting "yearstripes" and chemical analysis of its shell.

Members of the Sphaeriidae are widespread, whereas unionaceans usually have narrow ranges, and this can be related to important differences

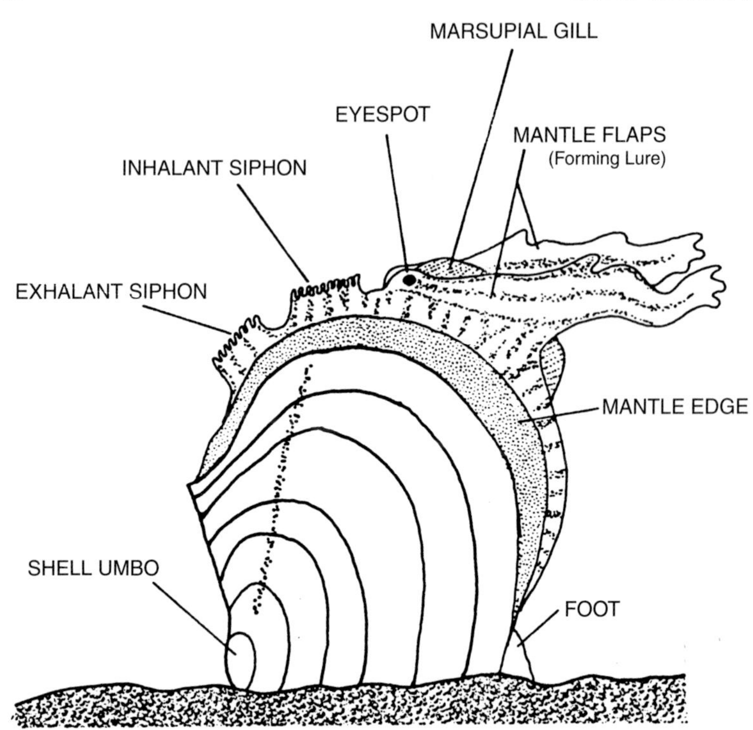

FIGURE 15.5 The mantle flaps of female *Lampsilis ventricosa* mimic a small fish with the eyespot and lateral pigmentation suggestive of a fish's lateral line. When the mantle flaps are disturbed by an attacking fish, glochidia are released. Mantle flap lures are characteristic of the unionid genus *Lampsilis*. (From McMahon, 1991.)

in their reproductive biology, which in turn affects dispersal ability. Unlike unionaceans, whose dispersal is determined by their fish host, juvenile sphaeriids clamp their shell valves onto creatures that can walk or fly to the next river basin, hitch-hiking on the limbs of beetles, feathers of birds, and limbs of salamanders. As a consequence they are less likely to be confined within a river basin.

The Invaders

Two nonnative bivalves are causing serious problems in North American freshwaters, affecting human activities and the native biota. The Asiatic clam was introduced to North America from Southeast Asia in the early 1900s, and now is widespread. Populations occur in West Coast drainages, the southern states, and throughout the region east of the Mississippi River, except for the most northerly states. It disperses effectively, grows rapidly, and seems well suited to unstable, disturbed river habitats. At present there is scant evidence that they pose a serious threat to native bivalves, but the economic damage from fouling of human structures, including the intake pipes used to cool power plants, is considerable.

The zebra mussel (Fig. 15.4f), in contrast, is truly a biological disaster. It first arrived, we think, in Lake St. Clair near Detroit, in the mid 1980s, when an oceangoing ship must have emptied its ballast water and released some zebra mussel larvae. Reported in 1988, it had since spread to at least 19 states by 1999, from Lake Champlain in the northeast, across the Great Lakes, west through the Mississippi River system, and south to the Gulf of Mexico (Fig. 15.6). Eventually, it is expected to occupy most of the continent from southern Canada south. It likely will not occur in soft waters, soft substrates, and the hottest regions of the southern United States. Its free-swimming larval stage likely is disadvantageous in small, fast-flowing streams. In Europe, where it is also a pest, zebra mussels do not occur in rivers less than 10 m wide. However, transformation of free-flowing streams into a series of impoundments works to this animal's advantage. Normally one might think of the adult stage as a poor disperser, but unfortunately humans provide a helping hand. Zebra mussels produce a very strong attachment thread, termed a *byssal* thread, and this allows them to attach to rocks, wood, or the shells of native bivalves that protrude above the substrate surface. Adults attached to the hulls of boats can withstand days of exposure to air, more than enough time to be transported across the country.

This small (up to 4 cm) invader, originally of the basins of the Caspian, Black, and Aral Seas, has profound economic effects. By fouling water intake pipes of power plants it forces costly maintenance and redesign. It affects the recreational value of beaches, cutting the feet of bathers. It has caused a remarkable increase in the clarity of Lake Erie, enabling 15 m visibility where recently divers found objects by touch, to the point where scuba divers are declaring it a mecca for exploring shipwrecks. But with so much of the lake's productivity turning into zebra mussel biomass, one has to worry about the consequences for other residents of the ecosystem. Freshwater mussels, already in serious trouble for reasons about to be discussed, may finally succumb to fouling by zebra mussels, which readily use native unionaceans for attachment. They can form very thick mats, effectively smothering as well as outcompeting native species, which live more within than on the surface of the substrate. In extreme circumstances, zebra mussel densities of 100,000 and a biomass of 15 kg have been reported on a square meter of substrate.

The Decline of Freshwater Mussels

Indian middens provide evidence that species once abundant enough to be harvested by Native Americans no longer can be found within the area of the midden. At least 32 unionaceans are known from middens near the upper Ohio River. A survey in the same area in 1921 found only 25 midden species, while a 1979 survey found only 13 midden species. Today, by every accounting, the freshwater mussels have the dubious honor of being the most imperiled members of the North American biota. The Endangered Species Act (ESA) protects 111 U.S. species of freshwater invertebrates; all but

FIGURE 15.6 Zebra mussel distribution in North America, as of 1999. (National Aquatic Nuisance Species Clearinghouse.)

25 are mollusks. The Nature Conservancy counts nearly 70% of the freshwater mussels in one of their "at risk" categories (vulnerable, imperiled, critically imperiled, and presumed/possibly extinct). An independent team of scientists as of 1993 considered 71% of freshwater mussels to be endangered, threatened, or of special concern (these correspond to categories of the ESA). How did freshwater mussels become so unlucky?

Reasons for the widespread decline and local extirpation of freshwater mollusks are many. At the beginning of the twentieth century mussels were harvested for the pearl and button industries, contributing to the decline of many larger species. In 1912 nearly 200 factories in 20 states in the south and southeast were engaged in pearl and shell products, mainly buttons, from mussels dredged from the larger rivers of the Mississippi system and the Gulf Coast. During peak harvests, 13 million kg of shells were taken from Illinois streams in a single year. Plastics contributed to the demise of the pearl button industry around the beginning of World War II, but another use was discovered. A small pellet of mussel shell, inserted into pearl oysters, is an excellent "seed" for cultured pearls. Since about 1950 and continuing today, mussel shells are shipped to the Orient, crushed, and inserted into pearl oysters, which then secrete layers of nacreas material around the foreign object, forming pearls. In 1999, one company paid a six-figure fine, and several collectors drew jail terms, for participating in illegal harvests that included endangered species of mollusks in several Midwestern states.

Habitat destruction has been another major factor in the decline of freshwater mollusks, many of which prefer shallow, swift-water habitats of coarse sand and gravel. Silt and sediments from agricultural activity have choked many formerly clean riffles, and dams have drowned many more beneath deep, slow water and the muddy depths of reservoirs. Pollution became a factor with the first clearing of the land, introducing silt into waterways, and the first industrial wastes, from tanneries, sawmills, and so on. Pollution from modern-day industries, municipal wastes, and farm chemicals continue to deal additional blows. In the Pearl River, Louisiana, five unionid species not encountered during the previous 20 years reappeared after modern sewage treatment was established. The bacterial breakdown of sewage effluent robs water and especially the surface of the streambed of oxygen, and this is particularly serious for sedentary, infaunal organisms like the bivalves.

Because they are localized, freshwater mussels can be vulnerable to chemical spills. A truck accident that dumped toxic materials into the Clinch River, Virginia, killed virtually all aquatic life in a 10 km stretch of river, including many endangered mussels. To date, this remains the largest "take" of imperiled species since the Endangered Species Act was authorized in 1973.

The status of the freshwater molluskan fauna, and its uncertain future, should concern us all. It might inject a note of optimism if we could say that only recently has this come to our attention, but sadly, that is not so. One can go back 20, 50, or more years into the scientific record and find accounts by malacologists that testify to an early recognition of the problem. Nor will culturing of these species—the aquatic equivalent of captive breeding—provide much help here, because the specific habitat needs and hosts for the glochidial stage demand too complicated a culture system for widespread use. Native species need natural habitat and intact ecosystems. Protecting habitat is the issue, and the highly endangered and species-rich river snails and mussels tell this story unambiguously.

MOLLUSKS AND THE RCC

Mollusks fit the RCC in at least one respect: mussels and clams reach their greatest diversity and abundance in the higher-order reaches of large rivers, where abundant FPOM is available for suspension feeding. However, FPOM is ubiquitous along the river continuum, and for this reason bivalves occur in smaller streams as well as larger rivers, although they are not abundant in headwater streams. River snails that feed on algae should be more abundant in mid-order and unshaded streams. We suspect this is true, but more research is needed to confirm this expectation.

Recommended Reading

Pennak, R. W. (1989). *Fresh-Water Invertebrates of the United States*. 3rd ed. John Wiley & Sons, New York.

Thorp, J. H., and Covich, A. P. (1991). *The Ecology and Classification of Freshwater Invertebrates*. Academic Press, New York.

Williams, J. D., Warren, M. L. Jr., Cummins, K. S. Harris, J. L. and Neves, R. J. (1993). Conservation status of the freshwater mussels of the United States and Canada. *Fisheries* 18(9):6–22.

Suggested Web Sites

<www.inhs.uiuc.edu/cbd/musselmanual> This site of the Illinois Natural History Museum provides a superb photo album of freshwater mussels.

<www.ansc.purdue.edu/sgnis> The Nonindigenous species site of the National Sea Grant College Program is an excellent source of information about zebra mussels and other invasive species.

CHAPTER 16

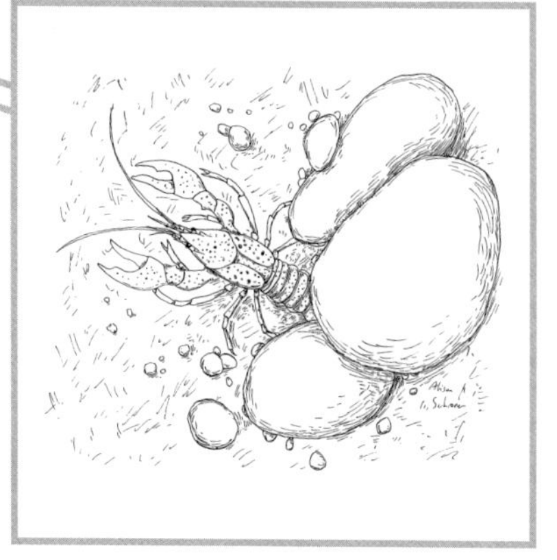

Crustaceans

INTRODUCTION

Of the three groups of common crustaceans found in rivers, two are small in size, but easily visible, and the third is easily seen and known to just about everybody who has explored freshwater habitats. The order Isopoda includes the small animals known as aquatic sow bugs; these have their familiar counterparts in the terrestrial environment, known as pill bugs and found in moist earth under rocks and other objects. Members of the order Amphipoda are somewhat similar in size and shape to the isopods and are commonly known as scuds, freshwater shrimp, or sideswimmers. The order Decapoda includes the larger and more conspicuous crayfish and shrimp. All three orders are in the class Malacostraca of the subphylum Crustacea.

Before getting into the characteristics and ecology of these three groups, some general information about crustaceans is in order. The great majority of the some 40,000 known species are marine; only about 10% live in freshwater. The body is covered by a hard *exoskeleton* made of chitin, and consists of three general regions—the head, thorax, and abdomen. The head, or *cephalic* region, consists of several fused segments and may further be fused to one or more segments of the *thorax* to form a *cephalothorax*. In many species the cephalothorax is completely or partly covered by a shieldlike *carapace*. The last segment of the *abdomen.* is referred to as the *telson*. Crustaceans differ from the insects in that they have two sets of antennae, rather than one.

The isopods and amphipods can be found in various habitats, including still pools, riffles, hot springs, and within the hyporheic zone. Crustaceans, in general, are omnivorous and carnivorous; we will have more to say about this later. They have well-developed chemosensory systems by which they are able to find food and mates. Other senses are also highly developed. Crustaceans generally require cold, clean water where dissolved oxygen concentrations are high. This, coupled with the fact that they feed on a wide variety of food material and concentrate chemicals within their bodies, makes them ideal test organisms for biomonitoring studies and for use as "indicator organisms" to foretell possible habitat degradation.

With the above as a general introduction to these three groups, let us now look at each in more detail.

Isopoda

About 130 freshwater species of sow bugs are found in the United States. They occur in springs, spring streams, rivers and streams, and in the interstitial waters of the hyporheic zone.

Isopods are flattened from top to bottom (dorsoventrally) (Fig. 16.1). The seven thoracic segments behind the cephalothorax are expanded laterally to form a plate over the basal segments of the seven walking legs. They move by a slow, crawling movement. Total length ranges from 5 to 20 mm, and they vary in color from blackish to yellowish and red. Isopods are omnivorous scavengers; they are known to feed on dead and injured aquatic animals and decaying aquatic vegetation. Juveniles feed largely on microbial food such as algae and bacteria. Adults shift to larger food items. After hatching and leaving the female, young isopods pass through at least 15 growth stages (instars) before becoming adults. Life span is about one year or less.

Freshwater isopods are secretive, remaining under rocks, debris, and vegetation and rarely venture into open waters. As with amphipods, they are essentially restricted to regions of constant, cold water. They may be numerous in the "recovery zone" of polluted streams where stream conditions are returning to normal. Although not as numerous, generally, as are the amphipods, large masses in the thousands of the isopod *Caecidotea* have been found in small streams, probably as a result of reactions to current velocity.

Amphipoda

There are about 150 freshwater species of amphipods in North America. They are found in most aquatic habitats—rivers, streams, spring streams, and subterranean waters.

Amphipods are flattened from side to side (laterally) and range from 5 to 20 mm in length. They have much in common with the isopods in

FIGURE 16.1 Line drawing of isopod (*Caecidotea*). (From Pennak, R. W. (1989). Fresh-Water Invertebrates of the United States, 3rd ed. Copyright © 1989 John Wiley & Sons, Inc.]

terms of body structure; the major difference is in direction of body compression—isopods are compressed top to bottom, amphipods from side to side (Fig. 16.2). Amphipods are more active at night, crawling and walking by flexing and extending their bodies among the debris and particles of the stream bottom. They also swim just above the bottom sediments, often turning over on their sides or back—the reason that they are often called "sideswimmers." More is known about the life history of amphipods than for isopods. Females of most species produce one brood per year, but some species produce multiple broods during an annual cycle. Newly hatched young have all of the adult appendages, and *Hyalella azteca* goes through at least nine instars. It may undergo as many as 15 to 20 molts during its adult period, each of which is completed in less than an hour.

Amphipods are omnivorous scavengers, feeding voraciously on all kinds of plants and animals, although they are not known to attack living animals. They also feed on the periphyton community. They have well-developed chemosensory ability, which helps them in locating their food.

Three common genera are *Hyalella, Gammarus*, and *Crangonyx*, which are widely distributed in clean, cold waters. Like isopods, they are secretive and prefer living under rocks and debris or in the hyporheic zone. They can be very abundant; reports of finding *Gammarus* populations exceeding

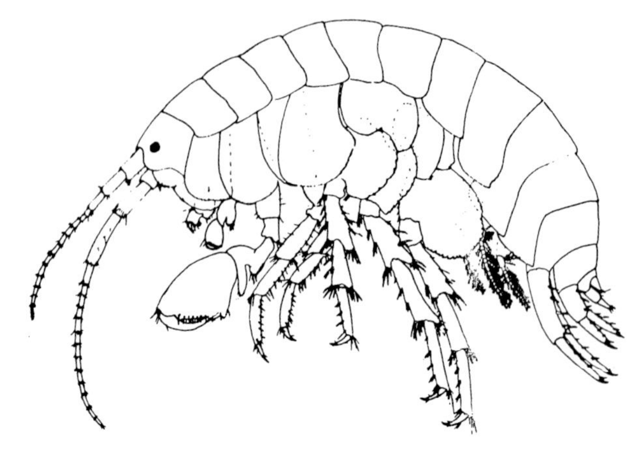

FIGURE 16.2 Line drawing of amphipod (*Hyalella* × 14). (From Pennak, R. W. (1989). Fresh-Water Invertebrates of the United States, 3rd ed. Copyright © 1989 John Wiley & Sons, Inc.]

10,000 per square meter are known, and one of us (CEC) has collected them with a small aquarium net in such high numbers that you could literally scoop them from the net by the handful! They are strictly shallow water organisms; only *Hyalella azteca* has been found in water deeper than 1 m.

Amphipods are important trout foods in many of the productive fisheries found in the tailwaters below dams. A good example is the Green River below Flaming Gorge Dam. The constant cool temperature of the water released from the dam has resulted in excellent populations of aquatic insects and scuds—and a large population of big trout.

It is obvious from the above that isopods and amphipods have much in common in terms of size, general body form, feeding habits, and habitat. This is particularly true in the fact that many species of both groups are found in subterranean stream systems in caves and other underground locations. Many of these forms have unique adaptive characteristics including lack of body pigmentation, reduction or absence of eyes, and elongation of some body parts.

Decapoda

Amphipods and isopods may be unfamiliar to those who visit streams on a casual basis, but certainly the decapods, crayfish and shrimp, are familiar to almost everybody. They go by many names—crayfish, crawfish, crawdads—and many people who have waded and turned over rocks in the shallows of ponds or streams have encountered them. Crayfish (families Astacidae and Cambaridae) (Fig. 16.3) are common in streams in the central, eastern, and southern United States. The freshwater shrimp (family Palaemonidae) (Fig. 16.4) have a more restricted distribution in the United States, but are important constituents of stream fauna in tropical rivers and

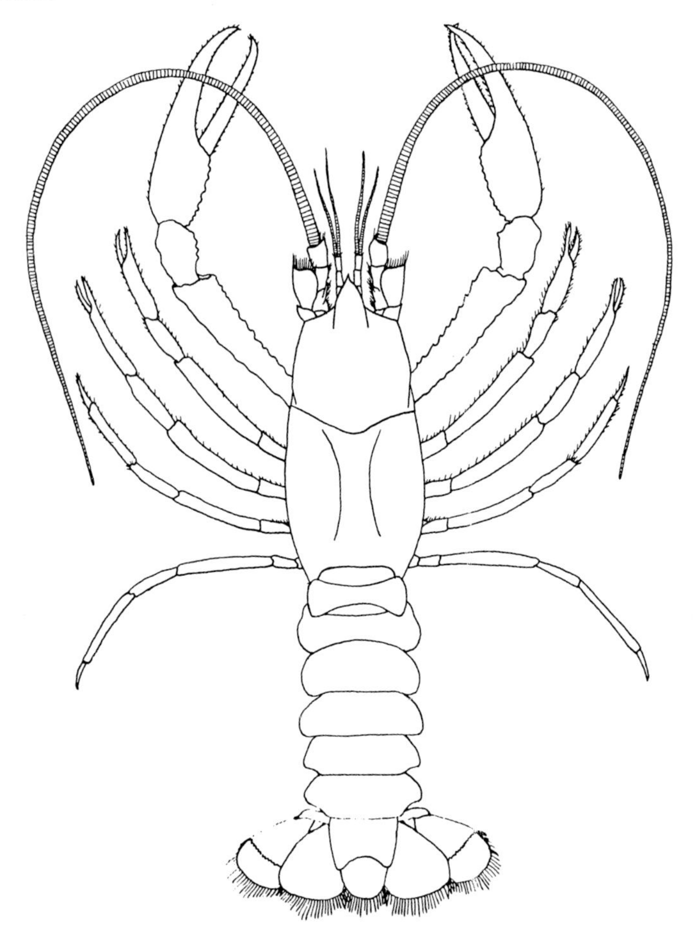

FIGURE 16.3 The crayfish *Procambarus acutus*. [From Pennak, R. W. (1989). *Fresh-Water Invertebrates of the United States*, 3rd ed. Copyright © 1989 John Wiley & Sons, Inc.]

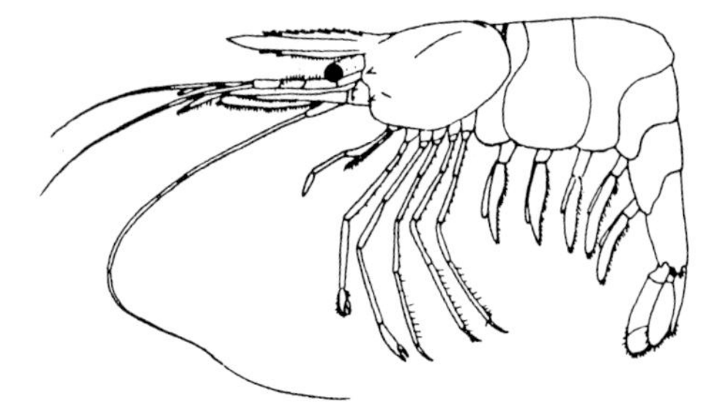

FIGURE 16.4 The shrimp *Atyoida bisculata*, an endemic atyid shrimp from Hawaii. [From Pennak, R. W. (1989). *Fresh-Water Invertebrates of the United States*, 3rd ed. Copyright © 1989 John Wiley & Sons, Inc.]

streams (see the description of tropical streams in Chapter 11). They are largely absent from greater portions of the Rocky Mountain region, although one species from the Pacific Northwest has migrated to headwaters of the Snake, Yellowstone, and Missouri rivers and into the Great Basin.

Body shape is mostly cylindrical and the animals have stalked, compound eyes which are movable. They range from 10 to 150 mm in length. Species found in fast-flowing waters may have compressed or depressed bodies. The head and thoracic segments are fused into a large cephalothorax covered with a carapace, which extends over the gills. The anterior end of the carapace terminates in a sharp point called a *rostrum*. The abdomen has six segments, the first five of which bear small leglike structures called *swimmerets*. In the females, they are used as places of attachment for the incubating eggs; in the males, the first two pairs are modified for clasping during mating. The last abdominal segment has a pair of flat, platelike projections called *uropods* and the telson. There are 19 pairs of appendages, the most noticeable of which are the two pairs of antennae and the five pairs of walking legs. In the crayfish, the first pair of walking legs is modified into large pincerlike grasping claws called *chela*; shrimp have small pincers at the end of their walking legs, but do not have these large, modified appendages.

Body coloration is variable, ranging from blackish and brown to bright colors such as red, blue, orange, and green. Coloration is often related to background color of the habitat and crayfish have the remarkable ability to change their pigmentation to match that of their surroundings.

Crayfish move about slowly by walking and climbing on the last four pairs of walking legs and may move forward, sideways, or backward. When alarmed, they move rapidly backward by flexing the abdomen and tail fan. They may also leave the stream bottom for short excursions into the water column using abdominal contractions to propel them.

Decapods are omnivorous, feeding by scavenging on a wide variety of food items, and play an important role in processing and transforming energy in rivers and streams. Food items include aquatic vegetation and carrion, and they have been known to prey upon snails, insects, and small fishes. The large chelate claws of crayfish are used to grasp, crush, and pick up food before passing it to smaller appendages, which strain and break up the food particles. Young crayfish are detritivores and herbivores, filtering suspended FPOM and grasping larger particles. Adults are more active predators and grazers on larger food items, but they, too, graze microbes from solid objects and shred vegetation. The presence of predators can alter the feeding strategy of crayfish, reducing both activity and extent of movement from cover. Crayfish also appear to be able to use their highly developed sense of chemical detection to identify and locate food. Shrimp function as grazers, utilizing the dense periphyton community found where light penetrates the stream canopy. Some interesting experiments were done in Puerto Rican streams to determine the magnitude of influence that the high densities of atyid shrimp had on the existing periphyton community. Shrimp were excluded using small (32 cm diameter by 2.5 cm high) electrified "fences," from areas on the top of rocks. When shrimp were absent, periphyton mass increased, as did the deposition of organic matter. Feeding as they do, crayfish and shrimp both act as decomposers, breaking down and altering the organic matter and releasing fecal material into the ecosystem. Both are actively fed upon by other predators, including fishes, snakes, etc.

Fertilized eggs are carried on the swimmerets in berrylike clusters. Newly hatched instars differ in appearance from adults; the cephalothorax is enlarged and rounded and the rostrum, too, is enlarged (Fig. 16.5). Young remain attached to the female's swimmerets through the second instar and leave during the third. Crayfish hatching in spring go through 6–10 molts by autumn; shrimp have 5 to 8 larval molts. Adult body characteristics develop during the larval molts. Growth rates of adults are highly variable, even among inhabitants of the same stream or river; this is probably related to activity and food availability. After the first mating season, most crayfish only undergo 2 to 4 molts before dying. Normal life span for crayfish is usually less than 24 months.

Crayfish remain hidden under rocks, stones, and other debris on the bottom of rivers and streams; some dig burrows in the bank near the shore-

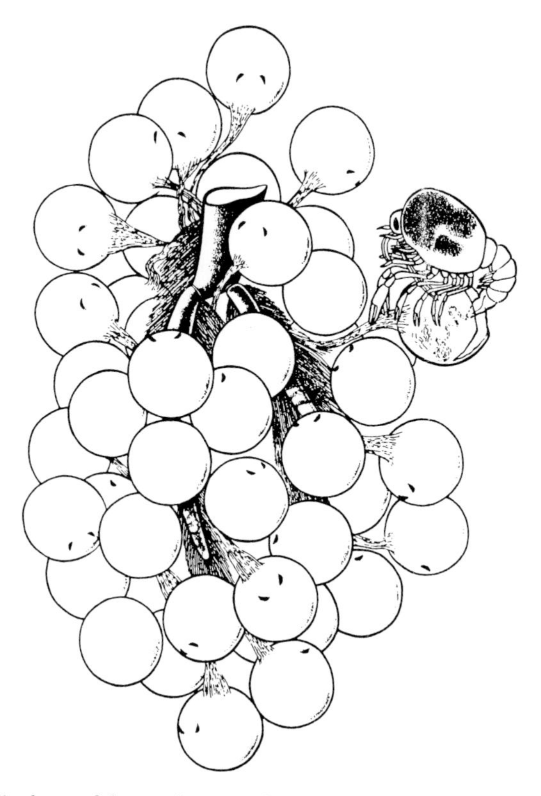

FIGURE 16.5 Single crayfish egg cluster with one hatched young. (From Pennak, R. W. (1989). Fresh-Water Invertebrates of the United States, 3rd ed. Copyright © 1989 John Wiley & Sons, Inc.]

line if the ground is suitable. They are more active from dusk to dawn, although immature animals regularly crawl about the bottom and in vegetation during the day. They are usually found in waters less than 1 m deep. They are found in habitats having a wide range of temperature, pH, and CO_2 concentrations, although river and stream species appear to be less tolerant than those living in standing waters. Home range is usually less than 30 m in streams. Population densities vary with habitat and species. Shrimp are found in habitats where the water is slow flowing and vegetative cover is dense.

The shrimp are represented by two genera in North America, *Palaemonetes* and *Macrobrachium*, with about 10 and 5 species, respectively. Crayfish, on the other hand, have about 319 species in North America north of Mexico. Common genera of crayfish include *Procambarus*, *Cambarus*, and *Orconectes*. It should be mentioned that many representatives of both the shrimp and the crayfish are found in subterranean caves.

Although neither crayfish nor shrimp are territorial, they are aggressive and will defend their location in coarse sediments. They demonstrate a range of reactions to potential competitors, including increased alertness, threats, combat, submission, avoidance, and escape. Some of these are reflected in movements of the antennae or other body parts. Crayfish also exhibit a variety of responses to predators other than hiding. They may assume a defensive posture, which varies with species, display the claw in various positions, and position the tail to be able to use it for rapid backward movement.

Decapods have a highly developed sense for detecting chemicals in the environment. Chemical cues are thought to be used to distinguish sexes and initiate mating when the proper chemical cues are released by the female. This sense is also likely involved in regulating population distribution and hybridization when encountering new populations.

ENVIRONMENTAL PROBLEMS

Several environmental problems have resulted from the widespread introduction of nonnative crayfish into various habitats. Intense competition for available resources has resulted in species displacement due to reproductive interference, alteration of aggression patterns, and changes of competitive dominance. Changes in community composition, obviously, can have potentially severe effects on the energy flow patterns within an ecosystem.

In many cases, these competitive interactions between native and nonnative crayfish populations have resulted in population reductions to the point that several species of crayfish have become endangered, and more will likely be added to the list. It has been estimated that nearly half of the 338 native crayfish species in North America are in need of serious conservation efforts. The Nature Conservancy estimates that 36% of the crayfish fauna of North America is extinct and an additional 26% is vulnerable to extinction. Much of this is related to the problem mentioned in the preceding paragraph. Despite this need of conservation efforts, there are currently only 4 species recognized as endangered under the Endangered Species Act of 1973, although several other species receive some level of state protection.

SOLUTIONS

Several things can be done to reduce the likelihood of endangered species of crayfish becoming extinct. Captive breeding and reintroduction of endangered species is a reasonable approach. Habitat protection programs can lessen the chances of losing species from destruction of their existing habitat and can protect habitat from invasion by more competitive nonnatives. A third way to reduce the number of possible competitive interactions is to prohibit the use of live crayfish as fish bait, thus lessening the potential for accidental introductions of nonnative species into habitat already occupied by a less aggressive competitor. Prohibiting the use of live crayfish as bait may not be possible. However, educating anglers as to the potential destructiveness of throwing away their unused live bait would be helpful.

CRUSTACEANS AND THE RCC

The crustaceans do not fit as neatly into the predictions of the RCC as do the insects. This is for several reasons, foremost among which are the facts that some of them shift their feeding habits at different life stages and that they are omnivorous.

For instance, juvenile crayfish function largely as filter feeders, but shift to larger diet items as they mature. Indeed, even adults often filter particles from the water. Thus, they also fit into more than one functional feeding group. The fact that crustaceans are omnivorous and scavenge extensive food resources ranging from vegetation to carrion to active predation means that they can capitalize on some kind of food resource almost anywhere along the stream continuum. Amphipods are important shredders in many eastern streams, but are found throughout the stream continuum, not just where CPOM is prevalent in the headwater reaches.

Information for this and the following chapter was drawn largely from two excellent books: *Fresh-Water Invertebrates of the United States*, 3rd ed. by R. W. Pennak (1989) and *Ecology and Classification of North American Freshwater Invertebrates* edited by J. H. Thorp and A. P. Covich (1991).

Recommended Reading

Pennak, R. W. (1989). *Fresh-Water Invertebrates of the United States*, 3rd ed. John Wiley & Sons.

Thorp, J. H. and Covich, A. P. (eds.). (1991). *Ecology and Classification of North American Freshwater Invertebrates*. Academic Press, San Diego.

CHAPTER 17

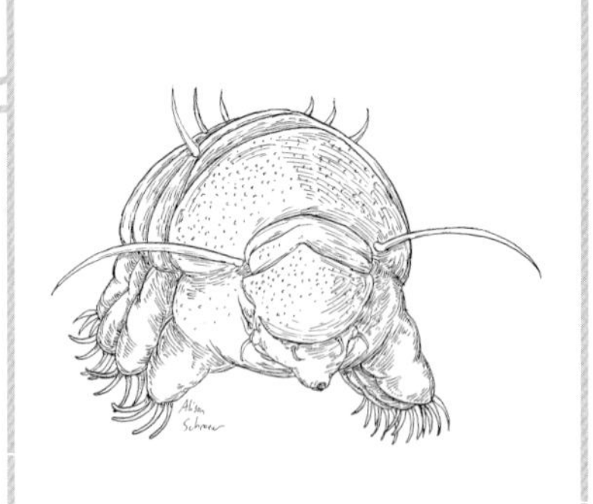

Other Invertebrates

INTRODUCTION

In previous chapters we described the macroinvertebrates, which includes insects, crustaceans, mollusks, and other groups found in rivers and streams and usually visible to the naked eye, at least in their later instars or growth stages. In this chapter we examine a diverse group of smaller invertebrates, collectively called the *meiofauna* (pronounced "my-oh-fauna"). Aquatic scientists usually define the meiofauna as any organisms that pass through a 0.5 mm sieve and the macrofauna as anything that does not. This is a useful working definition, although some early stages of macroinvertebrates change from meio- to macro- as they grow. The primary focus of studies of the meiofauna is on the several groups of organisms that are in this small-size category for all or most of their life cycle, and so constitute a functional category of tiny secondary producers. For the most part, they can be found within the biofilm and periphyton layers, within the interstitial waters of the stream bottom (where they are referred to as *hyporheic fauna*), in bottom muds, and free-floating in the water column. We also include one group which is small, but visible, doesn't conveniently fit into any of the other chapters, yet plays an important role in rivers and streams. These are the freshwater worms (phylum Annelida), that is, the oligochaetes that were mentioned previously in relation to the outbreak of whirling disease in trout streams.

This chapter also includes two groups of crustaceans that, because of their size and habitat, are usually included with the meiofauna in ecological studies rather than with their larger cousins. These are the harpactacoid copepods and ostracods, which are important members of the community dwelling among the interstitial waters of the stream bottom.

Probably one of the best places to find the greatest diversity of meiofauna is within the film of periphyton and microbes that coat rocks and other surfaces in streams. In previous chapters, we have described this community as consisting mainly of algae, and this is true, although bacteria, fungi, and cell exudates are combined in a matrix often referred to as a biofilm. Dwelling within this matlike community a wide range of meiofauna forms can be found—protozoans, rotifers, worms, and others—that function importantly in the transformation of energy within the stream ecosystem. A wide diversity of meiofauna also occurs within the interstitial waters among the rocks, gravels, and other materials composing the stream bottom. Many of the forms dwelling here are negatively phototropic, that is, they avoid light and prefer the darkness found deep within the stream bottom. As with those meiofauna found within the periphyton, organisms of the hyporheic regions perform important functions in terms of energy conversion and transfer. Certain meiofauna, such as the oligochaetes, are found in the muds of stream bottoms, and still others are free-floating in the water column.

Although members of the meiofauna are small, this does not mean that they are unimportant in terms of production of organic matter for other members of stream ecosystems. Many of them have extremely high reproductive rates, and this, coupled with the very high numbers that may be present, results in significant secondary production—food for other consumers. We will speak more of this when we describe the "microbial loop" below.

MAIN GROUPS OF THE MEIOFAUNA AND THEIR ECOLOGY

Although their are several taxonomic groups that make up the meiofauna, five are probably most important in terms of their importance in stream ecosystems. These are the Protozoa, Rotifera, Nematoda, Annelida, and microcrustacea. We will emphasize these five groups in our discussion and then mention briefly the minor taxa.

Protozoa

Protozoans are microscopic, single-celled organisms; however, several species may form colonies consisting of many to thousands of individuals. They occur in many places within flowing waters and may be present in surprisingly large numbers. The ciliates, those having short, hairlike structures on their surface, are among the most commonly found protozoans in

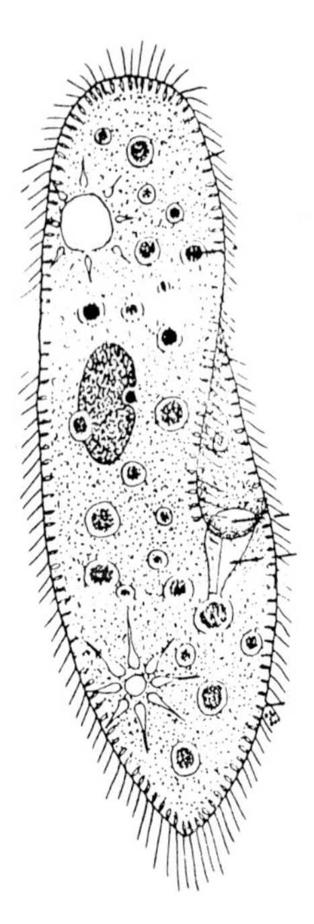

FIGURE 17.1 Line drawing of *Paramecium*. [From Pennak, R. W. (1989). *Fresh-Water Invertebrates of the United States*, 3rd ed. Copyright © 1989 John Wiley & Sons. Reprinted by permission of John Wiley & Sons, Inc.]

rivers and streams. A typical, well-known ciliate, *Paramecium*, is shown in Fig. 17.1; most readers were first introduced to this organism in high school biology class. Others may have flagella, hairlike structures longer than cilia, or no appendages at all.

Ecologically, protozoans are important constituents of river and stream food webs. They are heterotrophic, feeding on dissolved organic matter, bacteria, detritus, algae, rotifers, and other protozoans. Those feeding on bacteria and detritus are the most numerous in terms of number of genera, while those utilizing DOM are the fewest. Thus, protozoans function as both consumers and also release unassimilated organic material into the ecosystem. Protozoans are probably important contributors to the regeneration of nutrients, but the magnitude of their role in this respect is controversial and probably not as important as that of the bacteria. The great majority of protozoan species are widely distributed throughout the world due to their small size, production of cysts, and ease of distribution from one place to another. Protozoans are fed upon by larger members of the meiofauna, which, in turn, become food for macroscopic invertebrates.

Rotifera

Rotifers are perhaps the most numerous taxonomic group of animals characteristic of freshwaters. However, this statement is more applicable to freshwater habitats as a whole rather than to rapidly flowing rivers and streams. Nevertheless, they are important constituents of the meiofauna. They occur as free-living species, sessile species, and within colonies and clusters.

Rotifers are microscopic heterotrophic organisms which feed on bacteria, detritus, and algae; in fact, large rotifers may ingest smaller rotifers. Feeding occurs by the creation of a current produced by a ring of cilia which sweeps the suspended bits of organic matter into the region of the mouth. Figure 17.2 shows several typical rotifers and their cilia, which aid in feeding. Ingestion rates are quite high and rotifers convert a high percentage of the food they ingest into body mass, thus contributing significant amounts of food for consumption to the food web. As consumers of bacteria, algae, and other microscopic bits of organic matter, rotifers and protozoans are thus competitors for these food resources.

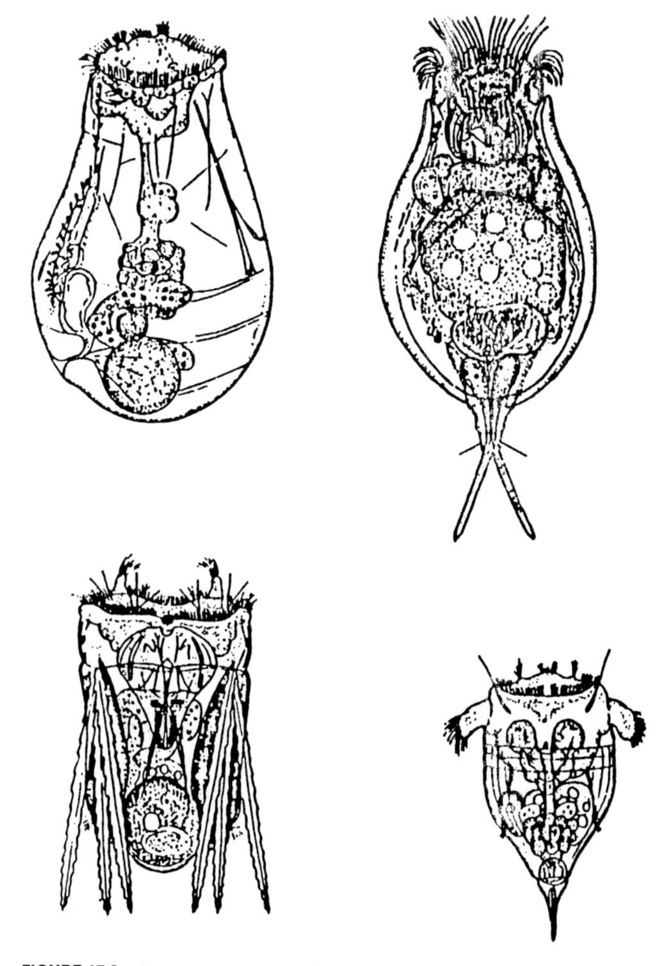

FIGURE 17.2 Representative rotifers. (From Wallace and Snell, 1991.)

Nematoda

Nematodes are round worms found in any sample of sand, mud, debris, or other vegetation collected from the bottom or wet margins of streams and rivers, living mainly in the upper two or three centimeters of sediment. They are often the most abundant benthic animal found in freshwater ecosystems. Nematodes are highly adaptable from both an ecological and physiological aspect; they are found in extreme conditions, including polar ice and hot springs. Nematodes occurring in rivers and streams are less than a centimeter long and move by whiplike movements of their bodies. They don't get far this way, but their motions tend to agitate the sediments where they live, stirring up food particles and mixing the soil. Figure 17.3 shows the typical body form of nematodes.

Nematodes feed mainly on bacteria, algae, protozoans, and other unicellular organisms. They are among the most numerous animals feeding on both primary decomposers (e.g., bacteria and fungi) and primary producers.

Because of their small size and difficulties with accurate identification, nematodes have been largely ignored by aquatic ecologists, an unfortunate oversight because their high numbers and high production rates indicate that they play an important role in the functioning of aquatic ecosystems.

Annelida

Many different groups of organisms compose the phylum Annelida, but the group most important to river and stream ecosystems is the class Oligochaeta, commonly called the aquatic earthworms or sludge worms

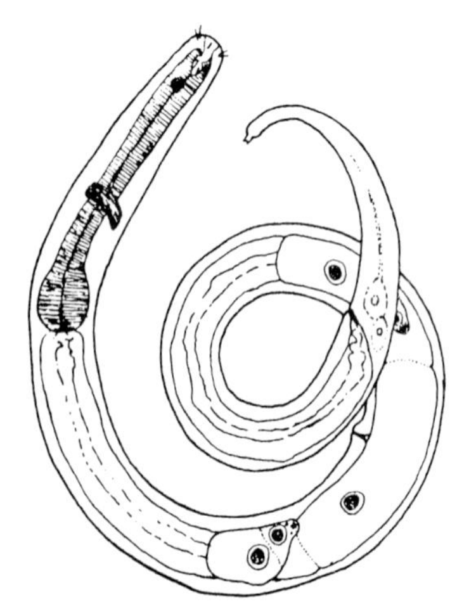

FIGURE 17.3 Line drawing of nematode (*Achromadora*). [From Pennak, R. W. (1989). *Fresh-Water Invertebrates of the United States*, 3rd ed. Copyright © 1989 John Wiley & Sons. Reprinted by permission of John Wiley & Sons, Inc.]

because of large, dense populations found where organic enrichment (pollution) of the sediments is high. Indeed, if you look at them closely under low magnification, they look much like common earthworms, although much, much smaller; in fact, their anatomy, physiology, and behavior is nearly identical to their larger terrestrial relations.

As mentioned, they are common inhabitants of the mud and debris of freshwater environments everywhere. Many species are tube-dwellers, living with their heads in tubes within the sediments where they feed by ingesting the sediments and with their tails waving in the water above the sediments (Fig. 17.4). In this fashion, they obtain oxygen from the water through the body wall of the tail. They are able to thrive under conditions of low oxygen concentrations in the water, indicated by their ability to form dense colonies under polluted, low-oxygen conditions.

Ecologically, oligochaetes perform similarly to their terrestrial cousins—agitating and ingesting the sediments to obtain the bacteria, algae, and other organic material that forms their basic food items. They, in turn, are fed upon by animals which essentially ingest periphyton or bottom sediments—places where the oligochaetes live. Thus, oligochaetes are ingested "by accident" rather than sought out specifically.

Microcrustacea

Two groups of microcrustaceans are commonly found inhabiting the interstitial water spaces between stream bottom particles—rocks, gravel, and

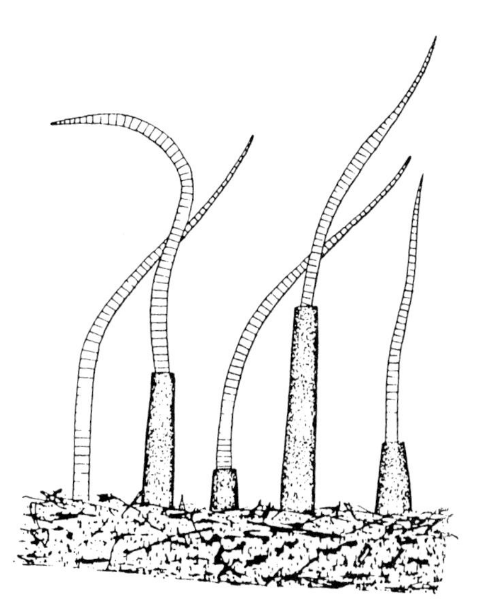

FIGURE 17.4 *Tubifex* colony. [From Pennak, R.W. (1989). *Fresh-Water Invertebrates of the United States*, 3rd ed. Copyright © 1989 John Wiley & Sons. Reprinted by permission of John Wiley & Sons, Inc.]

sand. The first of these are the harpactacoid copepods, small crustaceans with many affinities to the open-water dwelling calanoid and cyclopoid copepods. Some cyclopoid copepods may be found within the stream sediments, but the harpactacoids are prevalent in flowing waters and we will emphasize these here. The second group is the ostracods, commonly called "seed shrimps."

Harpactacoid copepods (Fig. 17.5) are small, usually less than 2 mm long, and can range in color from drab gray to brilliant red, orange, or purple. They are largely restricted to the benthic regions of rivers and streams. They move by crawling or swimming among the bottom sediments. Their mouth parts are adapted to scraping, seizing, and raking food from the sediments—mainly algae, bacteria, fungi, rotifers, other small animals, and organic debris. Thus, they function as herbivores, carnivores, and omnivores in terms of trophic position in the food web. They are also known to use DOM, and thus become important members of the "microbial loop" which will be described later.

Harpactacoids can be found in amazingly large numbers. For instance, one study reported up to 277 copepods in 10 cm^3 of sand, and they have been found in high numbers up to 6 cm deep in the sand. Numbers in a second-order North Carolina stream ranged from 3000 to 18,000 per square meter! Some common genera include *Attheyella, Canthocampus*, and *Bryocampus*, show in Fig. 17.5.

Fossil ostracods have been found in Cambrian marine sediments and are the oldest known microfauna. Ostracods (Fig. 17.6) are widely distributed and individual species are not restricted to any one kind of habitat, such as ponds or streams, but appear to be able to adapt to widely differing ecological con-

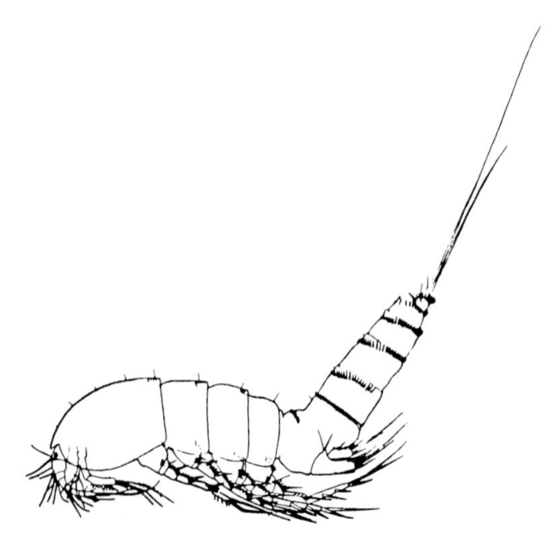

FIGURE 17.5 Line drawing of copepod (*Bryocampus*). [From Pennak, R. W. (1989). *Fresh-Water Invertebrates of the United States*, 3rd ed. Copyright © 1989 John Wiley & Sons. Reprinted by permission of John Wiley & Sons, Inc.]

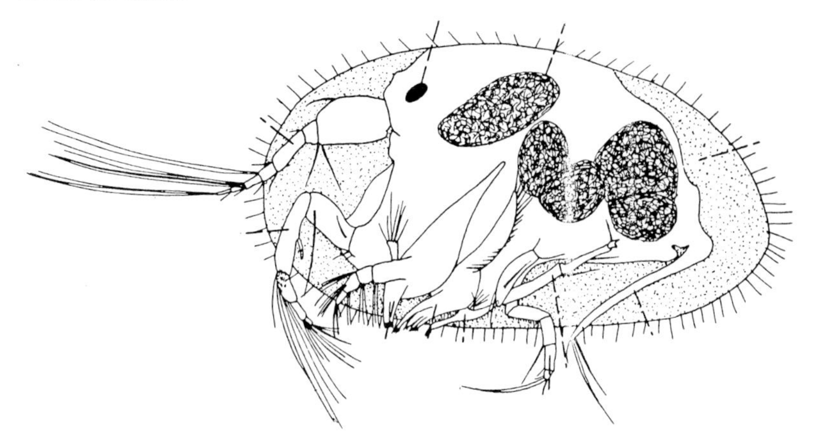

FIGURE 17.6 Line drawing of ostracod (*Cypricercus*). [From Pennak, R. W. (1989). *Fresh-Water Invertebrates of the United States*, 3rd ed. Copyright © 1989 John Wiley & Sons. Reprinted by permission of John Wiley & Sons, Inc.]

ditions. One species, *Cyprinotus incongruens*, has been found in sites ranging from muds to swift-flowing streams. Ostracods have not been as thoroughly studied as other microcrustacean forms; in fact, one of us (CEC) found a species in a spring stream in Washington State in 1963 whose previous known distribution was Yucatan and Trinidad! Obviously, this wasn't a rare occurrence; it just meant that nobody had looked very hard for them in other places. *Candona acuta, Potamocypris variegata, Ilyocypris braadyi, Ilyocypris gibba*, and *Cypria obesa* are considered to be primarily flowing water species.

Ostracods can be found in rooted vegetation, periphyton, mud, sand, and organic debris. They are small, usually less than 1 mm long, and have been given the common name of "seed shrimps" because they resemble small seeds to the unaided eye. Closer examination shows that they are similar to small clams in that they have two "valves" which look like the shells of clams. They range in color from clear to white to brighter colors such as red and green. They move about mainly by beating movements of the first and second antennae and crawling between sediment particles. Their main food includes bacteria, fungi, algae, and fine detritus, and thus, ecologically, they are omnivorous scavengers.

MINOR GROUPS

There are several other groups of organisms found in rivers and streams that play varying ecological roles, although usually minor ones. Under certain conditions, they can be significant contributors to either the total biomass present (e.g., sponges), but rarely are they important in the overall energy transfer within the food web.

Of the groups playing minor roles in river and stream ecosystems, five can be considered macroscopic in size, while the others are microscopic.

The macroscopic forms include the freshwater sponges (Porifera), freshwater hydroids and jellyfish (Coelenterata), flatworms (Turbellaria), the horsehair worms (Nematomorpha), and the mosslike animals (Bryozoa). Sponges can be found in large colonies attached to stones, sticks and other objects, and can be significant in terms of total community biomass. Living sponges are rarely fed upon by other consumers, so they do not play a major role in energy transfer through the food web despite their abundance. Free-living freshwater coelenterates include the well-known genus *Hydra* and the freshwater jellyfish *Craspedacusta sowerbyi*. The turbellarians are the common planarians that most readers became acquainted with in high school biology, the little "cross-eyed" worms that we cut in half and marveled at as each half became a whole worm. Horsehair worms resemble just what they are called—long, slender worms often found tightly coiled into a ball. Bryozoans are found as matlike communities resembling patches of moss or as large gelatinous masses attached to piers, rocks, or other solid objects.

Of the microscopic forms, gastrotrichs (Gastrotricha) closely resemble some rotifers and are found in association with them; they are true members of the meiofauna. Tardigrades (Tardigrada) are often called "water bears" because they have four pairs of "legs" and a rather "stumpy" body and give the appearance of a hunched over bear. They, too, live in close association with other members of the meiofauna.

GENERAL ECOLOGICAL CHARACTERISTICS

The meiofauna are often referred to as the "hidden dimension" of the benthos because of their small size and the fact that they dwell in the "unseen" habitats of rivers and streams. One of these habitats which we want to explore a bit further is that of the hyporheos. This is the region below the level of the stream bottom where the water fills the spaces between the stones, gravels, and sand. This is part of the vertical dimension of habitats, but there is another overlooked dimension of the hyporheos and this is the lateral dimension—the water-filled regions among stones and gravels extending laterally from the water's edge. Indeed, thriving populations of stonefly nymphs and other typical riverine biota were found existing up to 2 km laterally from the channel of the Flathead River in Montana, with indications that the hyporheic zone may exist 3 km laterally from the channel and 10 m in depth. Given that the depth to which organisms inhabited the hyporheic zone within the channel of the Flathead River was only about 0.25 m, the volume of the hyporheic zone extending laterally, far exceeded that existing under the river bed. Conceivably, the biomass of organisms inhabiting the lateral regions may be much greater than that within the river channel, although this may be the norm only in large gravel-bed rivers.

The food web of the hyporheos is most likely based on organic matter and biofilms as the primary food sources. This detritus comes from several sources including deposition of FPOM from the water column, particulates originating in the soil profile, and components of the "microbial loop," which will be described next. The organisms found in these hyporheic regions are quite varied, ranging from forms indistinguishable from river organisms to colorless, blind forms that spend their entire lives in the regions where light does not penetrate.

Organisms of the meiofauna, especially the microscopic ones, function in a unique system of food (energy) transfer called the "microbial loop." This cycling of material involves the uptake and transfer of carbon, beginning with DOM, by bacteria and its subsequent consumption by other microscopic organisms. Part of this is recycled within this community, and part of it escapes this loop when the meiofauna are consumed by macroscopic organisms, including macroinvertebrates and zooplankton, and enters the higher food webs of rivers and streams. The importance of this "microbial loop" is still debated among ecologists, but it appears to be of considerable significance in streams and rivers where large amounts of dissolved and particulate organic matter are present. Figure 17.7 depicts the "microbial loop," and a close study of it will help readers understand the above discussion.

An interesting aspect of stream ecology is the relationship between the scouring of the stream bottom by floods, and the rate at which the stream becomes repopulated by invertebrates. One popular theory relating to this repopulation by the meiofauna was that the organisms retreated deeper into the hyporheic zones and laterally from the channel when increasing flows were sensed, and they migrated back upward and toward the stream

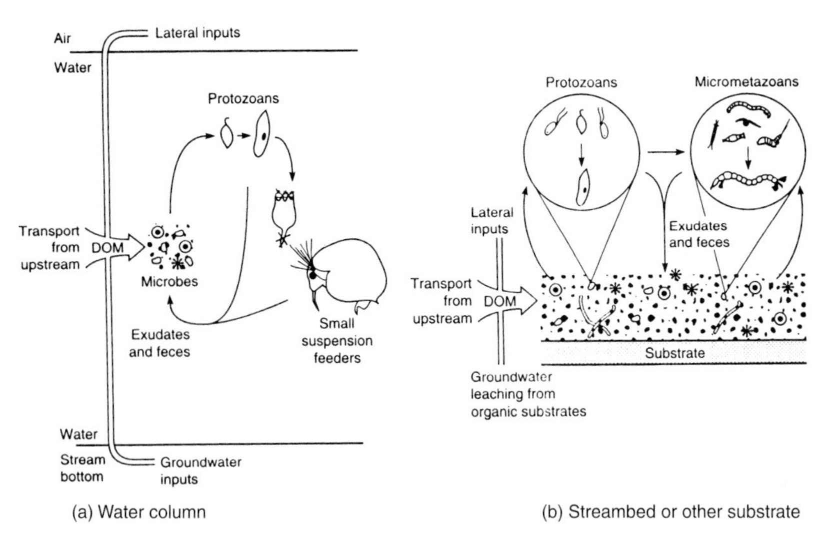

FIGURE 17.7 The "microbial loop." Microbial food webs within the water column of a large river (a) and on the streambed or other substrate of a small stream or larger river (b). Microbes within the water column, primarily bacteria, are consumed by flagellates and ciliates, which in turn are grazed by zooplankton such as rotifers and microcrustaceans. Exudates, waste products, and decomposing consumers are likely to be utilized by microbes, completing the loop. (From Allan, 1995, with permission.)

channel when normal flows returned. Some interesting experiments to test these theories were performed in the field and in laboratory flumes. Although several members of the meiofauna migrated a few centimeters into the hyporheic zone in response to increased flow, and thus might be a partial source of colonists after the disturbance, this movement alone was not sufficient to prevent significant loss of meiofauna during floods. So, where does recolonization originate? Most likely from downstream drift of organisms from upstream reaches which settle and repopulate the hyporheos following floods.

If you have access to a microscope, you may want to look at the amazing numbers and forms of animals described in this chapter. Simply take a scraping of the periphyton from the surface of a rock or some mud from the edge of a stream and make a smear of it on a slide. Some of the larger species can be seen readily under a low-power dissecting microscope, but a high-power compound microscope is needed to see some of the microscopic animals. You will be surprised at what you see!

Recommended Reading

Pennak, R. W. (1989). *Fresh-Water Invertebrates of the United States,* 3rd ed. John Wiley & Sons.

Thorp, J. H., and Covich, A. P. (eds.). (1991). *Ecology and Classification of North American Freshwater Invertebrates.* Academic Press, San Diego.

CHAPTER 18

Fishes

INTRODUCTION

With some 24,000 species distributed in 445 families, fishes are the dominant vertebrates of the world. North America has about 790 native species of freshwater fishes, in 36 families. It turns out not to be easy to report what fraction might be called "riverine." Under that term would we include only obligate river-dwellers or also river specialists that may occasionally be encountered in lakes? How should we count migratory species that use rivers only for reproduction? Even species that we might think of as primarily lake-dwellers will flourish in slow-moving sections of rivers, particularly if the river connects a chain of lakes. Under The Nature Conservancy's habitat categorization, most native fishes of North America qualify as riverine.[1] Figure 18.1 shows a state-by-state listing of numbers of native, riverine fishes as well as the number classified as at-risk of extinction. The number of species that might be considered river specialists is smaller, but still considerable. Cornell fisheries biologist Mark Bain suggests that we use the categories of fluvial specialist, river dependent, and habitat generalist to recognize differing degrees of river dependency.

[1]This statement is based on species counts provided by The Nature Conservancy and the International Network of Natural Heritage Programs and Conservation Data Centers.

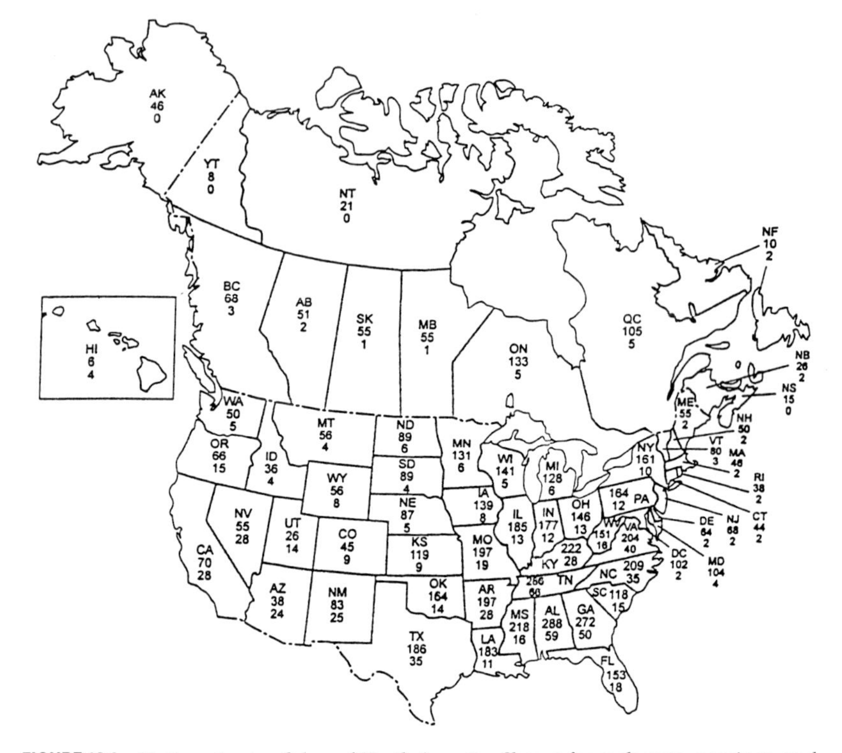

FIGURE 18.1 Native, riverine fishes of North America. Shown for each state, province, and territory are the number of river-dwelling fish species native to that place (top number) and the number of those species that are at risk (bottom number). The largest proportion of imperiled fish species inhabit the arid Southwest; the greatest number of imperiled riverine fishes is found in the species-rich Southeast. Source: The Nature Conservancy and the International Network of Natural Heritage programs and Conservation Data Centers. (From Karr et al., 2000.)

Although a species count according to these finer categories is not now available, it is clear that flowing waters are enormously important for fish populations and their conservation.

Numerous books have been written exclusively about fishes; three we find especially useful are *Fish: An Enthusiast's Guide* by P. B. Moyle (1993), *Fish Watching: An Outdoor Guide to Freshwater Fishes* by C. L. Smith (1994), and *A Field Guide to Freshwater Fishes* by L. M. Page and B. B. Burr (1991).

We have only one chapter to attempt some coverage of a very rich subject area. Our approach is to begin with a few general topics: the anatomy of a fish, their feeding roles, habitat and behavior, and some promising new efforts to develop bioassessment tools based on the diversity of the fish assemblage found at a site. Then we will examine many of the major families of freshwater fishes occurring in rivers and streams of North America. We will provide more detailed coverage of a half dozen or so families likely to be of particular interest to river enthusiasts and say only a few words about others.

THE ANATOMY OF A FISH

The external features of a fish are illustrated in Fig. 18.2, which portrays a generalized bony fish. The adipose fin is characteristic of several fish groups, but absent from most. The dorsal fin may be single or divided into two or three separate parts. In spiny-rayed fishes, the presence of spines associated with different fins is a useful diagnostic character. The tail may or may not be forked. A symmetrical tail is called *homocercal*, an asymmetrical (sharklike) tail is *heterocercal*. The narrow region between the anal fin and the tail is the *caudal peduncle*.

Peter Moyle (referenced in previous section) recognizes a number of fish body shapes. Rover-predators are streamlined, like a trout, and cruise about for prey. Lie-in-wait predators are more torpedo-shaped, like a pike, and capable of sudden bursts of speed from ambush. Surface-oriented fishes have an upward-directed mouth and large eyes, and usually are small species that capture prey from the water surface. Bottom fishes usually are flattened, and may have ventral mouths if they are bottom-feeders, like suckers and sturgeon, or they may be small crevice-dwellers. Deep-bodied fishes usually are flattened from side to side, spiny to defend against enemies, and good at maneuvering, often with their large pectoral fins. Eel-like fishes have long bodies, long dorsal and anal fins, and dwell in crevices and burrows. The sixth category, rattail shape, is seen in some deepwater marine fishes.

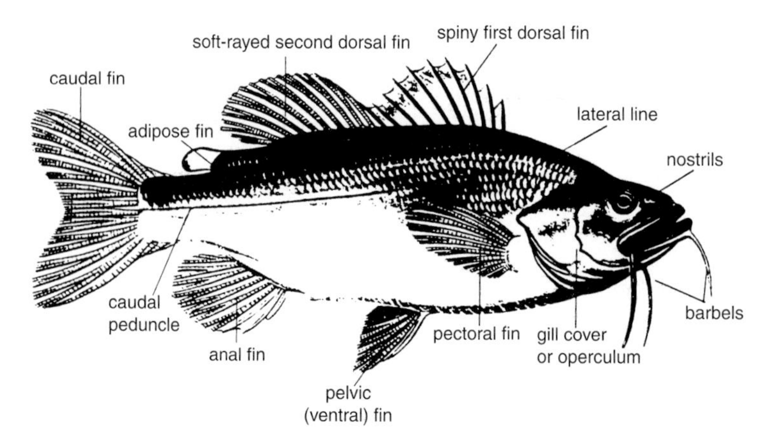

FIGURE 18.2 A generalized bony fish. Fishes vary widely in size, position, and arrangement of fins and other appendages, and no single species has all the features shown. (From Paxton and Eschmeyer, 1995.)

Brief mention of some additional aspects of internal and external anatomy will help us appreciate the capabilities of fishes. The *lateral line* refers to a row of mechanical receptors along each side of a fish's body. It provides "distance-touch" ability to detect prey and obstacles. Internally, many fishes possess a *swim bladder*, which ancestrally probably served as a lung, but in modern fishes conveys neutral buoyancy. This allows fishes with swim bladders to maneuver easily in three dimensions. Bottom-dwellers often have reduced swim bladders or have lost this adaptation entirely. Fishes breathe by extracting oxygen from water forced over their *gills*, but differ greatly in their activity and hence metabolic demands, in their tolerance of low-oxygen levels, and ability to supplement aquatic respiration by gulping air. Most fishes hear little, but in characins, carps, minnows, and drums, connections between the swim bladder, which acts as a sound amplifier, and an elaborate series of small bones in the head allow them to hear well, detecting predators and prey and communicating during mating.

Fishes are marvelous swimmers, as everyone knows, but how fishes swim might not be as well appreciated. They do indeed oscillate the tail fin, but in fact the entire body oscillates in a series of coordinated waves from head to tail, driven by contractions of the large, swimming muscles that make up 40 to 65% of a fish's weight (and all of the fish dinner we eat). Eels have reduced tail fins, and to watch an undulating eel swim is reassurance that the thrust comes from waves of body motion. At the other extreme, the fastest-swimming fishes of all, the tunas, do indeed drive their bodies with their tails. Fins add another dimension, maneuverability, both steering and propelling. Sunfish are adapted for maneuverability rather than speed, and use their pectorals as oars, to turn as well as move forward and back.

SOME GENERAL REMARKS ON FISH ECOLOGY

Feeding

The diet of fishes has received a great deal of study. Fishers, including Sir Isaac Walton and certainly others before him, often cut open a fish to see what it recently has eaten. Can there exist a fish biologist who has not done her or his share of this slimy duty to science? From all of this study comes a number of not necessarily compatible conclusions: fishes are generalists, specialists, and opportunists. Some generalizations do hold up, at a broad level, but fishes are individuals, capable of learning. Their environment and food supply changes from place to place and day to day, and fishes must eat to survive.

One can approach the question of feeding roles by asking, first, what do fishes eat? And second, how do they feed? In streams and rivers, the vast majority of fish species eat invertebrates at some point of their life cycle, and many derive most of their nutrition over their lifetime from invertebrates. Some become predaceous on other fishes as soon as they are large enough. A few feed on plant material, but this is much less common in temperate than in tropical streams, due to the seasonal availability of plant food (which also is difficult to digest). We've just described the food habits of *invertivores, piscivores,* and *herbivores,* respectively. *Omnivores* (suckers are good examples) feed on both plant and animal matter. How fishes feed adds another layer of complexity, particularly for invertivores. Some feed from the benthos (bottom), some from the water column, and some from the surface. Ooze or mud-feeding is uncommon in North American fishes, although common in tropical rivers, and is represented in U.S. waters by the introduced carp. Large rivers can provide enough plankton to support a large, *planktivorous* fish, the paddlefish.

Fishes display a fascinating variety of adaptations of mouth, teeth, and gut to match their feeding habits. Mouths are ventral on bottom-feeders, such as suckers, or angled downward (subterminal) as in some benthic-feeding minnows. Mouths at the midline (terminal) are useful for feeding from the water column, such as a trout on the lookout for drifting insects. An upward angle (super-terminal) signals a surface-feeding invertivore. Fishes can have flexible lips, useful for sorting prey, and can create powerful suction by forming the lips into a tube, and expanding the rear compartment of the mouth. Fish teeth occur in three places: the upper and lower jaws, on the tongue and roof of the mouth, and in the pharyngeal region. Pharyngeal teeth often are used for grinding and crushing, particularly in bottom-feeders, and spectacularly so in the freshwater drum, whose large molars allow it to crush good-sized mollusks. Gill arches can help to capture food as well, by becoming specialized to strain plankton from the water, as does the paddlefish.

The gut offers yet another opportunity for anatomical adaptation to diet. Herbivores, like the stoneroller, and mud-feeders, like the carp, have long, convoluted intestines to provide for a lengthy digestion. Invertivores usually have short, simple intestines and may lack a stomach entirely. Larger predators, such as smallmouth bass, have an expandable stomach to handle large prey.

Despite all of this specialization, fishes can be very opportunistic, providing exceptions to almost every rule. While not a stream fish you are likely to encounter in the United States, piranhas, noted for their specialized carnivory, will consume fruit when it is abundant.

Avoiding Predators

Eat, survive, reproduce: that's what life is about. Predators abound; only the larger individuals of the larger species are likely to be free from this risk, and not even these individuals are immune to human predators. Fishes have a variety of defenses, of which hiding probably comes first. Occupying crevices and cover, and being nocturnal, are good defenses. Spines and poison add another dimension to defense, as do schooling and speed of flight. Vision, the lateral line which detects turbulence, chemical detection of the fear scent released by an injured fish, and in some cases, sound, are means of assessing risk. Fishes seem to be marvelously attuned to intent of a lurking predator, perhaps never ignoring its presence, but sometimes seeming to display a heightened awareness. Like many fishes, bluegills will change their use of habitat according to their assessment of predation risk. Their prey may be more available in open water, but if a bass is nearby, better to consume a smaller lunch in weedy habitat, than venture outside and become lunch.

Habitat

From tiny headwater streams to great floodplain rivers with bays and backwaters; traversing high- and low-gradient terrain; flowing over bedrock, gravelly, or sandy substrates; meandering, cascading, and alternating between riffles, glides, and pools; riverine habitat is enormously diverse. Fishes occupy every type of habitat, and most are found principally in a limited subset. Characterization of habitat thus becomes an important activity for biologists. Usually we distinguish microhabitat—say, shallow, fast water flowing over pea-sized gravel—from macrohabitat—say, a 100-m reach of trout stream that includes riffles, pools, runs, undercut banks, debris jams of wood, and more. Indeed, one can continue up the spatial scale to a valley segment, perhaps 10 km in length, the stream itself, which might be a tributary of a larger river, and ultimately the river system. This hierarchical view of habitat is useful for a number of reasons. It makes clear that the variety of habitats increases as we expand the spatial scale, which probably explains why the number of fish species increases with river size. It reminds us that different habitats occur in different river reaches, and so conservation of biological diversity requires conservation of the many different types of habitat.

Stream scientists have developed a number of quick, useful methods to characterize habitat quality. Table 18.1 shows the scoring system now recommended by the EPA, which you can easily apply to streams in your area. To learn more of this approach, which needs to be tailored to the conditions in your part of the country, check the Web site given in the table. Generally speaking, a stream with a lot of silt on the stream bottom and embedded within the substrate is poor habitat for invertebrates and for fishes to spawn in, because water and the necessary oxygen will not freely circulate. Stable banks are a good sign, eroded banks are not. Cover in the form of overhanging vegetation and wood within the stream also is good. A diversity of currents and depths will support more species of fishes than will uniform habitat. And so on.

Of course, habitat is important because all fish species show some level of habitat specialization, and some are extremely specific. We will highlight a number of examples in the treatment of various families of fishes, later in this chapter. It is good to keep in mind that fishes, like any organism, exhibit an array of adaptations that "fit together" to make the total organism. The habitat of sculpins is fast, stony, clean streambeds. The loss of the swim bladder and huge pectoral fins adapt them to place-holding there. Their big mouths and eyes aid them in capturing the insects that inhabit riffles. Nest-guarding isn't perhaps a necessity of life in this habitat, but it certainly provides lots of flat rocks, just right for a female's eggs to adhere to the underside.

Table 18.1 An Abbreviated Version of the Habitat Assessment Field Protocol Used by the U.S. EPA[a]

Habitat parameter[b]	Description (for optimal condition)	Optimal	Suboptimal	Marginal	Poor
1. Epifaunal substrate/ Available cover	Extent of substrate suitable for invertebrates and fish. Mix of wood, cobbles, stable habitat, and undercut banks.	>70% 20→16	40–70% 15→11	20–40% 10→6	<20% 5→0
2. Embeddedness	Extent of fine sediment covering gravel and cobbles.	0–25% 20→16	25–50% 15→11	50–75% 10→6	>75% 5→0
3. Velocity/Depth regime	All four regimes present (slow-deep, slow-shallow, fast-deep, fast-shallow). Slow: <.3 m/s, deep: >.5 m	All four present 20→16	Only 3 of 4 present 15→11	Only 2 of 4 present 10→6	Only one present 5→0
4. Sediment deposition	Little enlargement of gravel bars, little of stream bottom affected by deposition.	<5% affected 20→16	5–30% affected 15→11	30–50% affected 10→6	>50% affected 5→0
5. Channel flow status	Water reaches base of both banks, and minimal amount of streambed is exposed.	100% satisfied 20→16	>75% of channel is filled 15→11	25–75% of channel is filled 10→6	Little water in channel 5→0
6. Channel alteration	Channelization and dredging absent, channel is natural.	100% satisfied 20→16	Minimal channelization 15→11	Extensive, affecting 40–80% 10→6	Channel greatly altered 5→0
7. Frequency of riffles	Riffles fairly frequent, in regular spacing with pools.	100% satisfied 20→16	Riffles infrequent 15→11	Few riffles 10→6	All flat water 5→0
8. Bank stability	Stable banks with few signs of erosion.	100% satisfied 20→16	Small areas of erosion 15→11	Moderately unstable 10→6	Unstable and eroded 5→0
9. Vegetative protection	Stream banks surfaces and immediate riparian covered with native vegetation.	>95% 20→16	70–90% 15→11	50–70% 10→6	<50% 5→0
10. Riparian zone	For 18 m (60 ft) width, riparian is natural, crops and built structures absent.	100% satisfied 20→16	Width 12–18 m, few impacts 15→11	Width 6–12 m, some impacts 10→6	Width <6 m heavily impacted 5→0

[a]See <<www.epa.gov/owow/monitoring/rbp/>> for a detailed version.

[b]As an example, under habitat parameter 1, an observer might assign a score of 19 if conditions were judged to lie at the high end of optimal; or a score of 11 if the habitat was judged to be barely within the limits of suboptimal. After assigning a rating for each of the ten habitat parameters, scores are totaled. Trained personnel can produce consistent habitat scores that will range from 0–200 and provide a useful assessment of habitat quality.

Life Histories and Reproduction

A fish's life history begins as a fertilized egg, sufficiently provisioned (except in some live-bearers, see later discussion) to develop into a free-swimming larva. Larvae develop into juveniles, who generally resemble the adults but may differ in shape and color, and then into adults. Fishes grow throughout their lives.

Fish reproduction is diverse, sometimes strange, and certainly fascinating. A classification of fish breeding recognizes five categories. Fishes that scatter their eggs, on the bottom, on plants, or in the water column, are termed *scatterers*. They offer no care or protection after spawning. Many suckers and minnows scatter their eggs on the bottom; freshwater drum scatter their eggs into the water column of large rivers. *Brood hiders* are fishes that hide their eggs, usually by burying them, but provide no further care. The redds constructed in gravel beds by salmon and trout are an example. *Guarders* exhibit a higher level of parental care, protecting their eggs and sometimes their young after hatching. Males guard the nest, driving off predators and removing debris. Sculpins and many darters, well adapted to benthic life, guard eggs deposited by females on the underside of rocks. *Nest-builders* offer another form of advanced parental involvement. The sunfishes are easily observed nest-builders, and in the bluegill nesting is colonial, which evidently helps to reduce egg predation. Nests are also used as a basis for courtship. The stickleback displays in front of his nest, and engages in an elaborate and stereotyped courtship sequence that makes it a favorite subject for animal behavior classes. *External bearers*, including mouth-brooders like some cichlids, and the seahorse, where young are carried in a pouch on the male, carry embryos externally. We have no examples of external bearers in North America. *Internal brooders* carry the embryo or young inside the body, and so they also have internal fertilization, effected with a modified spine of the anal fin. Guppies represent an advanced form of this breeding pattern, providing internal nutrition to the developing embryos.

Mating behavior also is diverse, and some examples will be given as part of the family-level synopses. Here we will simply highlight some of the variation that occurs. Most fishes have external fertilization, but as was just mentioned, internal fertilization occurs. Most fishes reproduce repeatedly, from maturity onward. The Pacific salmon invests all of its energy in one, spectacular reproductive outburst; spent, it dies. Small fishes live one to a few years, and so they might reproduce once, twice, or at most three times. Large fishes can live for many decades, and so they may reproduce dozens of times. Nest-building is a more advanced behavior. It suggests that females might benefit by choosing among males and provides ample opportunity for the evolution of elaborate courtship rituals. Males commonly battle for access to females. In both salmon and sunfish, "sneakers" have been reported. These are small males, unable to hold a territory or contest access to a female, who rush in to spawn at the moment when the successful male and his mate release sperm and eggs. Unsurprisingly, tattered fins and other wounds abound on these fellows. A safer strategy, practiced by some small bluegills, is to assume female coloration, and then be permitted by the larger male to join the spawning act.

The River Continuum and Fishes

Different fish species occur along a river's length, and so the RCC has relevance to fishes. As one proceeds from headwaters to river mouth, more big-bodied fishes are encountered, as well as small fishes, and so the overall range of body sizes increases. The number of species increases with river size, partly by addition of species not found in the headwaters, and partly by replacement. The type and volume of habitat and the available food change along the river continuum, and this has profound effects on which species are present. Fishes that prefer clean, swift gravel runs and feed on the aquatic insects that are abundant in such habitat are likely to be absent from low gradient, soft-bottomed lower river reaches. Instead one would expect to find fishes that feed on plankton, or suction up soft sediments for their organic content and occasional invertebrate, or that feed on large mollusks.

Fish and Bioassessment

An experienced stream ecologist can "read" a stream for signs of health or abuse and make a pretty fair assessment. A fish biologist can sample the fishes, and tell you whether the seine haul includes good diversity, including species requiring special habitat and intolerant of pollution, or whether the collection yields only hardy generalists. But our confidence increases if these assessments are quantitative, repeatable, and consistent across investigators. When a hard-fought legal battle ensues, it is essential to have standard, reliable assessment tools at our disposal. Earlier we described how habitat assessment works (Table 18.1). Biological assessment using fish collections also can be very effective.

The Index of Biotic Integrity, or IBI, developed nearly 20 years ago by Jim Karr, now at the University of Washington, is a widely used and effective bioassessment tool. A biologist samples the fishes from a stream by a standard protocol that specifies effort, distance covered (e.g., 35 times the average stream width, or three repeating sequence of riffle-run-pool), and collecting device (often an electroshocker, which stuns the fishes, but doesn't kill them). The IBI is a bit like a Dow Jones of river health—it combines 10 or 12 measures (Table 18.2) into a single number that can then be associated with descriptors like "poor," "good," or "excellent."

NORTH AMERICAN FAMILIES OF FRESHWATER FISHES[2,3]

Note: Representative photos appear at the end of the chapter.

[2]This account omits a few families represented only by introduced species or by one of a few species that barely make it into the southern United States.

[3]All line drawings in this section are reproduced with permission from C. L. Smith, *Fish watching: An Outdoor Guide to Freshwater Fishes*, © 1994 by Cornell University Press, except the drawing of a pupfish which is reproduced with permission from P. B. Moyle, *Fish: An Enthusiast's*, © 1993 by University of California Press, Berkeley.

Table 18.2 The Index of Biotic Integrity (IBI)[a]

Metric	Score		
	1	3	5
1. Total number of fish species (1st order)	<5	5–9	>9
(2nd order)	<6	6–12	>12
2. Number of darter species	0	1	>1
3. Number of sunfish species	0–1	2	>2
4. Number of sucker species	0	1	>1
5. Percent of individuals as carp, green sunfish, and white sucker	>25%	10–25%	<10%
6. Percent omnivores	>45%	20–45%	<20%
7. Percent insectivorous cyprinids	<20%	20–45%	>45%
8. Percent piscivores	<1%	1–5%	>5%
9. Fish density (number/m^2)	<0.05	0.05–0.11	>0.11
10. Number of intolerant species	0	1–2	>2

[a]This example uses ten metrics and allows scores to vary between 1 and 5. Variants of the IBI use up to twelve metrics and maximum scores up to 10. Application of this method remains challenging: maximum scores must be adjusted for stream size and region, and it can be difficult to determine which streams to use as "reference streams" in establishing maximum scores. Note that a high percentage of some "weedy" species (metric 5) and generalist species (metric 6) results in a lower score.

Lampreys (Family Petromyzontidae [19][4])

Silver lamprey, *Ichthyomyzon unicuspis*.

Eel-like fishes lacking jaws, scales, or paired fins, with a circular suckerlike mouth, and seven pair of external gill openings in the North American species, you're not going to mistake a lamprey for anything else. The sea lamprey is the most famous (or infamous) member of these jawless fishes, because of its catastrophic impact on the fishes of the Great Lakes. All are parasitic as adults.

Some lampreys live their entire lives in streams, and all spawn in swift, gravel streambeds. After hatching, the larvae drift downstream to a muddy or sandy place, burrow into the substrate, and subsist on organic matter and invertebrates. The larvae, termed *ammocoetes*, spend four to eight years in this way, then transform into adults and seek hosts.

Much effort has gone into lamprey control, including the development of barriers and a highly specific chemical lampricide that seems to do just what it is supposed to, kill larval lampreys and nothing else. Ironically, because lampreys reproduce in exactly the clean, fast-flowing streams that also make good trout habitat, it is the "best" streams, especially those without dams, which produce the most lampreys.

[4]The number in brackets after each family refers to the number of species in North America north of Mexico, according to Page and Burr's *Field Guide to Freshwater Fishes* (Peterson Field Guides).

Sturgeons (Family Acipenseridae [7])

Atlantic sturgeon, *Acipenser oxyrhynchus.*

Sturgeons are the descendents of the most primitive bony fishes. Biologists like to put quotation marks around "primitive" as a sign of scientific caution since, after all, today's sturgeons are not the same fishes that dominated freshwater some 300 million years ago, pre-dating the dinosaurs by quite a lot. But they look pretty darn primitive! Sturgeons have a cartilaginous skeleton. Their fins are attached to the body by fin rays, rather than lobes (and so these are ray-finned fishes), and the fins are set far back on the body. They have few scales, and these are modified into ridged, bony plates along the back and sides. The tail is heterocercal, as in sharks—the upper or dorsal lobe is longer than the lower. The mouth is on the underside of the head. Two pairs of barbels in front of the mouth provide chemosensory information (taste) to these bottom-feeding fishes. An array of electrosensory detectors around the mouth region add to their food detection capability.

About 23 species of sturgeons occur worldwide, and some are among the largest fishes occurring in freshwater. The white sturgeon of western North America may reach 4 m in length and almost 600 kg; the famed beluga sturgeon of Russia is even larger, with recorded specimens of over 8 m and weighing 1300 kg. Sturgeon occur in rivers, lakes, and the sea, and many species migrate between these habitats. All spawn in freshwater, and all are creatures of big rivers or big lakes. Two species of the genus *Scaphirhynchus* are found in North America, including the pallid sturgeon of the Mississippi system and the shovelnose sturgeon found primarily in the southern region of the Mississippi system. Two species have flattened, shovel-shaped heads. Another five sturgeon species are in the genus *Acipenser,* including three that live in freshwater and two that reproduce in freshwater but spend part of their life at sea. The recovering Atlantic sturgeon and endangered shortnose sturgeon, both found in the Hudson River, are of this genus.

Such big fishes grow slowly and live a long time. Like us, they mature somewhere between the ages of 10 and 20 years. Large females produce huge numbers of eggs—one 4-m long female was reported to provide 180 kg of caviar. They feed on the bottom, consuming a wide range of invertebrates and small fishes. They lack teeth but the mouth extends like a tube, sucking up prey and organic matter.

Many sturgeon are imperiled. Some, including the pallid sturgeon, are responding well to captive breeding in fish hatcheries.

Paddlefishes (Family Polyodontidae [1])

Paddlefish, *Polyodon spathula.*

Only two species of paddlefish occur in the world, one in China, and one in large rivers draining into the Gulf of Mexico. They are closely related to the sturgeons. Paddlefish have lost their bony plates and possess an elongated, flat snout for which they are named. The paddle-shaped snout, once thought to be used as a shovel, is in fact a tactile organ with electrosensory detectors, helping the fishes to locate swarms of its prey, minute animals in the water column, called *zooplankton*. Paddlefish are wonderfully adapted to feed on zooplankton. The mouth is large, and numerous long, slender gill rakers strain the tiny animals from the water. They often feed at night near the surface, a good time and place to locate their prey.

Paddlefish are fairly abundant in the Missouri River. A good place to view them is from Gavins Point dam, which forms Lake Lewis and Clark near Yankton, South Dakota. Individuals up to a meter in length can be seen swimming near the base of the dam, which prevents upstream passage. Fishing boats congregate just below the dam, and fishers use weighted treble hooks to snag these large planktivores.

Gars (Family Lepisosteidae [5])

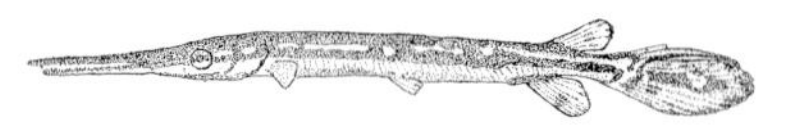

Longnose gar, *Lepisosteus osseus*.

Yet another group of primitive fishes, the gars include about seven species occurring in North and Central America and Cuba. Most common in the southern states, their distribution reaches into southern Canada. Gars are long, thin fishes with long, toothy jaws. They somewhat resemble pike, so much so that the great naturalist Linneaus grouped them together, and they continue to be referred to as gar-pike by some. The alligator gar of the Lower Mississippi Basin grows to about 3 m.

Gars tend to be found near the water surface, floating motionless, often in weedy areas of slow current. They stalk prey, which are attacked with a sideways slash of the head. Their prey includes small fishes, occasionally birds, and in estuaries, crabs.

Bowfin (Family Amiidae [1])

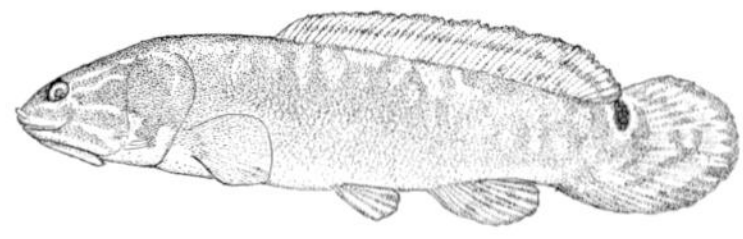

Bowfin, *Amia calva*.

Yet another primitive fish, only one species of bowfin exists in the world, and it occurs in North America. Most likely to be encountered in the south, they range as far north as the waters of Lake Michigan and Lake Ontario. Their habitat is weedy areas of little current, similar to the gar. If you encounter a bowfin, it will be easy to recognize by its long, wavy dorsal fin, nostrils at the end of short tubes, and a bony plate between the lower jaws. The tail is rounded and slightly upturned. Bowfins possess a swim bladder separated into chambers, and this allows them to gulp air, supplementing the oxygen extracted from the water by their gills. This adaptation allows bowfins to occupy water low in oxygen, such as areas of decaying vegetation, from which fishes lacking such an adaptation would be excluded.

Freshwater Eels (Family Anguillidae [1])

American eel, *Anguilla rostrata*.

The snakelike body and terminal mouth make freshwater eels instantly recognizable. Their well-developed jaws, paired pectoral fins, and single pair of gill openings easily distinguish these true bony fishes from the lamprey. The dorsal and anal fins extend half the length of the body, joining into a tail fin. Scales are tiny, embedded within the skin, and do not overlap.

Eels have an unusual life cycle. Adults spawn in the North Atlantic, including the deep water of the Sargasso Sea. The eggs hatch into a transparent, ribbon-like larva. Tarpon and bonefish have similar larvae, suggesting an affinity between eels and these prized saltwater gamefishes. Several years are spent at sea, where the young eels pass through distinctive larval stages, the first described as resembling a wet noodle; the second, a leaf. Maturing eels migrate into rivers and spend most of their lives in freshwater, returning to the sea to reproduce. Their reproductive behavior is virtually unknown, due to the difficulty of tracking and observing them in the ocean. This is known as a *catadromous* life cycle, the reverse of the *anadromous* life cycle for which salmon and shad (see later) are famous.

During their time in freshwater, including lakes and rivers of all sizes, their shape allows eels to make use of crevices and burrows. They have sharp teeth and prey upon a variety of invertebrates and other fishes. They reach from 35 cm to about 1.5 m after 6 to 12 years in freshwater, before migrating thousands of kilometers to spawn in the sea.

It is interesting that eels are not widely eaten in North America, but are highly prized elsewhere, including Europe and Japan. As a consequence there is a large but virtually unregulated commercial fishery for eels migrating from freshwater to the sea along the eastern seaboard. Lack of a domestic market for eels results in inattention to the health of their populations.

Mooneye and Goldeye (Family Hiodontidae [2])

Mooneye, *Hiodon tergisus*.

These two fishes are the only representatives of a group found mainly in Africa, Southeast Asia, and South America. Herringlike in appearance, they possess a large eye, forked tail, lack an adipose fin and fin spines, and the anal fin is longer than the dorsal fin. This family of fishes is notable for the large teeth on the roof and floor of the mouth and on the tongue. These are fishes of lakes and big rivers, including the Missouri, although they move into smaller streams to spawn. There is some debate concerning their value as human food: they have been used for dog food, but smoked goldeye has enjoyed enough popularity to support commercial fishing in Canada.

Herring and Shad (Family Clupeidae [8])

Gizzard shad, *Dorosoma cepedianum*

Herrings are laterally compressed, mostly small fishes with large, silvery scales and a row of sharp scales along the abdomen, a single dorsal fin a forked tail, a long anal fin, and no adipose fin. Most live in temperate seas, about five species are anadromous, entering rivers and lakes to spawn, and two, the gizzard and threadfin shads, are primarily freshwater. The alewife, famous for its massive die-offs in the early 1960s that littered Chicago's beaches, is usually anadromous but may become land-locked. The alewife gained entry to the Great Lakes via the Welland Canal, a bypass around Niagara Falls completed in 1824. It has become an important forage fish, supplanting native planktivores, and becoming the dietary mainstay of two more exotics, the coho and chinook salmon.

These fishes live in the open water, swim in schools, and feed on plankton, which they strain from the water column with their gill rakers. Because zooplankton are abundant in slow-moving water such as the embayments of large rivers, as well as reservoirs and lakes, but not in smaller or swift rivers, the gizzard and threadfin shads are found in big rivers and lakes. Anadromous shad enter rivers to spawn, and their young gradually move downstream, reaching the sea by the end of their first summer.

American shad, the biggest of the river shads, grow to more than 5 kg. Their roe makes them a valued commercial fish. They are gill-netted from the Hudson and other eastern seaboard rivers as they migrate upstream to spawn. In springtime when the spawning migrations enter eastern rivers, anglers enjoy excellent sports-fishing for shad. It is something of a surprise that a plankton-feeder on a spawning migration will strike at spoons, but strike they do. Several shad festivals are held along the Delaware River in New Jersey, New York, and Pennsylvania during April and May. Visitors can enjoy smoked shad, local shad specialties, and the premier shad delicacy, shad roe. In the mid-1800s, people by the hundreds came to fish for salmon and shad, carrying away bags of fishes on horseback. It is reported that shad were so plentiful they often were thrown back into the water as worthless; to eat them was an indication of poverty.

Trouts and Salmons (Family Salmonidae [38])

Brook trout, *Salvelinus fontinalis*.

For many readers of this chapter, this is the group of fishes that requires the least introduction. The salmonids also are rich in folklore and have received an enormous amount of scientific study. They are recognized by their graceful shape, which enables these fishes to be powerful swimmers, and their well-developed adipose fin. Salmon, trout, char, grayling, and whitefish comprise the salmonid family. The whitefish are primarily lake-dwellers, and the Pacific salmon spend much of their lives at sea. All the rest can be found in rivers and streams of various sizes. These are cold-water fishes, hence their distribution is northerly, but brown trout manage warmer waters up to about 25°C, as does the very adaptable rainbow trout.

Rainbow trout enthusiasts, of which there are many, may wonder how this species recently underwent a change in name, from *Salmo gairdneri* to *Oncorhynchus mykiss*. This application of scientific nomenclature made obsolete many a T-shirt that displayed a rainbow trout with its previous name. It also prevents us from saying, as we once could, that all the salmon are in the genus *Oncorhynchus* and all the trout are in the genus *Salmo*. Previously we had to acknowledge one exception: the Atlantic salmon, *Salmo salar*, is in the trout genus. But now the rainbow trout, *Oncorhynchus mykiss*, shares its genus name with the Pacific salmon. At least a char (*Salvelinus*) is still a char!

Scientific nomenclature, or the naming of species, recognizes evolutionary relatedness, and also historical precedent, in naming species. Close examination of the physical characteristics of the salmonids, along with powerful computer algorithms for comparing traits, led to the conclusion that rainbow, cutthroat, Apache, and golden trouts were most closely related to the Pacific salmon, and so they were transferred to the genus. Furthermore, the rainbow trout of Asia and North America were found to be the same, and the Asian trout was described first, so its name took precedence. And voila, *Salmo gairdneri* became *Oncorhynchus mykiss*.

Most anglers are keen students of the habitat and food of their prey, and no group attracts more study, excitement, and fish worship than the salmonids (with char near the bottom and the steelhead indisputably at the top, or so it seems to us). Thus, anything said about habitat and food habits is likely to bring to mind counterexamples in our readers. But here goes. In small streams, always cool and often stony, salmonids are usually the dominant fishes. They prefer regions of slow current, including pools, but will position themselves behind any obstruction. Because salmonids feed largely from organisms drifting in the current, fishes will select a position sheltered from flow but with a clear view of passing prey. They conserve energy by sitting out of the main current, but can easily make short forays for a tasty insect (or perhaps a well-presented artificial).

In general, salmonids less than about 30 cm feed mainly on aquatic invertebrates and also on the terrestrial infall which occasionally is very important (and explains why most fly fishers have a grasshopper and an ant imitation in their fly box). The chars, especially dolly varden, will feed from the bottom to a greater degree than do trout, although the latter also pick insects from the substrate. This is indicated by the lower proportion of "surface drift" in the stomachs of dollies and by their subterminal mouth. Once a trout is big enough, piscivory becomes an option. Most common in brown and rainbow trout, feeding on other fishes allows continued growth, producing fishes of 2–5 kg or more. Big trout get their pick of habitat, and usually they pick big water, such as a long pool formed by a river bend, with lots of cover from wood and undercut banks.

How much do resident trout move around? The conventional wisdom is, "not much." Examples abound of tagged fishes caught again and again at the same spot. Fish-watchers have observed brown trout to be very faithful to a site. Day after day, the same fish, recognizable by its markings, lies in the same position. This well-ordered world can be severely disrupted when a hatchery truck disgorges its cargo. One researcher, observing the disruptions that followed such an event, analogized these hatchery fishes to an ill-behaved motorcycle gang.

On the other hand, perhaps depending upon size and environmental circumstances, some movement does occur. Big trout can be radio-tracked with transmitters inserted into their belly cavity, and a study on Michigan's Au Sable River found that one fish rested in the same hole, day after day. At night, however, it ventured several kilometers up and downstream, presumably on the hunt for hapless little brookies. Because the transmission distance was limited, the graduate student earned his thesis by racing through the cedar swamps along this section of river all through the night, trying to keep the big brown within range. Another study in Colorado mountain streams marked enough brook trout to detect movement over time of at least 20 km, suggesting that in this instance, residency was not particularly strong.

Salmonids display a diversity of life cycles. Some are stream-dwellers all of their lives. Others are anadromous, spawning in freshwater but migrating to the sea (or a large lake) to feed and mature before returning to freshwater to spawn. All of the Pacific salmon die after spawning. Steelhead (which are sea-run rainbow), sea-run cutthroat, and Atlantic salmon are able to spawn repeatedly, returning to sea each time. They battle all of the odds of living long in the wild, as well as the arduousness of the journey, and these factors likely limit the number of spawnings. Finally, some anadromous species, notably the Atlantic salmon, have become land-locked. Like the Pacific salmon now established in the Great Lakes, a lake replaces the ocean as rearing habitat.

The spectacle of salmon migrating upstream to spawn takes the breath away. To observe 5- 10-, even 30-kg fishes, silvery and powerful, leaping up waterfalls, is a special experience. Many will have this experience looking through the window of a fish ladder that enables fishes to surpass a dam; it would be better to experience this thrill on a river without dams. The capacity of salmon to overcome such barriers is remarkable. But the decline of salmon in great rivers of the west (at this writing, every species of salmonid in the Snake River is federally listed) provides somber testimony to the cumulative impact of so many barriers (see Chapter 7).

Much more could be said of the life cycle of the Pacific salmon. Coho manage to reach very small headwater streams and mature for about one year in small streams, where they compete with dolly varden, young steelhead, and others for food. Chinook salmon spawn in large mainstem rivers, and also spend one to two years as juveniles in the river. Eggs hatch as fry, then they develop characteristic vertical bars and are known as parr for most of their juvenile life in rivers. Then, as they begin their migration to the sea, the young fishes become silvery and are referred to as smolts. They are undergoing a physiological change that will enable them to regulate their balance of internal water and salts as they leave a dilute environment and enter a salty one. At the other extreme, pink salmon fry hatch from the gravel and immediately drift to the sea, usually a short distance because pink salmon spawn mainly in small streams, not many kilometers above the estuary. Chum or dog salmon have a similar life cycle to the pink, although they spend longer (about 4 years vs 2 years) at sea. Sockeye salmon spawn in lakes, where they generally spend two years before migrating to sea, and so rivers are primarily migration corridors for this species. Sockeye can also become landlocked, and then they are known as kokanee.

Why the salmon evolved an anadromous life cycle is a matter of intense interest and informed speculation. The advantage to maturing at sea is said to be the greater food supply available there, and this does not seem to be contested. The advantage to returning to freshwater to spawn is less clear; reduced predation is the most plausible explanation. Although many fall prey, juvenile salmon certainly survive well in the river habitat. Migrating smolts run an impressive gauntlet of hungry fishes and waterbirds, but enormous numbers and synchrony ensure that many survive the journey. Indeed, reservoirs are by far the greatest threat. Migrating smolts lose direction, extend their travel time, and expend their energy reserves traversing slow waters. Then many are killed by turbines, the shock of going over a spillway, or the gas bubbles generated by the spillway, and the survivors face a crowd of predators waiting below the dam for the stunned and disoriented survivors. Amazingly, this has led to a combination of barge and truck transport of smolts first sucked out of the river, then squirted back in below the reservoir. Only because so many factors coalesce to bring about the salmon's demise, and because dams unquestionably bring economic benefits, could such an outlandish practice be viewed as sensible (see Chapter 7).

We'll close with an ecosystem story. You might not like dead fishes on your lawn, but they make wonderful fertilizer, and nowhere is this more apparent than in salmon streams. Because most of their biomass is accrued at sea, returning salmon transport great quantities of carbon, nitrogen, phosphorus, and other materials into rivers and streams. Migrating, spawning, and dying fishes are fed upon by eagles, bears, and the many birds and mammals that congregate during salmon runs. Animals and floods carry carcasses into the riparian region, where analysis of the leaves of trees and shrubs reveals the distinctive signature of marine chemicals. Microbes and invertebrates utilize the decaying organic matter, and recycled minerals stimulate algal growth. As Table 18.3 documents, salmon are the cornerstone of the stream and riparian food web, providing food for diverse consumers throughout their life cycle. In death, salmon provide one last, important contribution, fertilizing the stream and the streamside forest with great quantities of their carcasses.

Smelts (family Osmeridae [7])

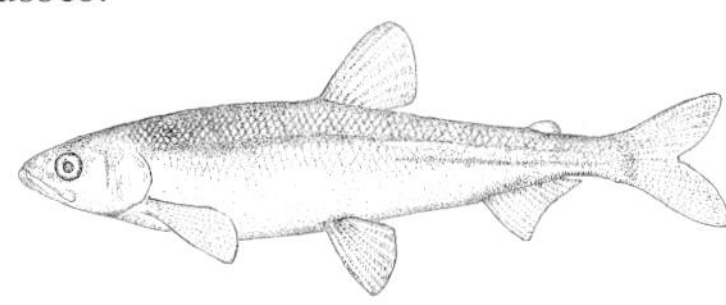

Rainbow smelt, *Osmerus mordax*.

Silvery, small, and torpedo-shaped, smelts have rough scales and a well-developed adipose fin. They live in deepwater of lakes or in the sea, and

Table 18.3 Wildlife Consumers of Salmon in or near Freshwaters of Southeast Alaska[a]

Consumers	Eggs	Juveniles	Adults and carcasses	
Mammals		River otter	Bears	Mink
		Mink	Wolverine	Wolf
			Coyote	Red fox
			Seals	Sea lions
			Deer mouse	Shrew
			Red squirrel	Flying squirrel
			Black-tailed deer	
Birds	Mallard	Loons	Bald eagle	
	Canada goose	Mergansers	Red-tailed hawk	
	Goldeneyes	Great blue heron	Northern harrier	
	Gulls (4 species)	Scaup	Gulls	
	American dipper	Gulls	Black-billed magpie	
	Robin	Arctic tern	Crow, Raven	
		Belted kingfisher	Steller's jay	
		Crow	Winter wren	
		Black-billed magpie	American dipper	
Fishes	Dolly vardin	Dolly vardin		
	Sculpins	Sculpins		
	Coho salmon	Coho, chinook salmon		
	Suckers	Rainbow trout/Steelhead		
	Grayling	Cutthroat trout		
		Walleye pollock		
		Pacific herring		

[a]From M. F. Willson K. C. Halupka, 1995. This table suggests that salmon are "keystone" species on which many other species depend.

they enter streams only to spawn. Nighttime migrations into streams of the Great Lakes can create quite a partylike atmosphere, as fishers with dip nets and lanterns scoop up migrating smelt. This rainbow smelt is an introduction, however, dating back to the 1920s. Eulachons, another migratory smelt, occupy nearshore marine environments from California to Alaska, entering streams to spawn, still in huge numbers in parts of Alaska. Indians used them for food, cooking oil, and when dried, as torches, which is why eulachons are called candlefishes.

Mudminnows (Family Umbridae [4])

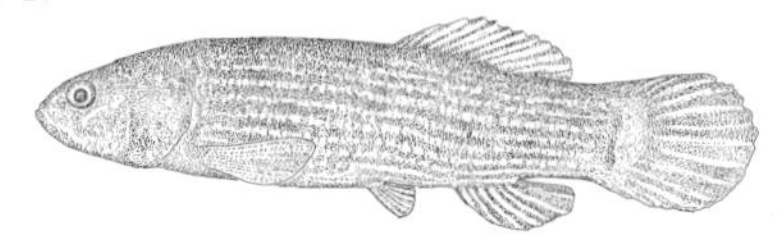

Eastern mudminnow, *Umbra pygmaea*.

Small, brownish, and cylindrical, with dorsal and anal fins set far back on the body, mudminnows are found in densely vegetated areas, and can tolerate low oxygen conditions. They are most likely to be encountered in a pond or a swamp, but they occur in slow-moving sections of rivers.

Characins (Family Characidae [1])

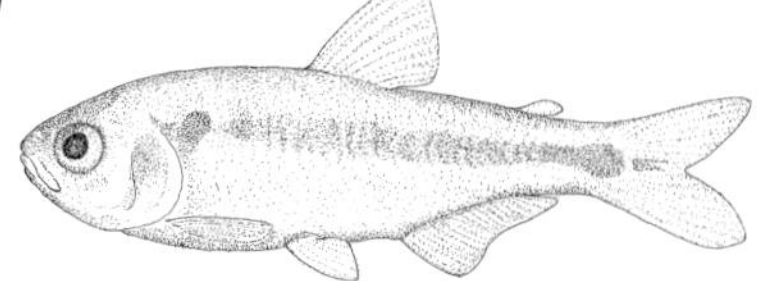

Mexican tetra, *Astyanax mexicanus.*

The small tetras, well known to aquarium fanciers, are represented by one species, the Mexican tetra, in the southern United States. Should you come upon a tetra in southern Florida, it is a visitor from South America, released by a well-meaning owner who probably has not thought about the consequences of releasing nonnative species into new habitats.

Pikes (Family Esocidae [4])

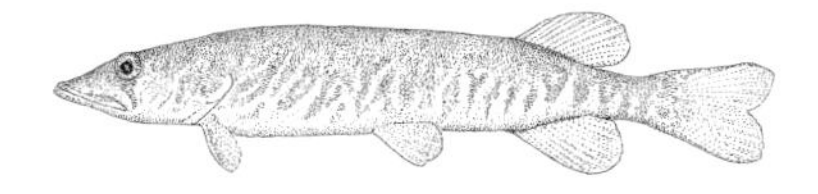

Redfin pickerel, *Esox americanus.*

Pikes, including the northern pike, the muskellunge or muskie, and the grass and redfin pickerels, prefer slow-moving water where they tend to be found in weed beds and other cover. They occur in North America and northern Eurasia, at northerly latitudes. Pikes are recognizable by their barracudalike shape, elongate jaws with large, sharp teeth, greenish color, and the position of their dorsal and anal fins, set far back on the body. Pikes, reaching 1.3 m in length and over 20 kg, and muskies, reaching 1.6 m and over 30 kg, are the giants of this family, occurring in larger rivers as well as in lakes. Grass and redfin pickerel reach a maximum length of about 40 cm, and small specimens are reasonably common in smaller streams of the eastern United States.

All are lie-in-wait predators, rushing from cover to strike at fishes, frogs, small rodents, and whatever else falls into their view. Prey are grabbed sideways, then turned around and swallowed head first. Due to their size and ferocious attack, pikes and muskies are popular with sport fishers, and also support a commercial fishery in some places. The successful angler, upon bringing a decent pike to shore, usually faces a considerable challenge in releasing the animal unharmed, as pike often swallow large, treble-hooked lures deep into their mouth cavity. Barbless hooks and a long, specially designed tool offer the best opportunity of releasing one of these large predators to continue its life at the top of the food chain. Muskie fishing has its own cult of dedicated specialists, who cast large (and expensive) lures using big fishing rods. Up to 100 hours of casting and then retrieving these lures with a jerking motion is the average effort per muskie, making patience and muscles two key requirements of this sport.

Bullhead Catfishes (Family Ictaluridae [40])

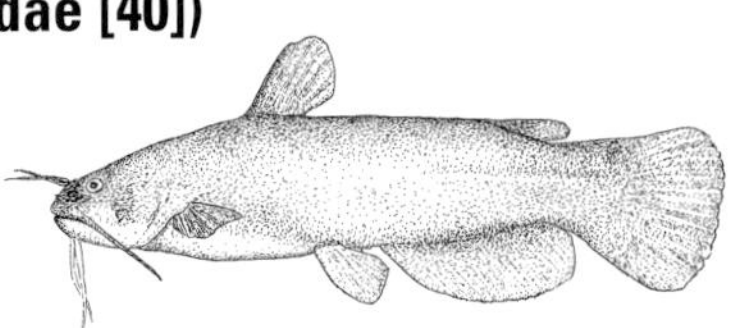

Yellow bullhead, *Ameiurus natalis.*

Of the thirty families of catfish worldwide, only the Ictaluridae is native to North America. Four pair of barbels on the head, scaleless skin, an adipose fin, and sharp spines at the front of the dorsal and pectoral fins are useful diagnostics. All can inflict pain with their spines. The diminutive (up to 10–15 cm) madtoms have a poison gland associated with their pectoral fins, and they pack a considerable wallop for the unwary fish handler. The larger catfish, including the bullheads, channel, blue, and flathead catfish, often reach 5 kg in weight but big channel cats can exceed 40 kg and flatheads can reach 60 kg. Sports fishers pursue channel cats especially, and there are commercial fisheries and aquaculture operations for channel catfish.

The big catfish typically are found in large, slow-moving rivers or lakes, often in cover or near the bottom, are mainly nocturnal, and so are seen infrequently. They feed on or near the bottom on invertebrates, algae, and organic matter, and rely on their barbels and a keen ability to discriminate odor, rather than their weak eyes, to locate prey. The madtoms are found in smaller eastern streams, in stony as well as weedy sections. Some have become specialized to cave life, giving rise to species that are pale and blind.

Minnows (Family Cyprinidae [238])

Creek chub, *Semotilus atromaculatus*.

With over 1600 species in the world, the minnows are the largest family of freshwater fishes. They come in a variety of shapes, but generally have a fusiform (trout like) body shape, large eyes, and small mouths with bony, flexible "lips" on a terminal mouth. Most are small (less than 10-cm long), corresponding to our notion of a minnow, but carp (introduced to North America from Europe, but originating in China) reach almost 1 m in length, and the Colorado pikeminnow reaches almost 2 m in length. The small, silvery shiners number about 200 species, and only with considerable experience can one assign names to the half-dozen or more species captured from a single seine haul in an eastern stream.

Cyprinids are a successful family within a successful superorder, the Ostariophysi, which includes the characins (familiar as the South American tetras and piranhas), suckers, catfishes, and others. Together these fishes dominate the freshwaters of the world. Their success can be attributed to several traits, including acute hearing, specialized pharyngeal teeth, and a chemical fear scent. Hearing is achieved by means of a small set of bones connecting the inner ear to the swim bladder, which vibrates with sound. Pharyngeal teeth are located behind the gills in the throat and, depending upon the species, are specialized for shredding, grinding, or chewing food. This placement allows this group of fishes to have specialized teeth, yet delicate, flexible mouths for prey capture. The fear scent is a chemical given off when a fish is injured. Others nearby, detecting this chemical, are forewarned of danger.

Most cyprinids are small, day-active predators, feeding on small invertebrates. Some are specialized for feeding on algae, like the stoneroller, and others for detritus. Unsurprisingly, large cyprinids prey on other fishes. The northern pikeminnow has a bad reputation with fishers because they prey on juvenile salmonids. They often congregate below dams and at strategic points along the downstream migratory path of salmon smolts. This reputation is accurate, although, as is usually the case, human activities (such as the existence of dams) create a favorable setting for these voracious predators.

The abundance and diversity of minnows make them promising subjects for ecological study. Workers in Oklahoma streams, where stonerollers can be very abundant, have shown that stonerollers avoid areas with bass, their predators. As a consequence, the presence of filamentous algae in a pool is a pretty good predictor that one or more largemouth bass are present. Noting that as many as half a dozen small, silvery minnow species can be captured in a single seine haul, other scientists have wondered whether each species truly occupies a distinct niche. Detailed study of feeding habits (water surface, water column, and bottom substrate), food (insects vs algae + detritus), and time (day vs night) seems to confirm that each species indeed is specialized to its own way of making a living. This partitioning of resources is thought to be a prerequisite for the coexistence of a number of similar-appearing fishes.

The astonishing success of carp and goldfish in new habitats is partly explained by their stout spines on dorsal and anal fins, which discourage predators. Carp are impressive practitioners of inducible defense, detecting the presence of pike from chemical cues, and changing shape to become more bulbous, and thus more difficult to swallow. As bottom feeders able to consume mud and ooze and extract its organic content during passage through their extraordinarily long digestive tract, carp are well suited to the degraded and polluted sections of mid-sized and larger rivers, and far too many such habitats have been created by human actions. Big carp also can be encountered in streams with barely enough water to cover their backs.

The minnow family exhibits a diversity of breeding styles. Males of many species have bright breeding colors and a distinctive pattern of *tubercles*, on the head, by the lips, or over the body. Some build nests, creating a pile of stones by moving each stone with their mouths. Eggs fall within the crevices, where they are protected from being eaten by other fishes.

Suckers (Family Catostomidae [63])

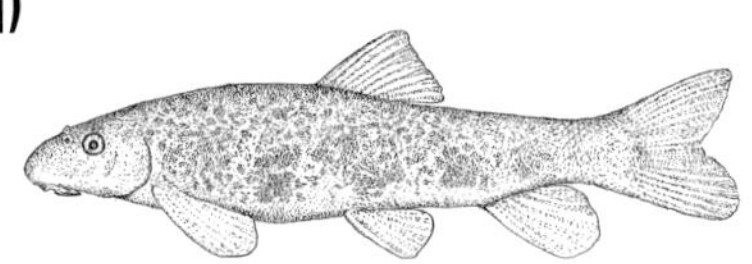

Longnose sucker, *Catostomus catostomus*.

Most suckers have a ventral mouth with thick, extensible lips, comblike pharyngeal teeth but none in the mouth itself, and a forked tail. The anal fin is farther back in suckers than in minnows. The body is relatively streamlined. Notable suckers include the buffalofishes, carp sucker, and blue sucker, which generally are big river fishes of the Missouri–Mississippi system; the colorful redhorses, which live in the deeper parts of large streams; the white sucker, very abundant in northeastern and Midwestern streams; and the hogsucker, a small fish of cool, fast, stony streams in the eastern and upper Midwestern regions of the United States. Adults of the bigger sucker species reach 40–80 cm in length.

All suckers are river-dwellers and many make spawning runs into small streams in spring, seeking gravel bars. Males develop horny tubercles, small pointed protrusions over their bodies, which apparently help maintain contact with the female during spawning.

As the mouth shape suggests, suckers suck or scrape algae, organic matter, and invertebrates from bottom habitats. Their pharyngeal teeth are effective at breaking down organic debris, which is digested in their elongated intestines.

Suckers have a bad reputation with trout fishers because they are suspected of increasing at the expense of trout. More likely, human actions that increase water temperatures and create siltier habitat shift conditions so that suckers fare better than trout.

Pirate Perch (Family Aphredoderidae [1])

Pirate perch, *Aphredoderus sayanus.*

A little fish with a stout, basslike body, the pirate perch is notable for its migrating anal and urogenital openings, which in young are found in the usual place, just in front of the anal fin. However, as the fish matures, these openings migrate forward, until they are in the throat region of the adult fish. The reason for this unusual development remains unknown. They live in weedy sections of slow-moving streams and rivers.

Trout Perch (Family Percopsidae [2])

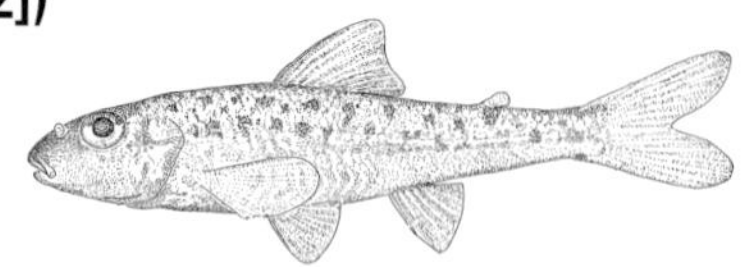

Trout perch, *Percopsis omiscomaycus.*

These are small, rather transparent fishes with rows of dark spots, a forked tail, and a blunt snout that extends over the mouth. Two species are known, one from the Pacific Northwest, and one that extends from the Atlantic seaboard to the Yukon. They occur in streams and lakes.

Cavefishes (Family Amblyopsidae [6])

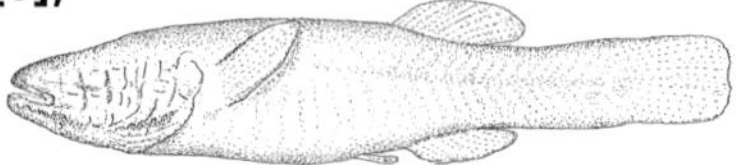

Northern cavefish, *Amblyopsis spelaea.*

These are small fishes dwelling in caves and springs of the Appalachians and the Ozarks. Most have reduced, nonfunctional eyes and little pigmentation, although two species live in surface waters and have small functional eyes and pigmentation. They share with the pirate perch the odd development of forward migrating anal and urogenital openings. Your best chance of seeing one of these fishes is while touring a cave.

Burbot (Family Gadidae [2])

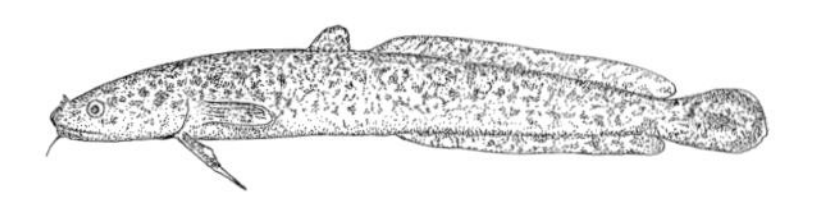

Burbot, *Lota lota*.

Only two codfish species are found in freshwater, and only the burbot or ling can spend its life in rivers. The burbot is an elongate fish with two dorsal fins, the second extending over the rear half of the body, as does the anal fin. The lower jaw has a single, prominent barbel. Though typically deepwater residents of large lakes and large rivers, stream-dwelling populations of burbot occur at scattered locations around the country.

Killifishes and Topminnows (Family Fundulidae [24])

Banded killifish, *Fundulus diaphanus*.

Small, minnowlike fishes with upturned mouths that live near the surface, killifish often have vertical bars, especially during breeding season. Most occur in shallow, slow-moving rivers, in marshes, and in estuaries. They are easily observable because they tend to swim near the water surface.

Pupfishes (Family Cyprinodontidae [13])

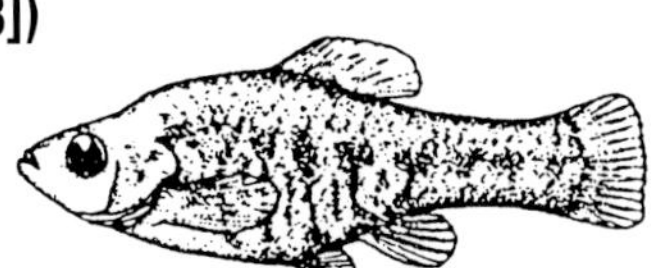

A pupfish, *Cyprinodon*.

Sometimes combined with the killifish and topminnows, pupfish are deeper-bodied and more commonly found in warmer waters. They survive extreme conditions, including high (42°C) temperatures, high salinity (142 parts per thousand—sea water is 35 ppt) and probably are the record holders for tolerating low oxygen conditions. Most are found in desert springs in the southwest, and a number are highly endangered. The Owens pupfish provides a dramatic example of a species saved from the brink of extinction. Thought possibly extinct, in 1969 one population survived in a single pond, which was drying up. Alerted by a student worker, Phil Pister, then of the California Department of Fish and Game, mounted a rescue effort that saved 800 individuals, and led eventually to a series of small reserves where the Owens pupfish lives on. Remarkably, there are other stories where a dedicated ichthyologist rushed to the rescue of the last few individuals of a fish species, sometimes driving off from destroyed habitat with one or two buckets of fish that represented all the individuals of that species in the world.

Livebearers (Family Poeciliidae [19])

Western mosquitofish, *Gambusia affinis*.

Familiar to aquarium hobbyists as guppies, mollies, and swordtails, most poecilids in North America are nonnative species that have been released and have established wild populations. Most are found in warm waters of the south, but the mosquitofish *Gambusia* makes it as far north as Chicago. They occur in slow streams, pool habitat, and various ditches and sloughs.

Silversides (Family Atherinidae [3])

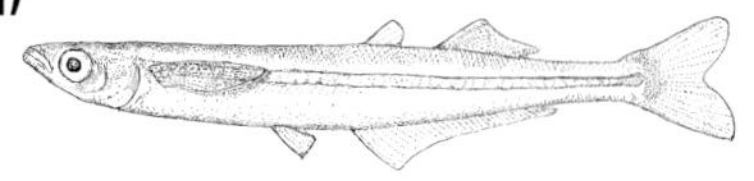

Brook silverside, *Labidesthes sicculus*.

These are slender, silvery surface dwellers and primarily marine, including the well-known grunion of California beaches. The brook silverside is widespread in vegetated streams and rivers of the eastern United States.

Sticklebacks (Family Gasterosteidae [4])

Brook stickleback, *Culaea inconstans*.

Small fishes of fresh, brackish, and salt water, sticklebacks lack scales, have a very slender caudal peduncle, and are distinctive because of their prominent and erect dorsal, anal, and pelvic spines. Their marine relatives include the seahorse, and probably the ancestor of the seahorse resembled a stickleback. The spines are an obvious defense against predators. The four species you may encounter in streams are the threespine, fourspine, ninespine, and brook sticklebacks. The brook is the only strictly freshwater stickleback.

Sticklebacks are excellent aquarium fishes, and their courtship has been extensively studied, even resulting in a Nobel Prize to the Danish behaviorist Tinbergen. After building a nest, males perform a courtship dance, which leads, step by ritualized step, to mating. In small lakes connected to the sea in western Canada, the threespine stickleback has repeatedly colonized from the sea, and repeatedly split into two types, effectively two species, one of which exhibits adaptations of the mouth to feed on plankton, while the other specializes on bottom feeding. Tiny in size, sticklebacks have given more than their share to scientific understanding of fish behavior and evolution.

Sculpins (Family Cottidae [27])

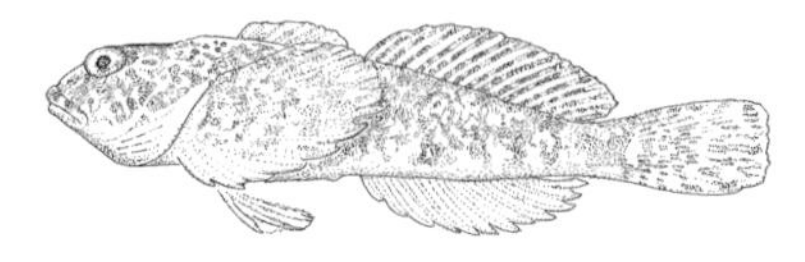

Slimy sculpin, *Cottus cognatus*.

Sculpins are small, bottom-dwelling fishes with a large mouth, prominent eyes, very large pectoral fins, and a tapering body. The first dorsal fin is small; the second is larger and matched by an equally large anal fin. They feed on bottom invertebrates in streams. Males guard nests under rocks and can be induced to mate and guard their eggs in small flower pots.

Temperate Basses (Family Moronidae [4])

White perch, *Morone americana*.

The four species in North America—striped, white, yellow bass and white perch (which is not a perch)—are silvery fishes with deep, laterally compressed bodies and two dorsal fins, the first of which is noticeably spiny. The striped bass is anadromous, entering rivers to spawn. It is a popular game fish reaching more than 40 kg in weight. None of the temperate basses is primarily a river-dweller, although all can be encountered in rivers, and all need flowing water in which to spawn.

Sunfishes (Family Centrarchidae [36])

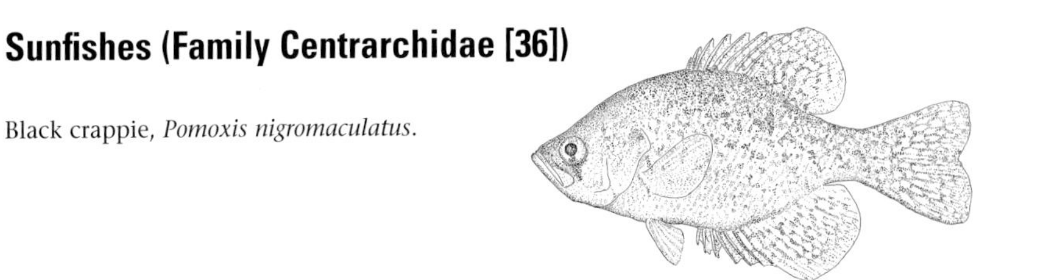

Black crappie, *Pomoxis nigromaculatus*.

The sunfishes, crappies, basses (black, rock, smallmouth, and largemouth), and pigmy sunfishes make up this distinctive family. The centrarchids are exclusively North American, and primarily eastern. Only one centrarchid, the Sacramento perch *Archoplites interruptus*, is native to the region west of the Rocky Mountains. Sunfish have deep, roundish bodies, generally a spot on the operculum, and they can be very colorful. Basses are more elongate, and their powerful bodies make them explosive hitters of lures almost as big as they are. Sunfish are better at maneuvering. Often thought of as lake-dwellers, sunfish can be found in slow rivers and smallmouth bass occur in relatively swift current, although in shelter. The sunfish feed mainly on invertebrates; basses are predators.

Male sunfish build nests, usually a shallow depression made by fanning the tail over soft substrate. Bluegills nest in colonies and can be seen vigorously defending their space against others, while hopefully awaiting a female's arrival. Small males can be very unlucky in such a system, but investigators have uncovered a rather neat trick. Small males take on the coloration of females, and so they are able to join a spawning pair, and join in the release of sperm. The larger male apparently "thinks" he is mating with two females.

Perches (Family Percidae [153])

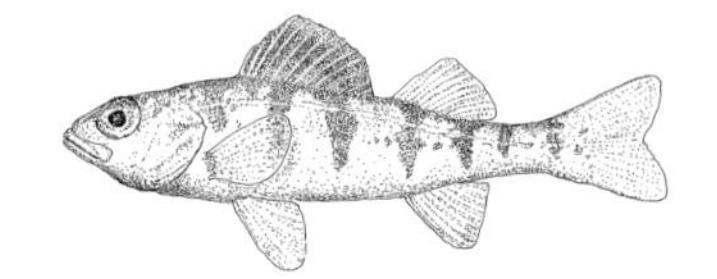

Yellow perch, *Perca flavescens*.

This large family includes the yellow perch, walleye, and sauger, and the many small darters which account for so much of the fish diversity of small eastern streams. The large-bodied species (and a sauger can weigh up to 10 kg) are mainly lake species, but also are plentiful in large, slow rivers of the north. But the darters (Fig. 18.3) really deserve the attention of stream enthusiasts, because they are the butterflies of the streams of the eastern United States—small, colorful, and diverse. Names like rainbow, leopard, green, amber, and spectacular darter are fitting for these colorful fishes. They tend to lie on the stream bottom, some preferring gravel, some sand. They occupy a range of swift to slow currents, and feed mainly on small invertebrates.

The snail darter is a famous name to many, especially in the south, for this is the little fish that almost stopped the construction of the Tellico Dam on the Little Tennessee River. It was an early battle involving the 1973 Endangered Species Act (ESA), and as it now appears, not the most edifying story from either side's perspective. Briefly, this darter was discovered

FIGURE 18.3 Some representatives of the darters, a diverse and colorful group of small fishes in the Percidae. (a) redfin darter (*Etheostoma whipplei*); sandy and rocky pools of headwaters, creeks, and small rivers. (b) tangerine darter (*Percina aurantiaca*); fairly deep rocky pools of creeks and small rivers. (c) rainbow darter (*Etheostoma ceuruleum*); fast gravel and rubble riffles of creeks and small to medium rivers. (d) orangebelly darter (*Etheostoma radiosum*); gravel and rubble riffles and runs of creeks and small to medium rivers. (Photos by W. Roston.)

FIGURE 18.3 *(Continued)*

FIGURE 18.3 *(Continued)*

and named in 1973 and at that time thought to inhabit only a small, remnant section of the free-flowing Little Tennessee River, just downstream of the dam that was already under construction. The snail darter was pitted against the dam and its powerful supporters. The battle ground on, and in 1978 the Supreme Court upheld a lower court decision in favor of the snail darter and against the dam, which was $100 million down the road to completion. Due to this conflict, Congress amended the ESA to form a special committee of agency heads and cabinet secretaries, which became known as "the God committee," because it had the power to decide the fate of a species. The committee ruled for the darter, but an exemption tacked onto a large water and energy bill made it through Congress. The dam was completed. Snail darters transferred to other habitats survived, however, and several additional populations subsequently were discovered.

If one could rewind the tape and play this story again, one would hope for a number of differences. With better inventories of aquatic resources, we might realize how much has been lost and how little remains. The snail darter stood for the scarce, remaining free-flowing rivers of that region, just as the spotted owl stands for the old growth forests of the Pacific Northwest and all they contain. If dams had to be justified on their overall merits, and not as proof of a member of congress's ability to bring jobs to his district, the discussion might have turned out differently.

Freshwater Drums (Family Sciaenidae [3])

Freshwater drum, *Aplodinotus grunniens*.

Drums (also known as croakers) are large fishes of large rivers, bays, and estuaries. Only the sheepshead is a regular inhabitant of freshwater, preferring the deeper waters of large rivers and lakes in the eastern and middle of the United States. Drums have strong pharyngeal teeth for crunching clams and other invertebrates. They vibrate their swim bladders during courtship to produce a drumming sound, which apparently serves to aggregate males and females.

This completes our tour of the 790 species of fishes found in North American freshwater, most of which will be found in a river at some time or another. We have excellent knowledge of the number of kinds of fishes in North America, although you'll still encounter different numbers from source to source. The count is over 1000 if one includes species and subspecies. Some 40 have gone extinct, many more have been introduced, and the lumping and splitting of subspecies further complicates things. By various estimates about one-third of the freshwater fishes of North America are imperiled, a sobering thought considering all of the marvelous diversity they encompass.

NORTH AMERICAN FAMILIES OF FRESHWATER FISHES—PHOTOS

Photos supplied courtesy of the American Fisheries Society and used with permission of the photographer. The photographer's name appears in parentheses.

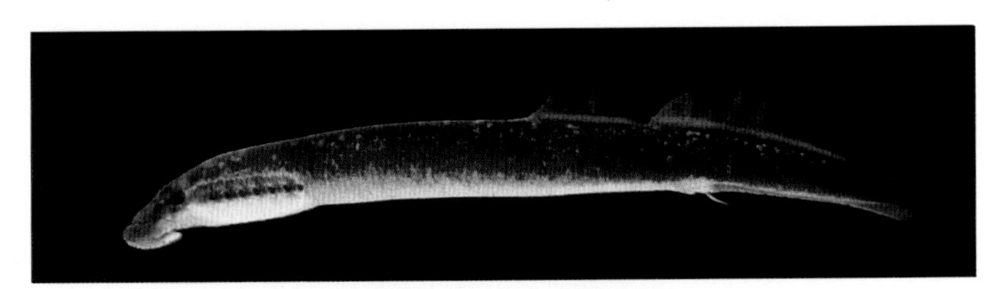

Least brook lamprey (Pflieger)

Shovelnose sturgeon (Pflieger)

Paddlefish (MacGregor)

Longnose gar (Pflieger)

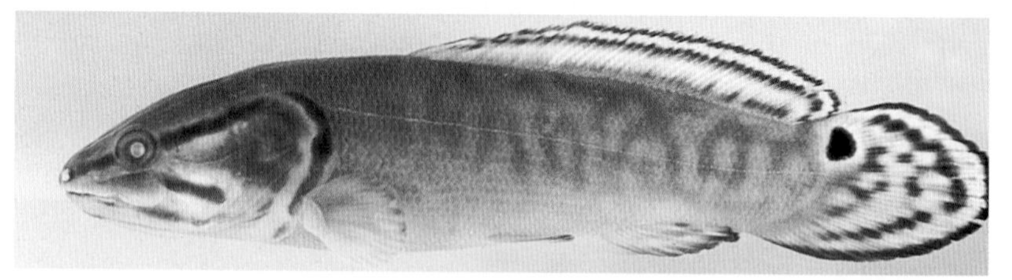

Bowfin (Bauer)

American eel (Scarola)

Goldeye (Pflieger)

American shad (Flescher)

Brown trout (Roston)

Rainbow smelt (Scarola)

Eastern mudminnow (Jenkins & Burkhead)

Mexican tetra (Johnson)

Grass pickerel (Pflieger)

Ozark madtom (Roston)

Hornyhead chub (Roston)

White sucker (Scarola)

Pirate perch (Pflieger)

Trout perch (Purkett)

Spring cavefish (Pflieger)

Burbot (Scarola)

Banded killifish (Scarola)

Pecos pupfish (Roston)

Pecos gambusia (Johnson)

Brook silverside (Roston)

Brook stickleback (Roston)

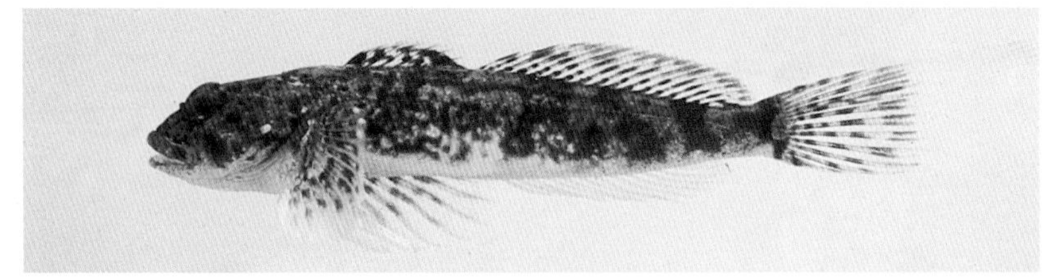

Slimy sculpin (Scarola)

White bass (Pflieger)

Orange-spotted sunfish (Roston)

Walleye (Scarola)

Freshwater drum (Purkett)

Recommended Reading

Moyle, P. B. (1993). *Fish An Enthusiast's Guide*. Univesiry of California Press, Berkeley.

Page, L. M. and Burr, B. B. (1991). *A Field Guide to Freshwater Fishes*. Peterson Field Guides, Houghton Mifflin, Boston.

Smith, C. L. (1994). *Fish Watching: An Outdoor Guide to Freshwater Fishes*. Comstock; Cornell University Press, Ithaca, New York.

CHAPTER 19

Reptiles and Amphibians

INTRODUCTION

Reptiles and amphibians are easily observed members of river and stream ecosystems—at least where they occur in significant numbers. Frogs chirping in marshy backwaters, turtles basking on floating logs, and alligators sunning themselves along the shorelines of slow-flowing warm-water rivers are familiar to most people who spend time near such rivers. Other members of this group may be less obvious to the casual visitor—water snakes and salamanders, for instance. But despite their secretive ways, they can be important members of the food web; a good example is that of the Pacific giant salamanders in the headwaters of Pacific Northwest rain forest streams, where they can be the dominant predator in terms of biomass; more about these later.

We will address the ecological role and some of the life history characteristics of five groups: three reptile groups (turtles, crocodilians, and snakes) and two groups of amphibians (salamanders and frogs and toads).

REPTILES

All reptiles are terrestrial organisms, despite the fact that many are closely associated with aquatic habitats and spend considerable time in the water. They breathe air, and most leave the water to bask and to reproduce, and so spend only a portion of their time in water. Turtles are omnivorous, but crocodilians and snakes are carnivorous; thus they function, as a group, as secondary or tertiary trophic consumers in aquatic food webs. Fertilization is internal, although development may be either external or internal. They also share one other common trait—all are heavily preyed upon by man for various reasons—some reasonable, such as for food, and some unreasonable, such as the almost universal fear of snakes. Now, some details about each of these three groups, emphasizing those that are found in rivers and streams.

Turtles (Order Chelonia)

General Ecology

Turtles, as a group, are omnivorous; they feed on a wide variety of animal and plant material. Some, such as the painted turtles, feed on plant material, and snapping turtles supplement their regular diet of vegetation and animals by feeding on carrion. Turtles mainly use their senses of vision and smell when locating food; they have been shown to have a keen sense of vision. Some turtles, such as those on the island of Trinidad, have adjusted from the usual diurnal feeding activities to one of feeding only at night; this is probably because that is the time most of their prey is active. Turtles, of course, serve as food to other, larger predators in river food webs and as such are in the secondary or tertiary trophic levels of the food web. They are fed upon by birds, mammals, other reptiles, and amphibians. A good share of the predation upon turtles occurs soon after hatching, when they are extremely vulnerable.

Freshwater turtles, as with other reptiles, are strictly terrestrial animals despite their being found closely associated with many aquatic habitats, including streams and rivers. Many species spend considerable time submerged in water, lying concealed on the bottom. They are less reliant on water than amphibians and have a number of adaptive characteristics. These include a scaly skin, which loses less moisture than the smooth skin of amphibians, and a parchmentlike cover for their eggs—again a moisture-retaining adaptation.

In flowing waters turtles are most often found in regions of slow-flowing currents; they are weak swimmers and not commonly found where swift currents prevail. The river cooter (*Pseudemys concinna*), found mostly along the southern states and in Mexico, prefers rivers with moderate current, plenty of aquatic vegetation, and rocky stream bottoms; it is, however, also found in a variety of standing-water habitats. This species is mainly herbivorous, but may eat animal foods or scavenge. It is wary and tends to dive when alarmed. Probably the best known turtle found in flowing waters is the painted turtle (*Chrysemys picta*), which is widely dis-

FIGURE 19.1 The common painted turtle *Chrysemys picta*. (Photo by W. Baker.)

tributed across the continent (Fig. 19.1). It prefers slow-flowing, shallow water, and is diurnal; it sleeps in vegetation on the bottom at night and forages or basks in the sun during the day. The American snapping turtle (*Chelydra serpentina*) is also widespread and may be found in flowing waters where there is abundant aquatic vegetation, submerged brush and tree trunks, and soft muddy bottoms. Several species of turtles are found in rivers and streams in Middle and South America. One, the cogswheel turtle (*Hosemys spinosa*) has coarsely serrated edges along the sides of the carapace, or shell, which are used as aids in moving in swift-flowing waters. In the southeastern United States, similar adaptations are found in stream-dwelling turtles of the genus *Graptemys*, except that the spines are found on the dorsal surface of the carapace. Several species of soft-shell turtles (*Amyda* sp.) are found in riverine systems, of mostly slow-flowing, silt-laden streams. The loss of a hard shell and the flattened body form appear to be adaptations to existence in flowing waters.

The breeding season of reptiles, in general, is largely related to temperature. Fertilization of eggs in all turtles is internal; they are *oviparous* and lay their eggs on land but mate in the water. Some species exhibit unusual courtship behavior. Eggs are laid in nests where they can be kept warm by the sun and/or decaying vegetation.

Problems

Turtle populations have declined worldwide in recent years. Several factors have contributed to this decline, including overcollection within the pet trade, decreases related to indiscriminate use of insecticides and herbicides, and high mortality of some species along highways. Extensive turtle fisheries as a food source have also contributed to the decline of many species, largely because the slow rate of maturation is not adequately considered when establishing fishing quotas. Add to these the continued encroachment upon their habitats by man and you have a classic example of declining populations.

Adequate conservation plans must be developed if turtles are to remain a conspicuous member of our environment. These include protection from property development and loss of habitat, better understanding of their biology to properly manage populations, especially where they are an important food resource, and better control of the pet trade. There are indications that such efforts are taking place. River turtles in South America have been overexploited for years, but long-range study programs, involving counting of eggs and hatchlings, branding of juveniles, and evaluation of breeding schedules, are underway.

Crocodilians (Order Crocodilia)

General Ecology

Crocodilians share a similar feeding ecology with turtles, except that crocodilians are strictly carnivorous and not known to feed on vegetation. They readily eat almost any animal—from insects and spiders to birds and mammals. They rely heavily on their sense of vision while hunting, and their sense of hearing may also be useful in locating food. As with other carnivorous reptiles, their first job is catching their prey. Crocodilians are both stalkers and ambushers, and very aggressive when attacking their prey. They use their large, powerful tails to sweep their prey to where it can be grasped by the jaws. Crocodilians have little competition for food because of their large size. However, size differences enable crocodiles of the Nile River to lessen competition for food; immature individuals feed on smaller prey such as insects, mollusks, and amphibians, whereas adults feed largely on fishes, other reptiles, and mammals.

Eggs and newly hatched crocodilians are heavily preyed upon by a variety of organisms including raccoons, fishes, herons, snakes, and larger crocodilians. Adults, however, are largely safe from predation except from man.

River-dwelling crocodilians are found where the water is slow-moving; they do not inhabit rapid currents. They are restricted to subtropical and tropical regions where temperatures are high enough to aid in the hatching of the eggs.

Except for the Nile crocodile (*Crocodylus niloticus*) and the American alligator (*Alligator mississippiensis*) (Fig. 19.2), little is known concerning the mating behavior of crocodilians. They are territorial and aggressively defend their basking sites and nearby area. Mating occurs in the water, although this has seldom been observed.

Crocodilians are oviparous, and females construct nests in which they lay their eggs. Eggs are deposited into pits dug into the sandy or pebbly soil

FIGURE 19.2 The American alligator *Alligator mississippiensis*. (Photo by J. Detterline.)

and then covered. The American alligator begins nest construction by using her body to trample down the vegetation in an area approximately 3 m in diameter. Loose vegetation is then piled into a mound, which may measure 2 m in diameter and 1 m high. Using her hind feet, she scoops out a hollow in the center of the mound and fills it with mud and vegetation. She then excavates a second hollow inside of the first one and deposits her eggs in this hollow. The eggs are then covered with mud and vegetation. She packs it all down, and this structure not only provides protection against predation, but also incubates the egg due to heat from fermentation of the nest materials. The female usually remains in the vicinity of the nest until the eggs hatch, and the females maintain parental care over the young, often transporting them in their mouths from the nest to the water.

There is no known instance where crocodilians have been identified as "keystone" species in any ecosystem, despite the fact that they are definitely predators. This certainly holds true for the American alligator, but further research would be needed to determine if the Nile crocodile might be a "keystone" species in terms of its heavy feeding on fishes in some streams. Although little is known of the ecological impact of crocodilians in terms of food web dynamics, it seems logical that in places where their populations are unexploited and they are present in large numbers, they must have a significant impact on fish populations.

Problems

The American alligator was an endangered species for several years, but it has recently been removed from this status due to a highly successful protection and conservation program. It was indiscriminately killed for both food and its hides—which produce a highly decorative and prized leather. In fact, it has recovered so successfully that it has become a nuisance species in some places. The American crocodile (*Crocodylus acutus*) inhabits the southern tip of Florida in both freshwater and brackish habitats. It was formerly classified as endangered, but its listing has been upgraded to vulnerable. The gavial (*Gavialis gangeticus*), a stream-dwelling crocodilian that inhabits streams in India, has had its range severely impacted by widespread human development.

Although alligators are potentially dangerous to man, there is little positive evidence of their attacking and killing humans. Only two Old World crocodiles, the Nile crocodile and the saltwater crocodile (*C. porosus*), have been shown to be dangerous to man. Several other large species, however, have no known records as man-eaters, despite their size.

Snakes (Order Squamata)

General Ecology

All snakes are carnivorous, including those living closely associated with streams. They feed on relatively large prey; their jaws and bone structure enable them to ingest whole food items considerably larger than the diameter of their bodies. Snakes use a combination of senses to locate prey: vision, smell, and, in the cases of venomous pit vipers such as the cottonmouth or water moccasin (*Agkistrodon piscivoros*), heat sensing through pits located in the sides of the head.

Snakes are stalkers, and once they locate their prey, they often kill it by constricting the victim in their muscular coils until it is asphyxiated. In the case of water moccasins, killing is by injection of poisonous venom. The prey is swallowed whole, and digestion is usually slow. Thus, snakes are known to feed quite irregularly, depending upon the size of the last meal.

The most likely snakes to be encountered closely associated with rivers and streams are the venomous cottonmouth and the nonvenomous water snakes (subfamily Natricinea). Cottonmouths (Fig. 19.3), as their specific name, *piscivoros*, implies, feed largely on fish and amphibians. Water snakes also feed on fishes and amphibians. The water snake *Nerodia sipedon* (Fig. 19.4), an opportunistic carnivore, is common along mid-sized streams in North America, and *N. septemvittata*, the queen snake which occurs in states from the Great Lakes south to the Gulf of Mexico, feeds mainly on soft-shelled crayfish. One study in Michigan showed that *N. sipedon* fed extensively on sculpins (*Cottus* sp.)—which, in turn, eat trout eggs. Three other genera of water snakes occur in rivers and streams around the world: *Natrix* in Europe, *Afronatrix* in Africa, and *Sinonatrix* in Asia. Garter snakes, racers, and a few other snakes are often encountered along rivers and streams, but cannot be considered to be as closely associated with these ecosystems as are the cottonmouths and water snakes.

Water snakes partition food resources by selectively feeding on different organisms, thus limiting pressure on some food organisms. One study in Louisiana, where four species of *Nerodia* occurred, revealed that all four

FIGURE 19.3 The water moccasin, *Agkistrodon piscivoros*, illustrating the white mouth lining that gives it its other common name, the cottonmouth. (Photo by J. Detterline.)

FIGURE 19.4 The water snake *Nerodia sipedon*. (Photo by J. Detterline.)

had several food types in common and that three ate mainly fish. *N. erythrogaster* fed mainly on toads and frogs, *N. cyclopion* ate large fish, *N. fasciata* fed on large fish and frogs, and *N. rhombifera* ate small and large fish. Temporal partitioning has also been reported, whereby some species feed during the day and others at night. In addition, the same scientists found that these five species of water snakes also partitioned their habitat. *N. erythrogaster* seldom were seen basking on branches or logs. *N. cyclopion* basked frequently during the day, whereas *N. fasciata* and *N. rhombifera* more commonly basked on emergent objects at night. *N. cyclopion, N. rhombifera,* and *N. erythrogaster* had similar microhabitat preferences, preferring gently sloping shórelines essentially devoid of vegetation, whereas *N. fasciata* showed no preference between vegetated and nonvegetated sites.

All snakes have been known to swim, but few can truly call rivers and streams their homes. Those that are closely associated with flowing water include the above-mentioned cottonmouths and water snakes. Cottonmouths occur in streams in warmer climates along the southern tier of U.S. states, while water snakes are distributed widely in the United States. The anaconda (*Eunectes murinus*), a large constrictor known to attain lengths of 12 m, inhabits slow-flowing streams in South America where it preys largely on mammals.

Snakes, as with other reptiles, have internal fertilization of the eggs; although most are oviparous and lay eggs, some give birth to living young (*viviparous*). Internal development is the rule for water snakes.

Problems

Snakes have taken a bum rap from man since the serpent in the Garden of Eden—they've always been the "bad guy" from ancient mythology right up to present times. They are presented as "slimy" creatures and inherently feared, hated, and indiscriminately killed by most humans. This phobia against snakes is hard to understand; it certainly is not genetic. Yet somehow, we develop this fear. Fear is not bad in itself; it is the manifestation of this fear that is unwarranted when people kill snakes just because they're snakes.

Snakes are not important members of river and stream food webs. Although they are predatory upon fishes, amphibians, and other members of aquatic food webs, it has not been shown that their presence or absence significantly influences the flow of energy through these ecosystems. So, despite the unwarranted killing of snakes by humans, it is not likely that this repugnant practice has significant repercussions on stream ecosystems.

AMPHIBIANS

Salamanders, frogs, and toads share many common traits. They live in environments rich in water, but poor in salt. They are carnivorous as adults and feed on live animals. They are sight-feeders and feed by both seeking out prey animals or ambushing them. They function at the secondary or tertiary trophic levels as adults; larvae feed by various means including grazing, filter-feeding, predation, and even cannibalism. Amphibians, in turn, are preyed upon by mammals, birds, other amphibians, reptiles, and man. Fertilization is either external or internal. Now, let's look at some details about stream- and river-dwelling salamanders and frogs.

Salamanders (Order Urodella)

General Ecology

Salamanders feed by a snapping behavior, usually on insects, invertebrates, and other aquatic vertebrates such as amphibians, fish, and reptiles—depending on size. The Pacific giant salamander (*Dicamptodon ensatus*) (Fig. 19.5) is the top predator in many small headwater streams in the rain forests of Oregon, sometimes making up as much as 99% of the total predator biomass. Of course, it takes a lot of predatory insect larvae to equal the weight of one salamander! Olympic salamanders (*Rhyacotriton olympicus*) are present in small numbers in some 1st-order streams. Free-swimming larval salamanders will eat arthropods and other soft-bodied invertebrates. Another large salamander that can often be found on the bottom of stony creeks and rivers is the hellbender (*Cryptobranchus alleganiensis*), which feeds on crayfish, insects, and worms.

One study of the Pacific giant salamander found higher densities of these animals in streams traversing open clear-cut regions rather than in shaded, forested sites. This was related to the increase of algae and other primary producers in these reaches where sunlight enhanced instream photosynthesis. Low-gradient streams had lower densities of salamanders and contained a greater quantity of fine sediment and organic matter.

Another study of the same species, in contrast, found that the abundance of salamanders was not affected by canopy type, but that densities were correlated with substrate composition. Salamanders in this study were found only at stream sites having high gradients and coarse sub-

FIGURE 19.5 The Pacific giant salamander *Dicamptodon ensatus*. (Photo by C. Hawkins.)

strates. Conflicting information such as this clearly indicates that much is yet to be learned concerning salamander ecology in rivers and streams; little ecological study of these organisms has been done.

Salamanders are essentially voiceless; thus, sound plays no part in their reproductive activities. Fertilization by stream-dwelling salamanders is external and mating takes place in the water. Males of *Ranodon sibiricus*, which live in mountain streams, deposit a sticky, elastic spermatophore and the female attaches her eggs to this. Although details vary in terms of courtship behavior, reproduction of aquatic salamanders basically involves the transfer of the spermatophore to the egg sacs.

Problems

Salamanders, indeed amphibians in general, suffer from some of the same man-related problems as do the reptiles—that of destruction of habitat by encroaching development. In addition, there appears to be a general disappearance of amphibians on a global scale. Theories as to why this is happening are many and varied, but one of the most credible is that of increasing exposure to harmful ultraviolet (UV) radiation because of the destruction of the ozone layer by man-made chemicals, such as chlorofluorocarbons. This layer shields the earth from sunlight and effectively screens out harmful UV radiation; without it, more UV radiation reaches the earth. While increased UV radiation appears to be part of the answer, other factors, including disease and climate change, might also play a role.

Frogs and Toads (Order Anura)

General Ecology

Tadpoles are omnivorous, as a group, feeding by grazing on the periphyton community and filter-feeding on bacteria, protozoans, and particulate organic matter. They also can be cannibalistic on other tadpoles. Adults are carnivorous, similar to the salamanders, feeding on a variety of insects, invertebrates, and fishes; the key being that they will take just about anything that is of appropriate size.

One of the few stream-dwelling frogs that has been studied is the tailed frog (*Ascaphus truei*) (Fig. 19.6), which inhabits cold streams in the northwestern United States and southern British Columbia. In these streams, the tadpoles may account for over 90% of the total herbivore biomass. The tadpole is well suited to life in swift streams. Its body is streamlined and ventrally flattened, and the mouth is modified into a powerful suctorial disc that is used to hold onto the stones and to scrape periphyton. Diatoms are the main food source for tadpoles. Densities of frogs were greater in streams in completely or partially forested drainages, where water temperatures were less than 18°C, and lowest in completely deforested basins where stream temperatures were near 20°C. Current speed, percent embeddedness of the stream bottom, and substrate size influence tadpole aggregations. Essentially, *A. truei* tadpoles are cold-water specialists

FIGURE 19.6 The tailed frog *Ascaphus truei*. (Photo by C. Hawkins.)

most often found in swift currents and coarse substrates. Tadpoles move actively over rock surfaces at night.

In contrast to salamanders, sound plays a very important role in the reproductive process of frogs. Males emit mating calls which are species-specific; thus, the females use this to recognize mating partners. Some studies have shown that this sound also helps the female become physiologically and psychologically prepared for mating. After the male has clasped the female, the eggs are released immediately into the water; they are fertilized as they emerge from the female's body. The characteristics of the egg masses vary depending upon the type of environment in which they are laid. Eggs which are deposited in heavy currents, such as those of the red-spotted toad (*Bufo punctatus*), are usually laid as single eggs, in short strands, or in loose, scattered masses.

The tailed frog, mentioned above, is the only New World species that exhibits internal fertilization. It inhabits rapidly flowing mountain streams; thus, external fertilization would be difficult since the current would sweep the sperm away. Eggs of the tailed frog hatch in late summer and tadpoles transform into adults two years later.

Problems

Frogs are thought to be suffering the same global declines in populations as are the salamanders, that is, because of increased exposure to UV radiation. In addition, spraying of DDT and other pesticides has resulted in severe mortality of amphibians where the practice is still common. This probably affects amphibians dwelling in still-water habitats adjacent to crops or in flooded crops such as rice, but it is obvious that general spraying practices cannot guarantee that the pesticides will not reach rivers or streams.

RELATIONSHIP TO THE RCC

For the most part, reptiles and amphibians do not play significant roles in the ecosystem dynamics of rivers and streams; thus, it is not easy to assess their fit into the predictions of the RCC. However, this generalization is not true in the headwater streams of the Pacific Northwest where, as mentioned above, predatory salamanders can dominate the biomass in these systems. Thus, the predicted relative populations of the various functional groups would be skewed from the expected 15% of the biomass, to a community where these predators compose up to 99% of the functional group biomass.

Conversely, in some streams in the same region, herbivorous tadpoles of the tailed frog may constitute 90% of the biomass, thus skewing the relative functional group relationships in favor of grazers. Remember, however, that these deviations are largely restricted to the 1st- and maybe 2nd-order reaches; thus, the predicted functional group relationships revert to those predicted in the downstream reaches of these rivers and streams.

Where predatory reptiles, such as snakes and crocodilians, are present in large numbers, it is surprising that they don't appear to play a bigger role in the structure of fish populations. Certainly they function as top-level carnivores in these food webs; perhaps, we simply have not studied these ecosystems thoroughly enough to document their role.

CHAPTER 20

Birds

INTRODUCTION

As with the mammals to be addressed in the following chapter, many species of birds, indeed probably most of them, at one time or another at least visit rivers and streams. One accounting lists 11 orders of birds that are associated with riverine food webs and the ecology of rivers and streams; however, little solid information exists on the ecological impacts that birds have on the dynamics of these ecosystems, and much that we do know is in generalities. We will not discuss reproduction, nesting, and general aspects of the biology; there are far too many species, and this information can be found in the numerous field guides devoted to birds. We will limit our observations to the ecology of birds in relationship to running water ecosystems. The relevant groups are the raptors, which feed largely on fishes; diving birds, which feed on fishes, underwater plants, and invertebrates; and a number of wading birds, which feed on invertebrates and other prey in the shallows of rivers and streams.

RAPTORS

The raptors (order Falconiformes) most closely associated with river and stream ecosystems in the United States are the bald eagle (*Haliaetus leucocephalus*), the golden eagle (*Aquila chrysaetos*), and the osprey (*Pandion haliaetus carolinensis*). Other fish-eagles perform similar ecological roles in other parts of the world.

Viewers have thrilled to see an eagle or osprey (Fig. 20.1), soaring over the water, suddenly fold its wings to plummet toward the surface and, just before hitting the water, put on the brakes, thrust its feet into the water, and rise with a struggling fish in its claws. The osprey is especially adapted for this task; the soles of its feet have scaly pads that help to grasp fishes, and the front toe is double-jointed so that it can swing backward to more efficiently grasp fishes from the water. Osprey are also known to train their young to catch fishes by dropping a fish in the air and allowing the young to attempt catching the fish before it hits the water. More often than not, the young have to retrieve the fish from the water, thus reinforcing the natural hunting instinct of snatching fishes from the water. Osprey nest near water and feed almost exclusively on fishes. Osprey were adversely impacted during the period when DDT was widely used, but have recovered and are common since elimination of the pesticide.

Despite the beauty and thrill of seeing a raptor dive and catch a live fish from the water, this is not truly indicative of how most raptors, particularly eagles, obtain their food from rivers and streams, especially

FIGURE 20.1 An osprey, *Pandion haliaetus carolinensis*, in flight. (Photo by J.-L. Cartron.)

from the salmon-rich streams of Alaska and the Pacific Coast. In these streams, a more common feeding pattern is to merely scavenge the spawned out carcasses of dead salmon (Fig. 20.2), which have completed their life cycle. This provide a rich harvest for the raptors, who can pick them up with a minimum of effort. Everyone who has spent time along streams in areas where eagles, especially bald eagles, live is familiar with the large concentrations of eagles that frequent the riparian zone of streams having large populations of salmonids. They perch in large numbers in trees (Fig. 20.3) along the stream, waiting to spot a dead or dying fish. As many as 3000–4000 birds have been counted along the Chilkat River in Alaska during the salmon run. They will dive and take live fishes, but scavenging is their main method of feeding. Bald eagle concentrations are also found in areas of extensive waterfowl nesting, where they may catch flying ducks and scavenge weak or dead birds on the ground.

One ecological impact of scavenging fish carcasses by eagles is that by removing the carcasses from the stream, they preclude nutrients and other chemicals from the decaying carcasses from entering the chemical mix of the water. However, given the huge numbers of fish bodies not removed by the birds, it is unlikely that this removal significantly alters the overall chemical regime of the water.

Other raptors, including hawks, are also known to prey on fishes and waterfowl, but do not play significant roles in the dynamics of river and stream ecosystems.

FIGURE 20.2 Bald eagles scavenging on salmon carcasses. (Photo by W. Baker.)

FIGURE 20.3 Bald eagles, *Haliaetus leucocephalus*, perching in tree alongside salmonid spawning stream. (Photo by W. Baker.)

DIVERS

Several orders of birds found associated with flowing waters fit into this general category: the loons (order Gaviformes), grebes (order Podicipitiformes), cormorants and pelicans (order Pelcaniformes), ducks and mergansers (order Anseriformes), and kingfishers (order Piciformes).

The loons, cormorants, pelicans, mergansers, and kingfishers feed primarily on fishes, while the grebes feed mainly on invertebrates and amphibians, although they may also take fishes and plant matter. Diving ducks consume significant numbers of invertebrates, especially mollusks, and may also feed on submerged macrophytes.

Loon (Fig. 20.4) populations have declined steeply, especially in the northeastern United States, and two reasons have been identified. First, the increase in recreational boating has increased the predation rate on loon eggs as the adults are frightened from their nests by passing boats. Second, the increase in acidification of lakes in this region has resulted in a decline of fish populations, the main diet of the loons. Although adults can readily fly to nearby lakes for feeding, the young cannot and so starve to death. These scenarios apply mainly to lake dwelling loon populations, but could apply to those nesting in the slower waters of river sloughs and backwaters.

Kingfishers are familiar to everyone who has spent much time along rivers and streams. They can be seen perched on branches extending out over the water (Fig. 20.5), where they scan the surface for likely meals. Although they are known to dive relatively deeply when pursuing a fish,

FIGURE 20.4 The common loon, *Gavia immer*. (Photo by E. Swenson.)

FIGURE 20.5 The belted kingfisher, *Ceryle alcyon*. (Photo by J. Kelly.)

most prey are taken at the water surface. Whereas fishes are the staple of their diets, kingfishers occasionally take other invertebrates as well. The young are taught to fish by the adults dropping dead meals into the water to be retrieved by the fledglings.

It is largely unknown, as mentioned above, what impact the feeding of these birds has on food web dynamics in river and stream ecosystems. Instances are known where mergansers and cormorants have consumed large numbers of fishes, and this certainly could impact populations. Cormorants on the eastern coast of the United States have been regarded as competitors by commercial fishermen. Still, until stream ecologists begin to include birds in their studies, we can only speculate on their full role in the ecology of stream ecosystems. Conversely, ornithologists might widen their observations to include aquatic habitats in their analyses—a mutually beneficial partnership appears to be in order.

There is one instance, however, where birds have a documented impact on other constituents of the stream ecosystem. A study of the Rio Frijoles in Panama revealed that little blue herons and green kingfishers preyed upon loricariid catfish in experimental open-topped enclosures located in shallow areas of the stream. These catfish graze algae from the surfaces of stones, and in the enclosures where the catfish were exposed to bird predation, algal growth exceeded that found where normal populations of the fishes were present in enclosures in deeper water. Because the fishes normally inhabit deeper water, thereby avoiding the predatory birds, algae proliferate in shallow areas. This represents an example of what ecologists call a "trophic cascade." The presence or absence of bird predation impacts the algae via their effect on the intervening trophic level, herbivorous fishes.

WADERS

Wading birds include the herons, egrets, bitterns, storks and ibises (order Ciconiiformes); the rails, coots, and waterhens (order Gruiformes); gulls (order Charadriiformes); and dippers and wagtails (order Passeriformes). For the most part, these birds are found along the periphery of rivers and streams, usually in backwaters where the current is not swift (Fig. 20.6). They take advantage of their long legs and long beaks to patiently stalk the shallow waters, spearing fishes and invertebrates. Some wading birds exhibit a type of commensal feeding whereby many smaller species, called "attendants," such as shorebirds and waterbirds, follow other species and take whatever comes their way after the water is stirred up by the "beaters." For example, Great and Snowy Egrets attend cormorants, and coots attend several species of ducks. Coots and waterhens feed largely on plant matter, but do ingest several types of invertebrates.

Shorebirds exhibit a wide range of physical attributes which are related to where and how they feed. They vary in the length of their feet and bills, and this allows a type of "partitioning" of the food resources. Species with long legs and bills (e.g., the long-billed curlew, *Numenius americanus*) probe for insects and bottom-dwelling crustaceans in deeper water. The semipalmated plover (*Charadrius semipalmatus*) has short legs and bills and feeds on similar organisms, but in shallow water or land just above the water's edge. Several species exhibit a range of bill and leg lengths intermediate between these two examples.

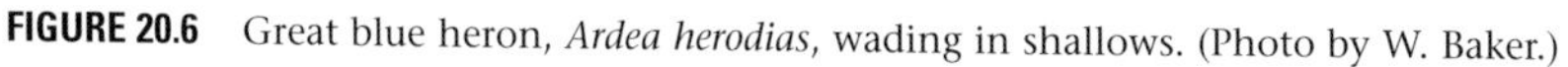

FIGURE 20.6 Great blue heron, *Ardea herodias*, wading in shallows. (Photo by W. Baker.)

Again, the impact of the feeding of these birds on the total productivity of river and stream ecosystems is largely unknown. Herons have been shown to have a significant impact on brown trout populations in Michigan's Au Sable River; mergansers and kingfishers also prey upon young trout. A great blue heron killed because it was feeding on tagged rainbow trout at an experimental facility in Colorado was found to have 26 tagged trout in its stomach—a significant number given the small size of the stream and the total number of fishes tagged. Gulls are notorious scavengers and are familiar sights gleaning the remains of fish carcasses left by bears feeding on spawning salmon.

The American dipper (Fig. 20.7), or water-ouzel, (*Cinclus mexicanus*) is one of the most intriguing birds found closely associated with stream ecosystems. In fact, dippers are completely dependent on the productivity of the rivers and streams where they dwell. The small, gray birds can be found along the sparkling, swift-flowing headwaters of most mountainous streams. The dipper perches on rocks at the edge of the water,

FIGURE 20.7 The American dipper, *Cinclus mexicanus*. (Photo by R. Ryder.)

where it constantly dips its body up and down every few seconds—looking much like it is doing a series of knee-bends. Then, with a quick leap, it jumps into the water to feed on aquatic invertebrates, which it picks from the rocks—mayflies, stoneflies, caddisflies, worms, or any other likely prey. It can actually walk across the rocks of the streambed, and maintains its position on the bottom by adjusting its wings so that the force of the current pushes it down. It has been known to "fly" as deep as 7 m in pursuit of food. As a predator of aquatic insects and other macroinvertebrates, it functions as a consumer in stream food webs; it is also known to feed on fish eggs, but the degree of competition between dippers and trout has yet to be documented.

GENERAL ECOLOGY

In general, birds have not been shown to play major roles in the food web dynamics and productivity in stream ecosystems. Instances have been shown, such as in the Panamanian streams, where predation by birds can impact the spatial distribution of the different components of the aquatic food web. Nevertheless, we are woefully ignorant of the ecological relationships of birds in river and stream ecosystems. What we do know is largely inferred from observations, and this indicates that for the most part, birds function as top predators, preying on everything from invertebrates to fishes. In some cases, as with those species that feed on macrophytes and other plant material, they function as primary consumers. Much remains to be learned—and quantified—in terms of birds and riverine ecosystems.

ENVIRONMENTAL PROBLEMS

Fish-eating birds, including eagles, ospreys, cormorants, terns, and seagulls, can be exposed to contaminant concentrations high enough to impair growth and reproduction. The pesticide DDT, dioxin produced from burning chlorinated wastes and as a by-product of some pulp processing, and PCBs used in various industrial processes, share the property of biomagnification. With each step in the food chain, as predators consume many prey, because these chemicals tend not to be excreted, the contaminant load accumulates. Eventually, predators at the top of the food chain experience contaminant levels that cause reproductive failure or impair growth. In the Great Lakes region, eagles and ospreys nesting along lakeshores reproduce below replacement levels and show high PCB levels in their eggs and greater frequency of birth defects compared to birds nesting on inland lakes and rivers.

In a surprising twist, these findings have led some biologists to argue for keeping dams on rivers to block fish movement from the Great Lakes into inland waters. According to the view that PCBs are the main culprit, dams protect inland populations of fish-eating birds, because fishes of rivers and inland lakes have lower PCB loads. Ironically, dams could be bad for the native biota of rivers, but good for eagles. An alternative explanation for higher reproductive success in inland waters is the possibility of greater food availability. Whichever argument ultimately wins out, one hopes that healthy rivers free of both dams and contaminants lie in our future. Such rivers make an important contribution to the well-being of fish-eating birds, including our national emblem.

CHAPTER 21

Mammals

INTRODUCTION

Probably there are few mammals that do not at least visit streams at some time—to drink, bathe, or any number of other activities. But, it is not our purpose to describe the ecological characteristics of all mammals that casually visit streams and rivers. Rather, we want to emphasize the mammals that play major roles in both food web and energy transfer processes—either directly, or indirectly via their influence through physical alterations of the habitat. Thus, we will concentrate on four species/groups that play significant roles in the ecology of rivers and streams, and mention a few others which are closely associated with flowing waters.

BEAVERS (FAMILY CASTORIDAE)

The mammal that by far exerts the most profound influence on river and stream ecosystems is the beaver, *Castor canadensis* (Fig. 21.1). The beaver's

FIGURE 21.1 The beaver, *Castor canadensis*. (Photo by W. Baker.)

most significant role is that of "ecological engineer." Its propensity for altering the physical habitat, especially in headwater regions, is legendary, and we will explore this in detail below because of its importance. Estimates of their historical abundance vary, but they once were sufficiently abundant to play a pivotal role in the exploration and settlement of North America, and especially Canada. Their original distribution ranged over most of Alaska, the Northwest Territories, Canada, and the United States except for Florida, Georgia, part of South Carolina, and southern California and Nevada.

Historically, beavers were the focal point of one of the most exciting and interesting chapters in our country's history. A worldwide demand for their fur, which produced an exceptionally fine felt for hats, resulted in extensive trapping efforts by the colorful mountain men in the 1840s, men like Jim Bridger, Hugh Glass, Black Harris, and Jedediah Smith, to name but a few. Mari Sandoz in her book, *The Beaver Men*, referred to beaver as "soft gold." Unfortunately, due to this demand beavers were so extensively trapped that even now they are just beginning to reappear in some of their former range. Eventually, silk replaced beaver as the material of choice for gentlemen's hats, and the fur companies and beaver trappers faded into history.

General Ecology

Beavers are most numerous in the headwater reaches of streams draining forested regions, and it is here that they exert the most profound ecological impact on flowing waters. You will not find it hard to recognize beaver country. Where habitat conditions are suitable, beavers significantly alter the environment. They construct series of dams from mud, sticks, stones, and branches to produce ponds which inundate the immediate riparian area of the stream (Fig. 21.2); we shall discuss the ecological effects of these actions below. Beavers don't necessarily just construct a single pond; by ingeniously building an intricate series of dams and canals, they can produce a series of ponds which are hydraulically connected and regulated so that the entire system of ponds and flowing reaches functions as an hydraulic entity, providing habitat for several families of beavers. The canals are used to transport food from trees distant from the ponds. Another indicator of beaver activity is the presence of tree stumps chewed by beaver—typically a pointed stump with chisel-like marks from their chewing teeth. One other way you know you are in beaver country is to unknowingly approach a beaver pond and surprise one or more beavers—the riflelike report as it slaps the water with its tail in alarm as it dives will announce their presence—as will your increased heartbeat!

In their ponds, the beavers construct domed lodges of sticks and mud with an underwater entrance that will be accessible after ice covers the pond during winter. Beavers are less active during the day, but highly busy during the evening and night hours. This is when they actively feed and work on their dams—repairing leaks, adding new dams, etc. The term "busy as a beaver" is a truism, not just a play on words. One of us (CEC) was involved in a project to drain an acre or two of land flooded by beavers. We started by trying to breach the dam using 2 or 3 sticks of dynamite at a time; this did little more than blow a few sticks and some mud into the air. Eventually, we had to dynamite the main dam using 27 sticks

FIGURE 21.2 Beaver dam and ponds, illustrating the extensive flooding that can be caused by extensive beaver activities. (Photo by W. Baker.)

of dynamite in one blast—it blew a hole in the dam approximately 3 m high by 1.5 m wide, right down to bedrock. Yet, by the following morning when we returned, the breach was patched up and water was again backing up behind the dam; we gave up on draining the land.

Beavers feed mainly on the inner bark of aspen (*Populus tremuloides*), willow (*Salix* sp.), and birch (*Betula* sp.) trees. As winter approaches, they begin to create a store of food which will be available when ice covers the pond and land cannot be reached. They cut down trees and float and tow the branches to the vicinity of their lodges where they poke them into the bottom muds in underwater caches.

Beavers apparently are monogamous, mating for life. Mating occurs in February and the litter, averaging about four in number but ranging from one to eight, are born in May. They can live on solid food within a month and are able to reproduce when two years old.

Beavers have significant impacts on the ecological processes in rivers and streams where populations are normal. Perhaps the most significant ecological studies of beavers and streams were done by a group of scientists from the Woods Hole Oceanographic Institution and Ecosystem Center in the 1980s. An extensive study of headwater streams of a boreal forest in Quebec found that beaver significantly influenced ecosystem processes by their dam building and resulting flooding. The construction of dams and resulting ponding of the water slows down the current, causing retention of sediments, organic matter, and water, and resulting in significant alteration of channel geomorphology and flow of water through the system. The buildup of soft sediments, which replace the normal rocky bottom of headwater streams, results in a shift of macroinvertebrate populations from the typical fast-flowing forms, such as mayflies, stoneflies, and caddisflies, to fauna characteristic of soft sediments, such as midge larvae (order Diptera). The studies in Quebec revealed that midge larvae made up 73% of the numbers and 47% of the biomass of emerging insects from these systems. It seems likely that overall aquatic production of insects would increase simply because of the increase in area of aquatic habitat. Retention of fine organic matter and a reduction in current results in an environment conducive to organisms that can thrive under such conditions. Flooding of the riparian areas by ponds behind the dams creates wetland areas where none existed—again an environment entirely different from that existing before the flooding. It is not unusual to find, in such situations, large areas of standing dead timber in beaver ponds—the result of the drowning of the root systems of these trees. In these wetlands created by the ponds, nutrient cycling and decomposition processes are profoundly altered through creation of wetted soils and anaerobic zones. For instance, studies of the nitrogen budget showed a reduction in allochthonous nitrogen and an increase in nitrogen fixation by sediment microbes. Overall, the beaver-modified section accumulated almost 1000 times more nitrogen than unaltered streams. The composition and characteristics of allochthonous inputs from the flooded wetlands are also qualitatively and quantitatively changed. One can easily imagine how these significant changes can ramify through food webs formerly associated with rapid-flowing brooks and small streams—now changed to a series of standing-water ponds and interconnected flowing reaches. In this respect, beaver function as "keystone" species—that is, they are a species whose presence significantly impacts the ecosystem and whose elimination would cause major change in the existing aquatic food webs.

Beavers are preyed upon by bears, wolverines, wolves, and lynx. Bears and wolverines are strong enough to dig through the beaver's lodge, and wolves and lynx capture beaver when they stray too far from the safety of the water.

Problems

As mentioned above, in many areas beavers were extirpated or reduced to very low numbers by concerted trapping efforts. Trappers finding rich beaver populations set up shop and didn't leave until returns dwindled—or Indians drove them out. Many times, the reduced populations they left were not sufficient to repopulate the area. Development by humans has also impacted beaver populations, although the beaver's propensity for dwelling in headwater reaches has minimized this to some extent. However, in these areas, their dams were often destroyed and ponds drained to produce richer bottomland habitat for the cultivation of crops.

Recent evidence suggests that beaver populations are reappearing in places where they had been eliminated. Using aerial photos since 1940, researchers have studied the reemergence of beavers in Voyageurs National Park in northern Minnesota. Currently at a density of about one colony per square kilometer, beavers again are a dominant force shaping this landscape. Selecting first those areas that created the largest ponds, within 20 years the area of ponds had increased ten-fold; thereafter, pond formation slowed as suitable sites were taken. This is encouraging news for many reasons—not the least of which is simply the fact that beavers are such an unusual and interesting animal, and their disappearance would indeed be a terrible loss.

OTTERS (FAMILY MUSTELIDAE)

General Ecology

River otters (*Lutra canadensis*) are some of the most fascinating creatures found in the wild—not just because they are found in rivers and streams, but because they seem to be having so much fun! Anthropomorphic?—perhaps, but we challenge anyone who watches a group of otters cavorting around a pool in a stream to deny that it doesn't look like they are having fun. So be it.

Otters are true aquatic mammals (Fig. 21.3), closely related to minks and weasels. They are well adapted to life along flowing waters; they have a distribution similar to that of the beaver, but also occur in the southeastern states. In the water, they are extremely agile and seem to delight in chasing one another, wrestling, and essentially acting like kids having a good time. One of us (CEC) had a most memorable experience watching three river otters swimming and chasing each other in a beautiful, crystal-green pool in Kelly Creek, Idaho, during a fly-fishing trip; it wasn't hard to forget cutthroat trout for awhile.

FIGURE 21.3 The river otter, *Lutra canadensis*. (Photo by T. Beck/Colorado Division of Wildlife.)

Otters feed on frogs, crayfish, and large insects, but their main food is fish, being able to outmaneuver even trout. It is not unusual for otters to prey upon birds and small mammals when the chance presents itself. Otters seize their prey underwater, but surface to eat. Curiously enough, otters almost always eat fishes from the head to the tail, bones and all, but stop short of eating the tail.

Females bear a litter of from one to five young in early spring. Mothers zealously guard their young, even from the father, for the first three months of their lives. Both parents take a role in their upbringing after this initial period.

Otters are not inhibited by winter weather. In fact, they seem to relish it by building slides in the snow and across the ice on frozen rivers. In summer, you can often tell if otters are in the vicinity by finding long mudslides on river banks. This habit of sliding, either on snow, ice, or on mud, contributes to the impression that these animals seem to enjoy life thoroughly.

Otters do not physically impact their habitat as do beavers, and thus do not have as significant an effect on river and stream ecosystem processes. Nevertheless, as carnivores and predators, they play significant roles in these food webs.

Problems

As with beavers, river otters were extensively harvested for their rich, dark furs to the point that their populations have been severely reduced in many areas. Fortunately, the listing of otters as an endangered species has resulted in conservation and protection programs, which have enabled their populations to begin to increase.

MINKS, MUSKRATS, AND OTHER FRIENDS

Here we introduce several closely related mammals, including minks, muskrats, and nutria. Minks are members of the family Mustelidae, which includes weasels, martin, fisher, wolverine, skunks, and otters. Muskrats and nutria are both rodents (family Muridae and Myocastoridae, respectively). As a group, they do not have the significant impacts on river and stream ecosystems as do beavers, but nevertheless are closely associated with these ecosystems.

General Ecology

Minks (*Mustela vison*) are carnivorous predators (Fig. 21.4a), feeding on a wide variety of small mammals up to and including muskrats, snowshoe hares, grouse, and even migratory waterfowl. Their diet also includes aquatic organisms such as crayfish, mollusks, and fishes. They are excellent swimmers. Minks are found in essentially all forested regions of Alaska, Canada, the Northwest Territories, and the United States. They dwell in holes in stream banks and range widely from their home territory. Mating season usually begins in February, and five or six young are born in April. The young remain with the mother through the summer. Minks function in river and stream food webs as top predators.

Muskrats (*Ondatra zibethica*) are large rodents (Fig. 21.4b) found closely associated with aquatic habitats with a distribution similar to that of the beaver. They are omnivorous, feeding on a wide variety of vegetation, such as the stalks and roots of flags, lilies, and reeds, and prey upon some aquatic animals—mainly clams, fishes, and insects. Where the water is sluggish, they construct conical houses similar to, but smaller than, that of beavers. The lodges are built of mud and vegetation and have an elaborate system of tunnels used for refuge, breeding, and escape. The mouths of these tunnels begin below the water surface and lead into the chambers, which are above the water level. The tunnels may extend well out into the open water. Along streams, muskrats live in dens burrowed into the banks; the tunnels to these dens begin underwater and lead upward to a safe, dry chamber. Muskrats mate in early spring, and savage fights may occur among males for a particular female. Four to twelve young are born in early May; blind and helpless, they grow quickly and can fend for themselves in about three weeks. This is necessary because a second or even third litter is produced during the summer; four to five litters are common in some regions. This high reproductive rate is necessary for muskrats to succeed against their many predators. These include large hawks and owls, wolves, foxes, lynx, weasels, wolverines, otters, and minks; even fishes such as pike and muskellunge will prey upon muskrats. Of these, the mink is the greatest enemy because of its ability to follow the muskrat into its lodge.

a

b

c

FIGURE 21.4 Three mammals commonly occurring in river and stream ecosystems: (a) the mink, *Mustela vison*; (b) the muskrat, *Ondatra zibethica*; and (c) the nutria, *Myocastor coypus*. [a and b from Zeveloff (1988), with permission; and c from *Palmer's Fieldbook of Mammals* by E. Laurence Palmer, copyright © 1957, renewed 1985 by E. Laurence Palmer. Used by permission of Dutton, a division of Penguin Putnam Inc.]

Nutria (*Myocastor coypus*), or coypu, are large rodents native to South America. They were introduced into the United States to be reared for their fur, but some escaped, and others were released due to a decline in demand. They occur in 40 states, primarily throughout the southern United States, but are found as far north as Michigan in the central states. The nutria looks much like a large muskrat (Fig. 21.4c), and feeds on aquatic vegetation, with a preference for reeds. Not only does it eat reeds, it uses them to construct a large nest where five young are usually born. They take to the water within 24 h. Breeding occurs year round and the gestation period is about five months. Nutria construct burrows, but these are short, seldom more than a few feet in length, and are open at both ends; the mouth is at water level.

Problems

As with other fur-bearing animals, minks and muskrats have been extensively trapped, although population numbers do not appear to be in jeopardy. To watch a mink hunting along a river bank is not too unusual, although in our years along rivers we cannot call this experience common. Muskrats are often viewed, however, espcially in ponds and in slow river sections. Muskrats have been pests in some areas where their burrowing activities have undermined dams and dikes.

The nutria is yet another example of a nonnative species that has proliferated to become a major pest. In Texas they cause economic damage to rice and sugar crops, in Louisiana, their burrows threaten dikes alongside bayous, and in coastal Maryland they are overrunning a wildlife refuge. Most nutria damage results from burrows that animals construct as underground dens. Burrows normally extend 1 to 2 m into canal banks but some are as long as 15 to 50 m. Burrows can undermine levees, banks, and roadways, increase bank erosion, and cause bank cave-ins. In coastal wetlands, their burrows allow saltwater to penetrate, killing native vegetation. At this writing, the State of Maryland is launching an ambitious program aimed at eliminating nutria from coastal marshes of a National Wildlife Refuge.

BEARS (FAMILY URSIDAE)

General Ecology

Many people are familiar with the sight of large numbers of grizzly bears (*Ursus arctos*) wading in Alaskan rivers catching and feeding upon the huge runs of salmon migrating upstream to spawn (Fig. 21.5). Further, it is common lore that all bears, both grizzlies and black bears (*Ursus americanus*) will feed on fishes, using their forepaws to scoop fishes from a stream. In reality, bears often pounce on salmon if the water is shallow and grasp the fishes in their mouths, and in deeper water they will submerge their heads to do so. In feeding of spawning runs of fishes, bears can be important predators in river and stream ecosystems. Bears, however, are

FIGURE 21.5 Alaskan grizzly bears feeding on upstream migrating salmon. (Photo by M. Margulies.)

omnivorous, feeding on a wide variety of plant roots, berries, carrion, and freshly killed animals. No animal is safe from the grizzly, truly the top carnivore in North America. In some areas of southeast Alaska, which has the world's highest densities of Alaskan brown bears, the large and aggressive variety of the grizzly, no biologist ventures out along streams during the time of fall salmon runs without considerable caution, and a firearm as a last resort.

Grizzly bears range throughout Alaska, the western Northwest Territories, British Columbia, western Alberta, and restricted areas in the United States, mainly Glacier and Yellowstone national parks and parts of Idaho. Black bears are more widely distributed, occurring in essentially all forested areas of Alaska, the Northwest Territories, Canada, and the United States. Both species enter a state of "winter lethargy" during the winter in dens, sometimes using the same den for more than one year. Although commonly referred to as "hibernation," this characteristic does not fit the true physiological definition of hibernation. Winter lethargy is preceded by a period of heavy feeding to build up fat reserves to see them through the long winter. Mating occurs in June or July and birth follows in January or February. One or two cubs are the norm, although three or four have been documented.

Probably the most significant role of bears in stream ecosystems is that of top predators on spawning salmon runs (see Table 18.3). Even though they gorge on the fishes and kill many individuals, it is questionable whether this significantly impacts fish populations or the overall ecological dynamics of the streams, since uncounted numbers of fishes escape the bears to successfully spawn. Bears likely have an indirect effect on the riparian zone, however, in leaving many carcasses above the banks, where decomposition provides nutrients to streamside plants.

Problems

The most significant problem associated with bear populations is that of man/livestock interactions with grizzly bears. Grizzlies are known to prey on livestock where their ranges coincide, and livestock owners campaign for the bears elimination in these situations. However, environmentalists are equally vociferous for their preservation, maintaining that they are the most obvious symbol of true wilderness. Compromise solutions have resulted where stockman who can demonstrate grizzly loss are reimbursed from funds set aside for this purpose.

OTHER MAMMALS

Many other mammals, both small and large, can be found associated with rivers and streams, and although their activities do not play significant roles in terms of ecosystem dynamics, they deserve mention.

The water shrew (*Sorex* sp.) is a mouse-sized mammal that can be found swimming and diving to the bottom of sluggish water to feed upon insects. Racoons (*Procyon lotor*) are commonly found associated with aquatic habitats, including rivers and streams. They feed on mollusks and crayfish, which they thoroughly wash before eating. Moose (*Alces americana*) feed extensively on aquatic macrophytes growing in the backwaters of rivers and streams. Bats feed on the swarms of emerging aquatic insects, thus playing a role as carnivores in aquatic food webs. The fish-eating bat of tropical America captures fishes which venture to the water surface at night. During the course of their activities, all of these mammals contribute feces to the water which, upon disintegration and dissolution, contribute both FPOM and dissolved nutrients to the water to become part of the food web dynamics.

Problems

Nutrients and FPOM, however, are not the only contributions that streams receive from the input of mammal feces. Along with these inputs comes a serious problem. Due to the presence of beavers, deer, marmots, and other mammals virtually everywhere in the headwater basins of streams, the protozoan *Giardia lamblia* is widely distributed to most flowing waters via mammalian feces If viable *Giardia* are ingested by humans by drinking unpurified water, the result is severe intestinal disorder, diarrhrea, and debilitation. It is no longer safe to lie on your stomach and get a fresh, cold drink from a mountain stream—you may get more than just your thirst quenched! Several commercial filters on the market will make drinking water in the backcountry safe; boiling the water for at least five minutes will also kill the giardia organism.

RELATIONSHIP TO THE RCC

Neither birds nor mammals were explicitly included when formulating the predictions postulated by the RCC. Mammals function largely as predators in river and stream food webs. They are essentially viewed as "non-aquatic," even though, as we have described above, some, such as beavers and otters, are truly aquatic mammals.

The one mammal that significantly affects traditional interpretations of the RCC is the beaver, whose activities can dramatically influence headwater regions of stream ecosystems. Here, their dam building activities change the habitat from that of rapidly flowing, boulder and rubble bottom streams to standing ponds with soft sediments. This, in turn, changes the functional groups of aquatic insects found, altering the usual mix of shredders, collectors, and grazers to one dominated by collecting, sediment-dwelling dipterans. No other mammals significantly alter the patterns predicted by the RCC.

Recommended Reading

Sandoz, M. (1978). *The Beaver Men*. Bison Books, University of Nebraska Press, Lincoln.

Part IV

Management, Conservation, and Restoration of Rivers

CHAPTER 22

Coping with the Threats to America's Rivers

INTRODUCTION

We appreciate rivers for many reasons. Their utilitarian values are many: drinking water, irrigation, waste dilution, transportation, fishes for harvest and sport, power generation, and more. Their aesthetic and recreational values also are many, although these may not translate into equivalent dollar values. From quiet contemplation to adventurous white-water rafting, from children catching crayfish to the most dedicated fly fisher, rivers provide pleasure, relaxation, and spiritual renewal. This book is about understanding and appreciating rivers as ecosystems, and in this chapter we focus on our concerns for the future. We hope to communicate two main ideas. First, river life is seriously imperiled, much more so than many people realize. Second, rivers have great powers of recuperation—we can conserve and restore them if we make the effort.

Rivers and streams harbor a great diversity of life, as featured in Chapters 12–21. It should be said that the biodiversity of U.S. freshwaters is exceptional on a global level (Table 22.1). The approximately 800 species of freshwater fishes place the United States seventh in world rankings, and far ahead of Europe (193 species) and Australia (188 species). The United States has three-fifths of the world's crayfish, 96% of which occur nowhere else, almost one-third of the world's freshwater mussels, and a high percentage of the described species of major groups of aquatic insects. While incomplete knowledge of the fauna of other regions must be acknowledged, particularly for aquatic insects, these figures reflect the genuinely high biological diversity of our rivers and streams.

The state of endangerment of freshwater biodiversity also is great. It may surprise readers to learn that, as a whole, freshwater species are much more imperiled than the terrestrial biota (see also Chapter 8). Birds, mammals, butterflies, and reptiles of the United States have between 10 and 20% of their species at risk (Fig. 22.1). In contrast, amphibians weigh in at about 40%, and the crayfishes and freshwater mussels are very highly imperiled. Table 22.2 provides further evidence of the alarming deterioration of rivers and their biota.

Clearly, human actions have wreaked havoc upon our rivers and streams. We have poisoned our rivers with contaminants, blocked rivers and inundated valleys with dams, and diverted water far outside its basin of origin. In converting forest, prairie, and wetland into productive agricultural land, and then converting it again into cities and suburbs, we alter flow regimes and introduce unprecedented quantities of sediments and nutrients into our streams. Inadvertent and often purposeful introductions of nonnative species have resulted in some benefits, at least from the viewpoint of anglers, but these exotics also have caused many declines of native species. Future climate change will bring additional stress to natural functions and native biota of river ecosystems. We turn now to a detailed examination of the stressors that must be addressed in order to conserve and restore our streams and rivers.

THE MULTIPLE THREATS TO U.S. RIVERS

Dams and Channel Modifications

Approximately 75,000 dams over two meters and 2.5 million smaller dams occur on U.S. rivers. So pervasive are dams that virtually all large rivers of the contiguous 48 states are highly modified and fragmented by dams, and only 42 rivers remain free-flowing for more than 200 km. These dams alter flow, temperature, and sediment regimes of rivers, impacting not just the biota but the physical system as well. Dams fragment river systems, often isolating previously continuous populations into unconnected subunits. And while one dam is bad enough, many rivers are subject to multiple dams, such that the river barely has a chance to behave like a river again, before it is once more impounded.

It is useful to realize that there are many types of dams and they vary greatly, not only in size but also in purpose and how they are operated. Impoundments constructed for water supply need to store enough water to

Table 22.1 Global Significance of U.S. Freshwater Species

Taxonomic group	Number of described U.S. species	Number of described species worldwide	Percentage of known species worldwide found in U.S.	U.S. ranking worldwide in species diversity
Fishes	801	8,400	10	7
Crayfishes	322	525	61	1
Freshwater mussels	300	1,000	30	1
Freshwater snails	600	4,000	15	1
Stoneflies	600	1,550	40	1
Mayflies	590	2,000	30	1
Caddisflies	1,400	10,564	13	1
Dragonflies & Damselflies	452	5,756	8	Uncertain
Stygobites	327	2,000	16	1

Source: Rivers of Life: Critical Watersheds for Protecting Freshwater Biodiversity (edited by L.L. Master, S.R. Flack and B.A. Stein for The Nature Conservancy, available online at <www.tnc.org>.

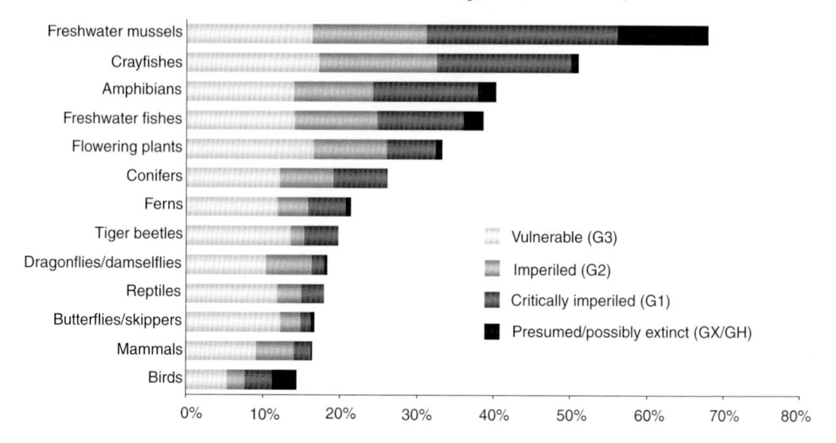

FIGURE 22.1 Extent of imperilment of major groups of U.S. species. Note that the most imperiled species groups are entirely or primarily freshwater species. (From Master *et al.*, 1998, with permission.)

meet projected needs and outlast droughts. Dams for irrigation must capture as much floodwater as possible, to release it later, during the growing season. Flood control reservoirs maintain only a small amount of water and draw down as soon as possible after a flood to maximize capacity for future flood events. Navigation dams maintain some needed depth for boat and barge traffic. Hydroelectric dams store water for release during times of peak energy demand, which may occur seasonally, due to heating and air conditioning demands, or even daily. Probably the worst form of dam operation, from the river's viewpoint, is a "peaking" dam that releases water during the day, as offices and industry demand more power during peak usage, and not at night, when energy demands are lower. Popular with energy utilities because a dam can come "on-line" much faster than a fossil fuel plant, the river is alternately in flood or at low flow, and sometimes even is dewatered. Probably the best form of dam operation, from the river's viewpoint, is a "run-of-the-river" dam that releases water from the dam in direct proportion to inflow to the reservoir. These differences are important because, while the removal of some dams is justifiable from all perspectives, in many instances the economic value of the dam is too great to counter, and so changes in dam operation offer the best opportunities for river restoration.

Dams alter the physical, chemical, and biological condition of the river below the dam, hinder fish passage, and may prevent interchange between upstream and downstream populations. Usually the river's flow is altered, often dramatically. A spring flood pulse, driven by snowmelt in the mountains, is a common feature of many great rivers of the United States. After Glen Canyon Dam was completed, forming Lake Powell and impounding the floodwaters of the upper Colorado River Basin, the river's downstream flow was highly regulated. The predam hydrograph, characterized by a highly predictable spring peak, became a flat line. Like many

Table 22.2 Conservation Status of Rivers of North America

State of the biota

- Since 1900, 123 freshwater animal species and subspecies have been recorded as extinct in North America, including 35 mussel species and 40 fish species.
- Scientific assessment of the fraction of U.S. species at risk of extinction includes:
  Two-thirds of the freshwater mussels,
  Half of the crayfish, and
  One-third of the fishes.
- At least 106 major populations of West Coast salmon and steelhead trout have been extirpated.
- An additional 214 West Coast salmonid stocks are at risk of extinction.
- Commercial fish harvests of U.S. rivers declined during the twentieth century by > 80% (Missouri and Delaware), > 95% (Columbia), and 100% (Illinois River).
- In 1910, more than 2,600 commercial mussel fishers operated on the Illinois River; virtually none do so today.

State of the habitat

- Of 5.2 million km of stream miles in lower 48 states, < 2% is federally protected and only about 42 rivers or river segments > 200 km long remain free-flowing.
- Almost all major river systems of North America outside of the far north are highly fragmented by dams. Exceptions include the Pascagoula, Yellowstone, and Salmon rivers.
- The U.S. has more than 75,000 dams, ranging in size from the Hoover Dam to small earthen dams. The federal government owns 2.7% of these.
- U.S. EPA reports 36% of surveyed river km to be impaired, with agriculture as the leading cause.
- So much of the Colorado River's water is appropriated and diverted that in wet years, a trickle reaches the sea; in normal years, none reaches the sea.

of the large Western dams, Glen Canyon was designed to release deep reservoir water, which is relatively cool and unchanging in temperature throughout the year. Like any reservoir, Lake Powell is a sediment trap, due to the settling of suspended particles because of reduced current velocities. And so the regulated Lower Colorado River has an altered flow, thermal, and sediment regime. Little wonder that many of its fish species are threatened, partly due to the onslaught of nonnative species, but also because altered river conditions disadvantage the native species.

Not everyone would object to some of the biological effects of dams. Wonderful "tailwater" sports fisheries exist just downstream of many large dams, where nonnative trophy trout can be pursued. However, scientific study reveals many changes to the community—loss of some invertebrate species and dominance by others, plant growth may be favored by less-turbid water and more constant flows, and the fish community usually is very different. Of the many biological consequences, blockage of fish passage surely is among the most important. Entire salmon runs have been eliminated by dams that prevented migrants from reaching upstream spawning areas. The Dworshak Dam on the Clearwater River in Idaho, a tributary of the Snake River, was completed in 1974, with the knowledge that spawning grounds for chinook and steelhead would be eliminated. Fish ladders and other fish passage devices designed to permit adults to migrate upstream have met with some success at some dams, although one has to wonder whether populations can survive even some small attrition associated with surpassing one dam, when a half-dozen more lie ahead. Not only must adults run the gauntlet of dams while swimming upstream, juveniles must run the same gauntlet in reverse. As also discussed in Chapter 7, dams pose serious risks for young salmon heading to sea. Without the current to direct and lend momentum to their journey, young fishes must swim the length of each reservoir, often wandering into blind alleys. Passage over the dam kills fishes by pressure changes and impact; passage through turbines can turn them into puree. Predators congregating below the dams find easy prey among the stunned, disoriented juveniles. For years, the solution has been to suck these juveniles into barges and trucks, transport them below the dam, and squirt them back into the river. Small wonder that many now believe it would be better to remove dams and let the river do the work. As anyone living in the West knows, we are entering an era of politically intense debate about "dams vs salmon." It will take a great deal of scientific knowledge, human ingenuity, and ability to compromise to find the best suite of solutions and to decide where and how to apply them.

An emerging consensus among river scientists might at least point toward the general form of a workable solution in this politically charged area. Recognizing that returning rivers like the Columbia or Colorado to their aboriginal state is out of the question, this consensus instead points to the features and functions of a natural river that are most critical to its ecological health. Wherever possible, we would hope to restore the natural flow regime, natural temperature regime, and natural sediment regime. Once the physical functions are restored, ecosystem functions and native biota have a better chance. Where conflicts with human uses limit our options, management objectives should attempt to move the river in the direction of its normal condition. In this way, the "normalized" river becomes a realistic goal.

Dams are the most obvious way that human actions directly impact rivers within their channels, but hardly the only way. According to one estimate made in 1977, more than 26,000 km of channelization had been completed and a further 16,000 km was then proposed. In water-poor Phoenix, canals now comprise a significant proportion of the flowing-water habitat of the area, formerly perennial rivers now flow only during periods of flooding and the native fauna has been decimated. Of the 15 original native species of the Salt River, only four remain, while 19 exotics have arrived. The evidence is clear that, when habitat and flows are altered, native species are disadvantaged. A study of the Kansas high plains reported that the distinctive local fish fauna was adapted to shallow streams subject to shifting sand beds. The absence of flood peaks has eliminated fishes dependent upon floods to trigger spawning, while increased water clarity has favored nonnative, piscivorous game fishes.

A unique study of historic changes in the Willamette River, Oregon, demonstrates how greatly the river channel has been modified by a century of settlement (Fig. 22.2). Early settlers encountered a brushy floodplain up to 3 km in width, with multiple, shifting channels filled with snags. It was wonderful biological habitat, but terrible for navigation. Beginning in 1868,

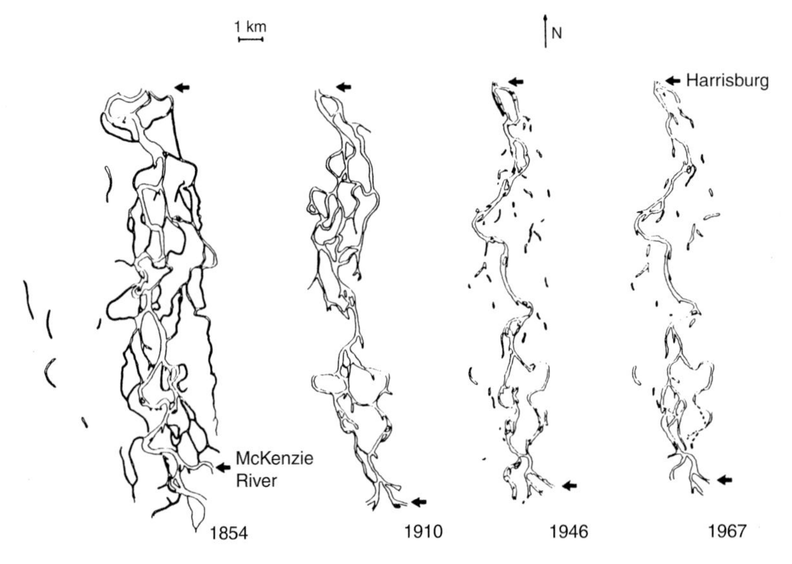

FIGURE 22.2 Transformation of the Willamette River, Oregon. In less than 150 years, a combination of snag removal, floodplain appropriation, channel deepening and dams have greatly changed this river. The estimated length of shoreline in a 25 km segment was 250 km in 1854, 120 km in 1910, 84 km in 1946, and 64 km in 1967. (From Allan, 1995, based on a study by Sedell and Frogatt, 1984.)

removal of snags using special riverboats fitted with cranes and deepening of the channel transformed the river, making it narrower, deeper, and more suitable for navigation and transport. The loss of riverine channel structure, floodplain area, and river–floodplain connectivity was great.

Virtually all of the large rivers of the contiguous 48 states have undergone similar transformation. The practical benefits are unquestioned. Floods are controlled, usually generating energy and storing water for irrigation and municipal use. Floodplains are converted to agricultural land. The river is navigable. The Missouri River, the "big muddy" that Lewis and Clark ascended in their dramatic effort to find a passage to the West by water, would be unrecognizable to those explorers today. Construction of wing dams along the Missouri River dramatically narrowed the channel. In sections used for barge traffic, controlled flow and stabilized river banks render the main river channel "self-cleaning," meaning that it transports enough sediment to maintain the navigable channel. If one visits an unchannelized section the differences are profound: the river is much wider and contains many shallow shoals and scattered islands, snags, and other natural habitat features. As Fig. 22.3 illustrates, human actions have dramatically modified the channel of this great river.

Nonpoint Source Pollution

Rivers receive a wide array of contaminants originating from domestic, industrial, and agricultural sources and as by-products of construction and development. The U.S. Clean Water Act of 1972, last reauthorized in 1987, has the stated objective "to restore and maintain the physical, chemical, and biological integrity of the nation's waters." The Environmental Protection Agency (U.S. EPA) is responsible for enforcement of this legislation in collaboration with the states, establishes minimum standards, and through its research and development activities works to improve environmental monitoring and evaluation. During the nearly three decades since this Act and Agency came into existence, substantial progress has been made in controlling "end-of-pipe" pollution. Municipal wastewater treatment plants have been upgraded in most of the country. A permit system for pollution discharge facilitates monitoring and enforcement, and pollution entering the stream at a point, from a defined source, usually can be reduced or eliminated—it is only a matter of enforcement, cost, and bringing the appropriate technology to bear. Point source pollution is, today, less of a problem that it was 30 years ago.

Unfortunately, some pollutants enter waterways from diffuse sources. Nonpoint source (NPS) pollution is particularly hard to control, because it emanates, often in small amounts, over a very wide area and enters the stream at many locations along its length. Agricultural lands deliver great quantities of sediments, nutrients, and chemicals into U.S. waters. It is estimated that soil erosion losses exceed two metric tons per hectare (5.6 tons per acre) per year in the United States, and that 65% of the sediment in our streams comes from croplands, pastures, and rangeland. Nutrient concentrations, including nitrogen and phosphorus, are elevated due to fertilizer application, resulting in eutrophication of lakes and coastal waters. Urban

FIGURE 22.3 Transformation of the Missouri River. Wing dikes constructed at Indian Cave Bend, Nebraska, in the 1930s have narrowed this section of the Missouri River to one-third of its original width. (Photos by U.S. Army Corps of Engineers.)

and residential areas also contribute to NPS pollution, as a result of the greater area of impermeable surfaces—roads, rooftops, and parking lots—and storm-drain systems. During a hard rain we have all seen water running along streets into a storm drain, but perhaps without realizing that lawn fertilizers, animal waste, automobile residues, and anything else that the water can carry as dissolved or suspended material is on its way to some waterway. In principle this is end-of-pipe pollution; only the source is diffuse. It receives no significant treatment, however. In municipalities where storm and waste sewers are combined, the wastewater treatment plant *might* be effective against animal waste and capture particulates in settling ponds. Often, however, a hard rain in such a combined sewer system sends too much volume into the plant, which then has to bypass some of that volume, including raw waste, directly into the stream. This is referred to as the combined sewer overflow (CSO) problem. By separating storm and waste sewer lines, municipalities can ensure that their wastewater never is bypassed due to a CSO event. But, to our knowledge, treatment of storm sewer water is virtually nonexistent. In effect, storm drains collect contaminants of diffuse origin and pipe them into rivers and streams.

NPS pollution is now recognized as the major source of pollution in most of the nation's rivers. According to a 1994 U.S. EPA report of impaired river miles in the United States, siltation, nutrients, and bacteria were the leading pollutants and agriculture was by far the leading source of river impairment. Fig. 9.3 graphically illustrates how high inputs of sediments are strongly associated with regions of high agricultural productivity.

Municipal point sources, and urban runoff/storm sewers, ranked lower as sources of river impairment, but of course the area of urban land is considerably less than the area of agricultural land. We suspect that urban streams are often more seriously degraded than agricultural streams; there are just fewer river kilometers in the urban category. Remarkably, we are not aware of any hard data to back up that speculation.

The consequences of NPS pollution are many; we will mention several that we think are especially important. Silt and sediments degrade habitat. Gravel riffles become covered with fine sediments, which can also become embedded among the coarser pebbles and stones, effectively eliminating any hyporheic habitat within the streambed and denying oxygen to any organisms or eggs within the substrate. Nutrients encourage plant growth, the effects of which are particularly serious in slow-moving streams, where aquatic macrophytes become so dense that mechanical harvest is necessary. Lakes and coastal waters are especially hard-hit, as nutrients cause algal blooms which in turn decay, consuming almost all oxygen and turning the deeper waters into an anoxic desert. Lakes are highly responsive to phosphorus. Thirty years ago, Lake Erie experienced severe eutrophication as a consequence of phosphorus inputs from untreated municipal waste. Reduction of point source phosphorus loading has been accomplished by improved wastewater treatment. Coastal areas including the Gulf of Mexico are highly responsive to nitrogen, almost all of which is NPS, mostly fertilizer from farmlands of the Upper Midwest. Wherever there is more impervious surface, and less vegetation, rainstorms will result in more surface runoff and a sharper floodpeak, which may result in scouring of stream banks and habitat.

Nonnative Species

At least 35 nonnative fishes have established populations in U.S. waters, and approximately 50 additional species have been recorded. The number of exotic species has risen sharply, from just six in the early years of the twentieth century, to nearly 100 today. Many biologists consider the proliferation of exotics to be an especially serious threat to native biological diversity. Through predation, competition, and habitat modification, nonnatives threaten natives. Once established, they are extremely difficult to control and virtually impossible to eradicate. Yet, in all likelihood the number of nonnative species will increase, aided in part by public agencies, in part by the public. How did this come to be?

Of the 35 well-established species, 12 were purposefully introduced as sports fishes, for aquaculture, and as agents of biological control. The remaining 23 are the product of unintentional releases of aquarium fishes. Not included in these numbers are what biologists call translocations: the transfer of a species outside of its range. Game fishes, along with small fishes in the bait industry, or just in some angler's bucket, are the main players in this form of introduction.

The introduction of species into new locations occurs for many reasons, which can be broadly grouped into purposeful and accidental. As Europeans colonized New Zealand, nostalgia led to determined efforts to introduce familiar species from "back home." Today, 20 of the 46 fish species found in New Zealand are exotic. In North America, stocking of the European brown trout reflects a similar motive, compounded, of course, by the desirability of the brown trout as a sports fish. Introduction, translocation, and stocking of sports fishes are major activities of government fish and game agencies, who believe—probably correctly—that they are doing what the public wants. Unfortunately, this practice is more often based on public pressure rather than on good science. Today, translocated and alien species comprise more than 25% of the recreational fisheries catch of freshwater fishes in the continental United States. And if the introduced species are unable to displace the native species on their own, fisheries agencies may lend a helping hand, by eradicating unwanted native, nongame species. This practice probably is less common than 20 years ago, which perhaps says that we have come far enough to (mostly) stop eradicating native species, but not far enough to stop the stocking of nonnative species.

The negative effects of exotics on native stream fauna are well-known. Predation, competition, habitat modification, hybridization, and introduction of diseases and parasites are the principal causes of declines in native biodiversity. Native species may not be well adapted to escape alien predators, and they may also be forced to occupy suboptimal habitat, which indirectly reduces their ability to maintain their populations. The Sacramento pikeminnow, a predatory minnow introduced into the Eel River in California, has forced a number of native species to abandon deepwater habitats. Hybridization can result in the loss of a species' unique identity, until it is as surely extinct as if the last individual was captured and killed. Perhaps more often a coup-de-grace, hybridization is most devastating once a species has been made rare. Wild Pacific salmon, seriously depleted by habitat loss and overfishing, interbreed with hatchery-reared fishes of unknown genetic ancestry, and in all probability this further weakens their populations. Native cutthroat trout in the Rockies, less hardy than the intensively reared and stocked rainbow trout, often form "cut-bow" hybrids. Diseases and parasites may "hitch-hike" into new locations in the bodies of alien species and then spread to natives. Asian tapeworm infections appeared in an endangered native species, the woundfin, in the Virgin River after the nonnative red shiner invaded the system.

The problems associated with nonnative species are vexing for a number of reasons. As we have learned from our efforts to control the sea lamprey in the Great Lakes, tremendous effort and expense are required to manage exotics, and the need to control the nonnative species never ends. We also must address some critical issues of what we value: native fishes, which the average person might never see, or sports fishes, which provide recreation to many and contribute to local economies. Neither the biological nor the social questions associated with species introductions will prove easy to resolve.

Climate Change

The rising level of CO_2 in the atmosphere, along with other "greenhouse gases," is widely believed to be the harbinger of global climate change.

These gases form a heat-trapping layer that will gradually increase the earth's mean temperature over the next century by an estimated 2–4°C. The change in temperature is expected to be greatest at higher latitudes and in winter. A warmer climate leads to more evapotranspiration, which feeds water vapor into the atmosphere, and more water vapor in the atmosphere means more precipitation. In at least some locations, the future climate will become warmer and wetter. The exact pattern of this climate change is very hard to predict, even at a relatively coarse spatial scale. We expect some areas to become drier, others to become wetter, and quite possibly storms will become more intense. Despite all of the uncertainty about the specifics, it seems highly likely that the rivers of North America will experience altered thermal and hydrologic regimes. An evaluation of the regional impacts of a doubling of CO_2 on aquatic ecosystems in North America was explored by scientists from several disciplines at a symposium held in 1994. Their assessments, though not always in agreement with one another, can be found in the book *Freshwater Ecosystems and Climate Change in North America*, edited by C.E. Cushing.

A U.S. climate that on average is warmer will see its river temperatures begin to rise earlier in the spring, reach higher summer temperatures, and cool later in the autumn. Biological production could be enhanced by a longer growing season. A study using latitudinal comparisons found the biological production of aquatic insects to increase from 3 to 30% for each 1°C increase in temperature. But because individual species are adapted to perform best at particular temperatures (Chapter 2), we should expect a change in species composition. It has been estimated that a 4°C warming would shift the center of distribution of aquatic insects some 640 km northward. Another study simulating the distribution of smallmouth bass and yellow perch estimated a 500 km northward shift. The angler can return to a favored stream, but may have to fish for a different species. Because range shifts in response to climate warming are a virtual certainty, biologists are concerned about the availability of dispersal routes. Where rivers run east–west, or basin divides act as barriers to poleward dispersal, we face the possibility that in those situations resident species will be badly stressed, and potential replacement species will not find natural entry routes.

Human activities can exacerbate or minimize the effects of climate warming. The presence or absence of riparian vegetation can influence summer maximum stream temperatures by shading, and this effect may be as great as the projected influence of climate warming. A chain of shallow reservoirs along a river will raise summer temperatures because the large expanse of shallow standing water will heat more than a shaded river. Steps to minimize these human-induced changes to stream temperatures could offset some of the expected impacts of climate change.

The flow regime of rivers also is expected to change under future climate scenarios. Some potential changes are easier to forecast than others. Rivers that receive much of their precipitation as winter snowpack, and have a predictable spring runoff fed by snowmelt, will become more like rivers governed by a winter rainy season. Erratic flow peaks will occur during winter, the spring peak will be reduced, and summer flows may also be lower because of reduced groundwater recharge during the gradual spring melt. A number of other changes are likely, but at present we can't be very specific about their severity or specific location. As some areas become wetter and others become drier, both the amount and the seasonal pattern of runoff will change. A warmer climate should increase both evapotranspiration (ET) and precipitation (P) (see Chapter 1 for a discussion of the hydrologic cycle), and therein lies the forecasting challenge. If ET increases more than P, perhaps because winds move the atmospheric moisture and resulting rain elsewhere, the location becomes drier; if the opposite occurs, the location becomes wetter. Greater intensity of storms and droughts are likely, and so "flashiness" and seasonality of flow are likely to be altered. Our understanding of the natural flow regime and the importance of flow as a master variable suggest we should be seriously concerned.

As is true for temperature, human actions modify streamflow, and we can exacerbate or reduce the projected impact of future climate change. Conversion of forested land to agriculture, and especially any increase in the amount of impervious surface, will produce flashier flows. Our drains and channelized river systems, designed to get water off the land and away from developed areas, reduce the opportunity for rainwater to recharge the groundwater. Then, during dry periods, base flow is reduced, sometimes to the point where streams dry up. Dams, discussed earlier in this chapter, also profoundly alter riverflow. By paying attention to the influence of land use and dams, we have some opportunity to counter the hydrologic effects of a warming climate.

A BLUEPRINT TO PROTECT AND RESTORE OUR RIVERS

Rivers have great recuperative powers. No longer can we live by the "pollution–dilution rule" that a river cleanses itself some small distance below an upstream town. Valid in frontier times, today our population is too dense, our ecological footprint too large. But rivers can recover relatively quickly, if we simply change many of our ways. We don't know enough yet about recovery to say with certainty how rapid it will be, in particular situations, with a particular history of abuse. The legacy of some abuses will be much longer than others. Large amounts of sediments deposited in a river no longer permitted to flood will have a long legacy. Streams that have lost both their woody debris and their riparian forests will not see naturally generated woody debris for a long time. Channels that have been modified and stabilized, and their flow regulated, will not become dynamic again without help. On the other hand, streams have recovered from chemical spills in only a year or two, depending on such things as percent of stream impacted, presence of upstream or nearby sources for recolonizers, and life cycles of surviving organisms. Farm streams planted with fast-growing woody vegetation look remarkably improved after only five years. Small, outdated dams, often removed as safety hazards, can be replaced with a free-flowing river section quite easily.

No single solution and no "one-size-fits-all" recipe exist for river protection and recovery. But a lot of sound science points the way, and a great many local projects are demonstrating success. We think that the following principles form the basis for river restoration. First, we need inventory and assessment, to know where we are and whether conditions are getting better or worse. Second, we need conservation planning directed at rivers and watersheds, which rarely get specific attention. Third, we need to expand upon and employ the existing protection and restoration techniques ranging from river-friendly farming to improved urban and suburban design. Fourth, we need to continue to improve our ability to apply our scientific knowledge to achieve sound river management. Fifth, we need to build citizen involvement and strengthen the nongovernmental organizations (NGOs) that work on behalf of protecting our natural heritage, so that rivers will have a voice in the political landscape. Here are some details on these five principles, based on an article by Jack Stanford and colleagues.

Inventory and Assessment

We are woefully short of a detailed understanding of the state of the nation's rivers. Various state and federal agencies, particularly the U.S. EPA, are now moving in the right direction. The goal is to develop a suite of indicators, such as the biological and habitat indicators described in Chapter 18 (see Tables 18.1 and 18.2), and then monitor rivers over time so that we can detect trends. At this writing, there aren't many places in the United States where government biologists can argue convincingly that a particular river has gotten better or worse.

Two stories will illustrate that more work is needed. Earlier in this chapter we said that "only about 42 rivers or river segments > 200 km long remain free-flowing." This widely quoted statement comes from our colleague Art Benke at the University of Alabama, who made an independent study from archived government data obtained by American Rivers, an NGO that specializes in river conservation. No official government figure of this sort exists. Also earlier in this chapter we reported on impaired river miles, citing the U.S. EPA, which compiles the data from individual state agencies. Individuals close to the process have warned us that these data are extremely uneven, because the individual states vary in the methods used, quality of their staff, and degree of commitment (i.e., funding). Those data appear to be no more than a very rough approximation.

Your state agency has responsibility for monitoring surface-water quality and publicly reporting the data. Your political representatives vote on the acts that regulate the health of the nation's rivers. Ask them what they are doing.

Conservation Planning

Rivers receive surprisingly little direct protection in the form of conservation reserves. The Clean Water Act has had some success in reducing pollution. The Endangered Species Act protects an increasing number of stream fishes (and one stream insect, a beetle in northern Michigan), but only once circumstances are dire. The Wild & Scenic River Act, and a number of similar state acts, protect rivers' segments but usually not entire river systems, and may permit a very high level of recreational use. Canoeists and fly fishers scowl at one another on the crowded Au Sable River in Michigan, and thousands raft down the New River in Virginia on a busy weekend—both Wild & Scenic stretches. Where are the reserves that protect whole rivers, along with their catchments? Some exist, but only as small streams within land areas protected as terrestrial rather than as aquatic reserves. For the most part, we lack a coordinated program of river protection centered around river reserves. But some good ideas are on the drawing board; here are two that we find especially promising.

Peter Moyle and his colleagues have developed a framework for river conservation reserves based on their studies of California aquatic resources. A key idea is that differing levels of protection would be applied in different situations. Class 1 reserves should protect entire catchments from ridgeline to ridgeline and have a nearly complete native biota. These will be few, and in remote areas. Class 2 reserves would permit some human presence and would often include public lands now managed for multiple use. Less pristine that Class 1 reserves, they are candidates for upgrading. Class 3 waters might look like a higher category, but are so altered by nonnative species and dams that they can never become Class 1. Moyle's plan includes six classes and draws upon ecological principles of reserve design. It is a blueprint waiting to be implemented.

The Nature Conservancy recently completed a national assessment of the imperilment of freshwater biodiversity (see Table 22.1). After mapping the at-risk populations of fish and mussel species throughout the more than 2000 small watersheds of the contiguous United States, they concluded that protection of 327 watersheds—15% of the total—would conserve populations of all at-risk populations. This is a powerful demonstration of how scientific data and information technology can help us focus our efforts where they will do the most good. Notably, this is the work of an NGO, supported by public and foundation donations. Valuable and valid, we should nonetheless point out that the data shine a spotlight on the southeastern United States, and would direct little attention to many other regions of the country. This results from the high biodiversity of the region (see Chapter 8 and Fig. 18.1) and the high state of imperilment of fishes and mollusks (Table 22.1), which are especially diverse in the southeast.

Protection/Restoration Techniques

We group protection and restoration methods together because often they serve the same purpose. In addition, because most rivers are not pristine, we usually want to protect them and also to make them better than they currently are. Most rivers cannot become "Class 1," but most can be racheted up a notch. The science of river restoration is young, but the scientific underpinnings are solid. We will describe some approaches that we think are especially important.

Protecting the riparian, the zone of land along both sides of the stream, may be the single most influential means of protection and recovery. Discussed in Chapter 9, riparian protection is widely championed by scientists. The required width of riparian buffer differs from place to place, sometimes for scientifically justified reasons, and also because of how the process of translating science into law plays out. Riparian protection is not yet uniformly practiced, but it should be.

Earlier we referred to silt and sediments as the single largest cause of impaired river kilometers and agriculture as the source. Clearly we need river-friendly farming. Agricultural best management practices (BMPs) include riparian buffer strips, planting in contours that parallel hillsides, reduced or no-till methods, leaving a vegetative cover on the land during winter, and many more techniques. Today, a modern farmer might use a global positioning system (GPS) and soil maps to tell him, meter-by-meter, how much fertilizer to apply. Wider adoption of BMPs could do much to improve the condition of U.S. rivers. Farmers, usually good stewards of the land, need an effective combination of methods and incentives to improve their practices.

Moving people off the floodplains, and allowing floodplains to once again be connected to the river, would be particularly helpful in the restoration of large rivers. Figure 23.3 illustrates how a former floodplain has been transformed into farmland and often settled with houses. We should tailor government incentives and insurance so that after each flood, fewer attempts to resettle in the river's historic path would be made, and more effort would be expended to encourage relocation outside of the floodplain.

Can we do anything about dams? Actually, quite a lot. Dam removal might sound far-fetched, and in some cases it is clearly an ambitious goal (see the discussion of salmon and the Snake River dams in Chapter 7). But in many instances the value of the dam is small compared to the value of a free-flowing river. Edwards Dam on the Kennebec River in Maine blocked passage for at least 10 species of migratory fishes that were denied access to a 27 km stretch of river above the dam site. Decommissioning of this dam, over the objections of its owner, not only is a triumph for migratory fishes, but it puts the interests of many river recreationists, and the merchants they patronize, over those of a small dam manager.

Most of the dams that could be removed won't be particularly controversial. Many are small, old, dysfunctional, and hazardous. Their safe removal and restoration of the river carry some cost, but often there is no party to oppose the action.

Big dams that generate considerable economic returns are not likely to be ripped out any time soon (but see Chapter 7). But these dams can be operated more in accord with the natural flow regime, in an effort to normalize river function. However, energy deregulation is an important complication for advocates of river-friendly dam operations. Utilities now can market directly to the consumer, offering us the same choices we now enjoy with providers of telephone services. Recognizing the public's concern for global warming, which is driven by energy consumption based on fossil fuels, utilities are beginning to market "green" energy. This might mean windmills or solar, but usually it means hydropower from dams. Some utilities, recognizing that not all consumers think that giant dams are "green," market energy from small dams. Green Hydropower, a coalition between some utilities and environmental organizations, provides a sensible procedure for certifying green hydropower (visit <www.amrivers.org/hydropowerdamreform>). It could also help you decide on which energy supplier you wish to buy from.

Better Science and Management

Application of the conceptual foundations of river ecology to the protection and restoration of streams offers challenges and opportunities to scientists and managers alike. Sometimes we just have to think about how to apply what we already know, and sometimes we have the basic principles but haven't yet worked out all of the problems of their practical application. Fish hatcheries are a good example of the former. Thanks to decades of rearing of fishes of sport and commercial value, the United States has facilities, trained personnel, and a wealth of knowledge on how to grow fishes in hatcheries. Today, that knowledge is increasingly being directed at captive breeding of endangered species, such as the pallid sturgeon or razorback sucker. Of course these species may require modifying old techniques, but consider what a headstart we have. The plight of freshwater mussels provides a counterexample. With their complicated life history (Chapter 15), and lack of a history of hatchery rearing, much scientific work is needed to develop feasible methods of captive breeding.

We suggest that five principles of river ecology serve as a useful framework for better river science and management. First, it makes sense to formalize the problem at the catchment scale. Our knowledge of the dynamic nature of rivers (Chapter 1), the river continuum (Chapter 5), and of the integrated nature of river networks dictates that we focus on the river system in a holistic fashion. Second, our knowledge of the detailed needs of particular species and of the complexity of habitat found in intact systems tells us that restoring environmental heterogeneity is crucial. Flow, naturally variable so that rivers and regions have characteristic flow regimes, can act as a master variable. If one can restore or normalize flows, the river will do much of the work of restoring habitat and the biota will follow. Third, it is important to retain the natural connectivity of the river, particularly upstream with downstream segments, and channel with floodplain. Populations can become isolated from habitat needed during some phase of the life history, or fragmented to the point where each is small and at risk. From fish passage of dams to proper siting of road culverts to restoring flood peaks, much can be done to restore connectivity. Fourth, we should minimize interference of all kinds. Stocking of nonnative species is a practice that should receive intense scrutiny, and we think will rarely be justified. Fifth, we should practice adaptive ecosystem management, which simply means that every management action should be viewed as an experiment to be monitored, and a lesson to be learned. It's hard to imagine a simpler idea, but close inspection of management actions shows that time after time, no one followed up a particular action to see what happened. Even big projects, such as done by the Army Corps, are governed by a project approach that, unsurprisingly, wants a nice, tidy ending, so that attention can turn to a new project. Without long-term monitoring to see the effects of a particular management experiment, valuable knowledge is lost. Typically, only when it is obvious to everyone that the project created a disaster, do the consequences become known; this happened within five years in the case of the Kissimmee River channelization (see Chapter 8). We need a more orderly form of learning from our mistakes.

Citizen Involvement

An educated and concerned citizenry that will advocate river protection could make a tremendous difference over the long term. Citizens can support public policies that benefit rivers and oppose policies that harm rivers. They can make their views known to their elected representatives and to representatives of government agencies. Individuals can support the good works of the large and small NGOs that work on river issues, and contribute directly through their individual actions. Fortunately, the number of concerned citizens and committed NGOs seems to be growing daily.

Rivers are wonderful settings for personal enjoyment, self-education and renewal of commitment. Almost all of us find some time to enjoy rivers, visually, recreationally, and perhaps adventurously. Few of us, can devote the time to explore many of America's wonderful rivers directly, but we can take advantage of a growing environmental literature that celebrates the beauty and diversity of our waterways. Tim Palmer's America by Rivers is an excellent entry into this vibrant literature.

Rivers provide great opportunity for hands-on involvement. Volunteer monitoring of river health has become sufficiently sophisticated that many state monitoring programs involve volunteers, and some, including Colorado and Kentucky, rely heavily on volunteers. The Izaak Walton League and Trout Unlimited both provide well-crafted monitoring protocols for volunteers that have well-structured quality-control guidelines. School children, elders, and just about anyone can make a contribution in this way.

Environmental education about rivers and the problems they face comes in many forms. Experiential learning from volunteer monitoring, government agency fact sheets and exhibits, the newsletters, literature, and public campaigns of NGOs, and we hope this book, comprise some of the ways that the public can become better informed. Surveys show that the public has a very high concern for water-quality issues, particularly with respect to drinking water. Local and regional issues, such as the plight of salmon in the Pacific Northwest, are well known to the public of that area. However, we suspect that the severity of the problems facing rivers and their biota is under-appreciated.

Many opportunities exist for those who would like to become involved. Table 22.3 gives the names, addresses, and web addresses of some organizations that are doing good work or serve as valuable sources of information. There are approximately 3000 organizations and agencies in the United States whose missions directly involve river or watershed conservation. Depending on one's preferences, one can work at the local,

Table 22.3 A Brief Listing of Some Non-governmental Organizations Active in River Conservation and Some Government Agencies with Informative Literature and Web Sites

Non-Government Organizations	
American Rivers	www.amrivers.org
River Revival	www.riverrivival.org
International River Network	www.irn.org
Izaak Walton League	www.IWLA.org
Pacific Rivers Council	www.pacrivers.org
River Network	www.rivernetwork.org
The Nature Conservancy	www.tnc.org
The Sierra Club	www.sierraclub.org
Trout Unlimited	www.tu.org
Federal Agencies	
U.S. EPA	www.epa.gov.owow
U.S. Geological Survey	www.usgs.gov
USDA Forest Service	www.fs.fed.us
USDA Natural Resources Conservation Service	www.nrcs.usda.gov

The River and Watershed Conservation Directory, published by River Network in 1996, is a guide to the 3000 + U.S. organizations and agencies active in river conservation.

state, or national level, and in all likelihood find an organization whose goals mesh with yours.

If we could have two wishes for this book, they would be to instill even greater enjoyment of rivers through a deeper understanding of their ecology and to motivate each reader to find a personal role in helping to protect and conserve rivers.

Recommended Reading

Master, L. L., Flack, S. R. and Stein, B. A. (eds). (1998). Rivers of life: Critical watersheds for protecting freshwater biodiversity. The Nature Conservancy, Arlington, Virginia.

Palmer, T. (1996). *America by Rivers*. Island Press, Washington D.C.

Palmer, T. (1996). *Lifelines: The Case for River Conservation*. Island Press, Washington D.C.

Conversion Factors

To change	Into	Multiply by	Definitions
millimeters (mm)	inches (in)	0.03937	ppb = parts per billion
centimeters (cm)	inches (in)	0.3937	ppm = parts per million
meters (m)	feet (ft)	3.281	ppt = parts per thousand
kilometers (km)	miles (mi)	0.6214	mg/L = milligrams per liter
grams (g)	ounces (oz)	0.03527	
kilograms (kg)	pounds (lbs)	2.205	
square meters (m^2)	square ft (ft^2)	10.76	
hectares (ha)	acres	2.471	
cubic meters/sec (m^3s)	cubic feet/sec (cfs)	35.31	
kilograms/hectare (kg/ha)	pounds/acres (lbs/acre)	0.8922	
degrees Celsius (°C)	degrees Fahrenheit (°F)	(9/5 C) + 32	

Suppliers of Sampling and Collection Equipment

Aquatic Eco-Systems, Inc.
1767 Benbow Court
Apopka, FL 32703
1-877-347-4788 (phone order)
1-407-886-6787 (fax order)
e-mail: aes@aquaticeco.com

Ben Meadows Company
3589 Broad Street
Atlanta, GA 30341
1-800-628-2068 (fax order)
1-800-241-6401 (phone order)
e-mail: mail@benmeadows.com

BioQuip Products
17803 LaSalle Avenue
Gardena, CA 90248-3602
310-324-0620 (phone)
310-324-7931 (fax)
e-mail: bioquip@aol.com

Forestry Suppliers, Inc.
205 W. Rankin Street
Jackson, MS 39284-8397
1-800-647-5368 (phone order)
1-800-543-4208 (fax order)
1-800-360-7788 (catalog request)

Wildlife Supply Company (Wildco)
301 Cass Street
Saginaw, MI 48602-2097
517-799-8100 (phone)
1-800-799-8103 (phone USA)
517-799-8115 (fax)
1-800-799-8115 (fax USA)
e-mail: goto@wildco.com

References

Note: The following publications either have been cited directly in the text of this book or were used as reference material. We are providing this list to aid readers who may wish to pursue different subjects in more detail.

Chapter 1. Rivers as Dynamic Physical Entities

Allan, J. D. (1995). *Stream Ecology*. Kluwer Academic. Dordrecht, The Netherlands.

Dunne, T. and Leopold, L. B. (1978). *Water in Environmental Planning*. Freeman, San Francisco.

Frissell, C. A., Liss, W. L. Warren, C. E. and Hurley, M. D. (1986). A hierarchical framework for stream habitat classification: Viewing streams in a watershed context. *Environmental Management* 10:199–214.

Leopold, L. B. (1994). *A View of the River*. Harvard University Press, Cambridge.

Leopold, L. B., Wolman, M. G. and Miller, J. P. (1964). *Fluvial Processes in Geomorphology*. Freeman, San Francisco.

Poff, N. L. and Ward, J. V. (1989). Implications of streamflow variability and predictability for lotic community structure: A regional analysis of streamflow patterns. *Canadian Journal of Fisheries and Aquatic Sciences* 46:1805–1818.

Postel, S. L., Daily, G. C. and Ehrlich, P. R. (1996). Human appropriation of renewable freshwater. *Science* 271:785–787.

Chapter 2. Abiotic Factors

Allan, J. D. (1995). *Stream Ecology*. Kluwer Academic. Dordrecht, The Netherlands.

Chapter 3. Energy Resources

Allan, J. D. (1995). *Stream Ecology*. Kluwer Academic. Dordrecht, The Netherlands.

Hynes, H. B. N. (1970). *The Ecology of Running Waters*. University of Toronto Press, Toronto.

Hynes, H. B. N. (1975). The stream and its valley. *Verhandlungen der Internationalen Vereinigung für theoretische und angewandte Limnologie* 19:1–15.

Kaplan, L. A. and Bott, T. L. (1982). Diel fluctuations of DOC generated by algae in a piedmont stream. *Limnology and Oceanography* 27:1091–1100.

Chapter 4. Feeding Roles and Food Webs

Allan, J. D. (1995). *Stream Ecology*. Kluwer Academic. Dordrecht, The Netherlands.
Cummins, K. W. (1973). Trophic relations of aquatic insects. *Annual Review of Entomology* 18:183–206.
Cummins, K. W. and Klug, M. J. (1979). Feeding ecology of stream invertebrates. *Annual Review of Ecology and Systematics* 10:147–172.
Cummins, K. W. and Merritt, R. W. (1996). Ecology and distribution of aquatic insects. *In* Merritt, R. W. and Cummins, K. W. (eds.), *An Introduction to the Aquatic Insects of North America*, 3rd ed., pp. 74–86. Kendall/Hunt, Dubuque, Iowa.
Wiggins, G. B. (1996). *Larvae of the North American Caddisfly Genera (Trichoptera)*, 2nd ed. University of Toronto Press, Toronto.

Chapter 5. Ecology: The Structure and Function of Riverine Ecosystems

Allan, J. D. (1995). *Stream Ecology*. Kluwer Academic. Dordrecht, The Netherlands.
Junk, W. J., Bayley, P. B. and Sparks, R. E. (1989). The flood pulse concept in river-flood plain systems. *In* Dodge, D. P. (ed.), *Proceedings of the International Large Rivers Symposium*. Canadian Special Publication in Fisheries and Aquatic Sciences No. 106, pp. 110–127. Ottawa.
McCafferty, W. P. (1981). *Aquatic Entomology*. Jones & Bartlett, Boston.
Minshall, G. W. (1978). Autotrophy in stream ecosystems. *BioScience* 28:767–771.
Stanford, J. A. and Gaufin, A. R. (1974). Hyporheic communities of two Montana rivers. *Science* 185:700–702.
Townsend, C. R. (1989). The patch dynamic concept of stream community ecology. *Journal of the North American Benthological Society* 8:36–50.
Vannote, R. L.,. Minshall, G. W., Cummins, K. W., Sedell, J. R. and Cushing, C. E. (1980). The river continuum concept. *Canadian Journal of Fisheries and Aquatic Sciences* 37:130–137.
Ward, J. V. (1989). The four-dimensional nature of lotic ecosystems. *Journal of the North American Benthological Society* 8:2–8.
Ward, J. V. and Stanford, J. A. (1983). The serial discontinuity concept of lotic ecosystems. *In* Fontaine, T. D. and Bartell, S. M. (eds.), *Dynamics of Lotic ecosystems*, pp. 29–42. Ann Arbor Science Publishers, Ann Arbor, Michigan.
Webster, J. R. and B. C. Patten. (1979). Effects of watershed perturbation on stream potassium and calcium dynamics. *Ecological Monographs* **49**, 51–72.

Chapter 6. Trout Streams

Bode, R. W., Novak, M. A. and Abele, L. E. (1993). Biological stream assessment: Beaver Kill, Sullivan and Delaware Counties, New York. NY State Dept. of Environ. Conser., Albany, New York, 34 pp.
Hendrickson, G. E. (1994). *The Angler's Guide to Twelve Classic Trout Streams in Michigan*. University of Michigan Press, Ann Arbor.
Van Put, E. (1981). The Beaverkill. *Trout* 22:48–54.
Waters, T. F. (1994). Productivity of streams—Can we increase it? *The FFF Quill*, Spring 1994, pp. 14–15.

Chapter 7. Large Rivers of the West

Anon. (1987). *River and Dam Management*. National Academy Press, Washington, D.C.

Anon. (n.d.) The 1996–1997 River and Watershed Conservation Directory. River Network and the Department of the Interior National Park Service, Rivers Trails and Conservation Assistance Program. Portland, Oregon.

Independent Scientific Group. (1999). Return to the River: Scientific issues in the restoration of salmonid fishes in the Columbia River. *Fisheries* 24:10–19.

Neal, V. T. (1972). Physical aspects of the Columbia River and its estuary. *In* Pruter, A. T. and Alverson, D. L. (eds.), *The Columbia River Estuary and Adjacent Ocean Waters*, pp. 19–40. University of Washington Press, Seattle.

Palmer, T. (1991). *The Snake River: A Window to the West*. Island Press, Washington D.C.

Reisner, M. (1993). *Cadillac Desert*. Penguin Books, New York.

Robeck, G. G., Henderson, C. and Palange, R. C. (1954). Water quality studies on the Columbia River. U.S. Dept. of Health, Education, and Welfare, Public Health Service, Cincinnati, Ohio.

Schmidt, J. C., Webb, R. H., Valdez, R. A., Marzolf, G. R. and Stevens, L. E. (1998). Science and values in river restoration in the Grand Canyon. *BioScience* 48: 735–747.

Stanford, J. A. and Ward, J. V. (1986). The Colorado River system. *In* Davies, B. R. and Walker, K. F. (eds.), *The Ecology of River Systems*, pp. 353–374. Dr. W. Junk Publishers, Boston.

Chapter 8. Diverse Rivers of the Southeast

Benke, A. C. and Jacobi, D. I. (1994). Production dynamics and resource utilization of snag-dwelling mayflies in a blackwater river. *Ecology* 75:1219–1232.

Benke, A. C., Van Arsdall, T. C. Jr., and Gillespie, D. M. (1984). Invertebrate productivity in a subtropical blackwater river: The importance of habitat and life history. *Ecological Monographs* 54:25–63.

Benke, A. C., Hunter, R. J. and Parrish, F. K. (1986). Invertebrate drift dynamics in a subtropical blackwater river. *Journal of the North American Benthological Society* 5:173–190.

Dahm, C. N. (ed.). (1995). *Restoration Ecology*, Vol. 3, No. 3, Blackwell Science, Oxford.

Lydeard, C. and Mayden, R. L. (1995). A diverse and endangered aquatic ecosystem of the southeast United States. *Conservation Biology* 9:800–805.

Merritt, R. W. and Lawson, D. L. (1992). The role of leaf litter macroinvertebrates in stream-floodplain dynamics. *Hydrobiologia* 248:65–77.

Merritt, R. W., Wallace, J. R., Higgins, M. J., Alexander, M. K., Berg, M. B., Morgan, W. T., Cummins, K. W. and Vandeneeden, B. (1996). Procedures for the functional analysis of invertebrate communities of the Kissimmee River-floodplain ecosystem. *Florida Scientist* 59:216–274.

Meyer, J. L. (1990). A blackwater perspective on riverine ecosystems. *BioScience* 40:643–651.

Meyer, J. L. (1992). Seasonal patterns of water quality in blackwater rivers of the coastal plain, southeastern United States. *In* Becker, C. D. and Neitzel, D. A. (eds.), *Water Quality in North American River Systems*, pp. 250–276. Battelle Press, Columbus, Ohio.

Meyer, J. L., Edwards, R. T. and Risley, R. (1987). Bacterial growth on dissolved organic carbon from a blackwater river. *Microbial Ecology* 13:13–29.

Smock, L. A. and Roeding, C. E. (1986). The trophic basis of production of the macroinvertebrate community of a southeastern U.S.A. blackwater stream. *Holarctic Ecology* 9:165–174.

Smock, L. A., Gilinsky, E., and Stoneburner, D. L. (1985). Macroinvertebrate production in a southeastern United States blackwater stream. *Ecology* 66:1491–1503.

Wallace, J. B. and Benke, A. C. (1984). Quantification of wood habitats in subtropical Coastal Plain streams. *Canadian Journal of Fisheries and Aquatic Sciences* 41:1643–1652.

Wallace, J.B., Benke, A. C., Lingle, A. H. and Parsons, K. (1987). Trophic pathways of macroinvertebrate primary consumers in subtropical blackwater streams. *Archive für Hydrobiologia/Supplement* 74(4):423–451.

Ward, A. K., Ward, G. M. and Harris, S. C. (1992). Water quality and biological communities of the Mobile River rainage, eastern Gulf of Mexico region. *In* Becker, C. D. and Neitzel, D. A. (eds.), *Water Quality in North American River Systems*, pp. 277–304. Battelle Press, Columbus, Ohio.

Chapter 9. Warm-Water Rivers of the Midwest

Andersen, O., Crow, T. R., Lietz, S. M. and Stearns, F. (1996). Transformation of a landscape in the upper mid-west, U.S.A.: The history of the lower St. Croix river valley, 1830 to present. *Landscape and Urban Planning* 35:247–267.

Hynes, H. B. N. (1975). The stream and its valley. *Verhandlungen der Internationalen Vereinigung für theoretische and angewandte Limnologie* 19:1–15.

Karr, J. R., Toth, L. A. and Dudley, D. R. (1985). Fish communities of midwestern rivers: A history of degradation. *BioScience* 35:90–95.

Sparks, R. E., Nelson, J. C. and Yin, Y. (1998). Naturalization of the flood regime in regulated rivers. *BioScience* 48:706–720.

U.S. Department of Agriculture Natural Resource Conservation Service "The Geography of Hope" <www.nhq.nrcs.usda.gov/land/meta/m2087.html>

Chapter 10. Desert Rivers of the Southwest

Busch, D. E. and Fisher, S. G. (1981). Metabolism of a desert stream. *Freshwater Biology* 11:301–307.

Crawford, C. S., Cully, A. C., Leutheuser R., Sifuentes, M. S., White, L. H. and Wilber, J. P. (1993). Middle Rio Grande ecosystem: Bosque biological management plan. Biological Interagency Team, USFWS, Albuquerque, New Mexico.

Deacon, J. E. and Minckley, W. L. (1974). Desert fishes. *In Desert Biology II, Special Topics on the Physical and Biological Aspects of Arid Regions*. Academic Press, New York.

Fisher, S. G. (1986). Structure and dynamics of desert streams. *In* W.G. Whitford (ed.), *Pattern and Process in Desert Ecosystems*, pp. 119–139. University of New Mexico Press, Albuquerque.

Fisher, S. G., Gray, L. J., Grimm N. B., and Busch, D. E. (1982). Temporal succession in a desert stream ecosystem following flash flooding. *Ecological Monographs* 52:93–110.

Grimm, N. B., Fisher, S. G. and Minckley, W. L. (1981). Nitrogen and phosphorus dynamics in hot desert streams of the southwestern U.S.A. *Hydrobiologia* 83:303–312.

Welsh, F. (1985). *How to Create a Water Crisis*. Johnson Books, Boulder, Colorado.

Chapter 11. Special Riverine Systems

Anon. (1996). Metolius watershed analysis, Sisters Ranger District. U.S. Forest Service. 186 pp. + appendixes.

Cushing, C. E. (1996). The ecology of cold desert spring-streams. *Archive für Hydrobiologie* 135:499–522.

Gaines, W. L., Cushing, C. E. and Smith, S. D. (1992). Secondary production estimates of benthic insects in three cold desert streams. *Great Basin Naturalist* 52:11–24.

Usinger, R. L. (1967). *The Life of Rivers and Streams*. McGraw-Hill, New York.

Chapter 12. Algae

South, G. R. and Whittick, A. (1987). *Introduction to Phycology*. Blackwell Scientific, London.

Sze, P. (1986). *A Biology of the Algae*. Brown, Dubuque, Iowa.

Chapter 13. Higher Plants: The Macrophytes

Fassett, N. C. (1940). *A Manual of Aquatic Plants*. McGraw-Hill, New York.

Hynes, H. B. N. (1970). *The Ecology of Running Waters*. University of Toronto Press, Toronto.

Muenscher, W. C. (1944). *Aquatic Plants of the United States*. Comstock, Ithaca, New York.

Chapter 14. Insects

Cummins, K. W. and. Merritt, R. W (1996). Ecology and distribution of aquatic insects. *In* Merritt, R. W. and Cummins, K. W. (eds.), *An Introduction to the Aquatic Insects of North America*, 3rd ed., pp. 74–86. Kendall/Hunt, Dubuque, Iowa.

Daly, H. V. (1996). General classification and key to the orders of aquatic and semiaquatic insects. *In* Merritt, R. W. and Cummins, K. W. (eds.), *An Introduction to the Aquatic Insects of North America*, 3rd ed., pp. 108–112. Kendall/Hunt, Dubuque, Iowa.

Gaines, W. L., Cushing, C. E. and Smith, S. D. (1992). Secondary production estimates of benthic insects in three cold desert streams. *Great Basin Naturalist* 52:11–24.

Grubaugh, J. W. and Wallace, J. B. (1995). Functional structure and production of the benthic community in a Piedmont river: 1956–1957 and 1991–1992. *Limnology and Oceanography* 40:490–501.

Hynes, H. B. N. (1970). *The Ecology of Running Waters*. University of Toronto Press.

McCafferty, W. P. (1981). *Aquatic Entomology*. Jones & Bartlett, Boston.

Merritt, R. W. and. Cummins, K. W (eds.). (1996). *An Introduction to the Aquatic Insects of North America*, 3rd ed. Kendall/Hunt, Dubuque, Iowa.

Minshall, G. W. (1981). Structure and temporal variations of the benthic macroinvertebrate community inhabiting Mink Creek, Idaho, U.S.A., a 3rd-order Rocky Mountain stream. *Journal of Freshwater Ecology* 1:13–26.

Müller, K. (1954). Die drift in fliessenden Gewässern. *Archive für Hydrobiologie* 49:539–545.

Stewart, K. W. and Stark, B. P. (1988). *Nymphs of North American Stonefly Genera (Plecoptera)*. University of North Texas Press, Denton.

Stetzer, R. S. (1992). *Flies, the Best One Thousand*. Frank Amato. Portland, Oregon.

Ward, J. V. (1986). Altitudinal zonation in a Rocky Mountain stream. *Archive für Hydrobiologie/Supplement* 74:133–199.

Ward, J. V. (1996). *Aquatic Insect Ecology. 1. Biology and Habitat*. John Wiley & Sons, New York.

Ward, J. V. and Stanford, J. A. (1991). Benthic faunal patterns along the longitudinal gradient of a Rocky Mountain river system. *Verhandlungen der Internationalen Vereinigung für theoretische und angewandte Limnologie* 24:3087–3094.

Waters, T. P. (1965). Interpretation of invertebrate drift in streams. *Ecology* 46:327–334.

Wiggins, G. B. (1996). *Larvae of the North American Caddisfly Genera (Trichoptera)*, 2nd ed. University of Toronto Press, Toronto.

Wiggins, G. B. (1996). Trichoptera families. *In* R.W. Merritt and K. W. Cummins (eds.), *An Introduction to the Aquatic Insects of North America*, 3rd ed., pp. 309–349. Kendall/Hunt, Dubuque, Iowa.

Wiggins, G. B. and Mackay, R. J. (1978). Some relationships between systematics and trophic ecology in Nearctic aquatic insects, with special reference to Trichoptera. *Ecology* 59:1211–1220.

Winget, R. N. (1993). Habitat partitioning among three species of Ephemerelloidea. *Journal of Freshwater Ecology* 8:227–233.

Chapter 15. Mollusks

Brown, K. M. (1991). Mollusca: gastropoda. *In* Thorp, J. H. and Covich, A. P. (eds.), *Ecology and Classification of North American Freshwater Invertebrates*, pp. 285–314. Academic Press, San Diego.

McMahon, R. F. (1991). Mollusca: Bivalvia. *In* Thorp, J. H. and Covich, A. P. (eds.), *Ecology and Classification of North American Freshwater Invertebrates*, pp. 315–399. Academic Press, San Diego.

Pennak, R. W. (1989). *Fresh-Water Invertebrates of the United States*, 3rd ed. John Wiley & Sons, New York.

Thorpe, J. H. and Covich, A. P. (1991). *The Ecology and Classification of Freshwater Inventebrates*. Academic Press, New York.

Williams, J. D., Warren, M. L. Jr., Cummins, K. S. Harris, J. L. and Neves, R. J. (1992). Conservation status of freshwater mussels of the United States and Canada. *Fisheries* 18:6–22.

Chapter 16. Crustaceans

Pennak, R. W. (1989). *Fresh-Water Invertebrates of the United States*, 3rd ed. John Wiley & Sons, New York.

Pringle, C. M. and Blake, G. A. (1994). Quantitative effects of atyid shrimp (Decapoda:Atyidae) on the depositional environment in a tropical stream: Use of electricity for experimental exclusion. *Canadian Journal of Fisheries and Aquatic Sciences* 51:1443–1450.

Thorp, J. H. and Covich, A. P. (eds.) (1991). *Ecology and Classification of North American Freshwater Invertebrates*. Academic Press, San Diego.

Chapter 17. Other Invertebrates

Allan, J. D. (1995). *Stream Ecology*. Kluwer Academic. Dordrecht, The Netherlands.

Palmer, M. A., Bely, A. E. and Berg, K. E. (1992). Response of invertebrates to lotic disturbance: A test of the hyporheic refuge hypothesis. *Oecologia* 89:182–194.

Pennak, R. W. (1989). *Fresh-Water Invertebrates of the United States*, 3rd ed. John Wiley & Sons, New York.

Stanford, J. A. and Ward, J. V. (1988). The hyporheic habitat of river ecosystems. *Nature* 335:64–66.

Thorp, J. H. and Covich, A. P. (eds.) (1991). *Ecology and Classification of North American Freshwater Invertebrates*. Academic Press, San Diego.

Wallace, R. L. and Snell, T. W. (1991). Rotifera. *In* Thorp, J. H. and Covich, A. P. (eds.), *Ecology and Classification of North American Freshwater Invertebrates*, pp. 187–248. Academic Press, San Diego.

Chapter 18. Fishes

Karr, J. R. (1991). Biological integrity: A long-neglected aspect of water resource management. *Ecological Applications* 1:66–84.

Karr, J. A., Allan, J. D. and Benke, A. (2001). River conservation in the United States and Canada: Science, policy, and practice. *In* Boon, P. J. Davies, B. R. and Petts, G. E. (eds.), *Global perspectives on river conservation*. pp. 3–40. John Wiley & Sons, New York.

Paxton, J. R. and Eschmeyer, W. N. (1995). *Encyclopedia of Fishes*. Academic Press, New York.

Willson, M. F. and Halupka, K. C. (1995). Anadromous fishes as keystone species in vertebrate communities. *Conservation Biology* 9:489–497.

Chapter 19. Reptiles and Amphibians

Hawkins, C. P. Murphy, M. L. Anderson, N. H. and Wilzbach, M. A. (1983). Density of fish and salamanders in relation to riparian canopy and physical habitat in streams of the northwestern United States. *Canadian Journal of Fisheries and Aquatic Sciences* 40:1173–1185.

Hawkins, C. P. Gottschal, L. J. and Brown, S. S. (1988). Densities and habitat of tailed frog tadpoles in small streams near Mt. St. Helens following the 1980 eruption. *Journal of the North American Benthological Society* 7:246–252.

Hebrard, J. J. and Mushinsky, H. R. (1978). Habitat use by five sympatric water snakes in a Louisiana swamp. *Herpetologica* 34:306–311.

Murphy, M. L. and Hall, J. D. (1981). Varied effects of clear-cut logging on predators and their habitat in small streams of the Cascade Mountains, Oregon. *Canadian Journal of Fisheries and Aquatic Sciences* 38:137–145.

Murphy, M. L., Hawkins, C. P. and Anderson, N. H. (1981). Effects of canopy modifications and accumulated sediment on stream communities. *Transactions of the American Fisheries Society* 110:469–478.

Mushinsky, H. R. and Hebrard, J. J. (1977). Food partitioning by five species of water snakes in Louisiana. *Herpetologica* 33:162–166.

Chapter 20. Birds

Ehrlich, P. R., Dobkin, D. S. and Wheye, D. (1988). *The Birder's Handbook*. Simon & Schuster, New York.

Power, M. E., Dudley, T. L. and Cooper, S. D. (1989). Grazing catfish, fishing birds, and attached algae in a Panamanian stream. *Environmental Biology* 26:285–294.

Usinger, R. L. (1967). *The Life of Rivers and Streams*. McGraw-Hill, New York.

Chapter 21. Mammals

Naiman, R. J. and Melillo, J. M. (1984). Nitrogen budget of a subarctic stream altered by beaver (*Castor canadensis*). *Oecologia* 62:150–155.

Naiman, R. J., McDowell, D. M. and Farr, B. S. (1984). The influence of beaver (*Castor canadensis*) on the production dynamics of aquatic insects. *Verhandlungen der Internationalen Vereinigung für theoretische und angewandte Limnologie* 22:1801–1810.

Palmer, E. L. (1957). *Fieldbook of Mammals*. Dutton, New York.

Sandoz, M. (1978). *The Beaver Men*. Bison Books, University of Nebraska Press, Lincoln.

Zeveloff, S. I. (1988). *Mammals of the Intermountain West*. University of Utah Press, Salt Lake City.

Chapter 22. Coping with the Threats to America's Rivers

Allan, J. D. (1995). *Stream Ecology*. Kluwer Academic. Dordrecht, The Netherlands.

Cushing, C. E. (ed.). (1997). *Freshwater Ecosystems and Climate Change in North America*. John Wiley & Sons, New York.

Master, L. L., Flack, S. R. and Stein, B. A. (eds.). (1998). Rivers of life: Critical watersheds for protecting freshwater biodiversity. The Nature Conservancy, Arlington, Virginia.

Moyle, P. B. (1995). Conservation of native freshwater fishes in the Mediterranean-type climate of California, U.S.A.: A review. *Biological Conservation* 72:271–279.

Stanford, J. A., Ward, J. V., Liss, W. J. Frissell, C. A., Williams, R. N., Lichatowich, J. A. and Coutant, C. C. (1996). A general protocol for restoration of regulated rivers. *Regulated Rivers* 12:391–413.

Sedell, J. R. and Froggatt, J. L. (1984). Importance of streamside forests to large rivers: The isolation of the Willamette River, Oregon, U.S.A., from its floodplain by snagging and streamside forest removal. *Verhandlungen der Internationalen Vereinigung für theoretische und angewandte Limnologie* 22:1828–1934.

The 1996–1997 River and Watershed Conservation Directory, published by River Network and the Department of the Interior National Park Service, Rivers Trails and Conservation Assistance program. (1996). River Network, P.O. Box 8787, Portland, Oregon 97207.

Index

Note: In this index, page numbers are given under common names; Latin names are also given, but the reader will be referred to appropriate page numbers under the common names.